Advances in Chemical Crystallography: A Themed Issue Honoring Professor Alexandra M. Z. Slawin on the Occasion of Her 60th Birthday

Advances in Chemical Crystallography: A Themed Issue Honoring Professor Alexandra M. Z. Slawin on the Occasion of Her 60th Birthday

Editors

William T. A. Harrison
R. Alan Aitken
Paul Waddell

MDPI • Basel • Beijing • Wuhan • Barcelona • Belgrade • Manchester • Tokyo • Cluj • Tianjin

Editors

William T. A. Harrison
School of Natural and
Computing Sciences
University of Aberdeen
Aberdeen
United Kingdom

R. Alan Aitken
Department of Chemistry
University of St Andrews
St Andrews
United Kingdom

Paul Waddell
School of Natural and
Environmental Sciences
University of Newcastle
Newcastle upon Tyne
United Kingdom

Editorial Office
MDPI
St. Alban-Anlage 66
4052 Basel, Switzerland

This is a reprint of articles from the Special Issue published online in the open access journal *Molecules* (ISSN 1420-3049) (available at: www.mdpi.com/journal/molecules/special_issues/Alexandra_Slawin).

For citation purposes, cite each article independently as indicated on the article page online and as indicated below:

LastName, A.A.; LastName, B.B.; LastName, C.C. Article Title. *Journal Name* **Year**, *Volume Number*, Page Range.

ISBN 978-3-0365-4196-9 (Hbk)
ISBN 978-3-0365-4195-2 (PDF)

© 2022 by the authors. Articles in this book are Open Access and distributed under the Creative Commons Attribution (CC BY) license, which allows users to download, copy and build upon published articles, as long as the author and publisher are properly credited, which ensures maximum dissemination and a wider impact of our publications.

The book as a whole is distributed by MDPI under the terms and conditions of the Creative Commons license CC BY-NC-ND.

Contents

Helen M. O'Connor, Sergio Sanz, Aaron J. Scott, Mateusz B. Pitak, Wim T. Klooster and Simon J. Coles et al.
$[Cr^{III}{}_8Ni^{II}{}_6]^{n+}$ Heterometallic Coordination Cubes
Reprinted from: 2021, 26, 757, doi:10.3390/molecules26030757 1

Guoxiong Hua, Cameron L. Carpenter-Warren, David B. Cordes, Alexandra M. Z. Slawin and J. Derek Woollins
Synthesis and Single Crystal Structures of N-Substituted Benzamides and Their Chemoselective Selenation/Reduction Derivatives
Reprinted from: 2021, 26, 2367, doi:10.3390/molecules26082367 11

Alister S. Goodfellow and Michael Bühl
Hydricity of 3d Transition Metal Complexes from Density Functional Theory: A Benchmarking Study
Reprinted from: 2021, 26, 4072, doi:10.3390/molecules26134072 21

Kaushik Naskar, Suvendu Maity, Himadri Sekhar Maity and Chittaranjan Sinha
A Reusable Efficient Green Catalyst of 2D Cu-MOF for the Click and Knoevenagel Reaction
Reprinted from: 2021, 26, 5296, doi:10.3390/molecules26175296 35

Stamatis S. Passadis, Sofia Hadjithoma, Panagiota Siafarika, Angelos G. Kalampounias, Anastasios D. Keramidas and Haralampos N. Miras et al.
Synthesis, Structural and Physicochemical Characterization of a Titanium(IV) Compound with the Hydroxamate Ligand N,2-Dihydroxybenzamide
Reprinted from: 2021, 26, 5588, doi:10.3390/molecules26185588 49

Craig A. Johnston, David B. Cordes, Tomas Lebl, Alexandra M. Z. Slawin and Nicholas J. Westwood
Synthesis of Indoloquinolines: An Intramolecular Cyclization Leading to Advanced Perophoramidine-Relevant Intermediates
Reprinted from: 2021, 26, 6039, doi:10.3390/molecules26196039 69

Jude N. Arokianathar, Will C. Hartley, Calum McLaughlin, Mark D. Greenhalgh, Darren Stead and Sean Ng et al.
Isothiourea-Catalyzed Enantioselective -Alkylation of Esters via 1,6-Conjugate Addition to *para*-Quinone Methides
Reprinted from: 2021, 26, 6333, doi:10.3390/molecules26216333 89

Jiyu Tian, Eli Zysman-Colman and Finlay D. Morrison
Azetidinium Lead Halide Ruddlesden–Popper Phases
Reprinted from: 2021, 26, 6474, doi:10.3390/molecules26216474 103

Peter De'Ath, Mark R. J. Elsegood, Noelia M. Sanchez-Ballester and Martin B. Smith
Low-Dimensional Architectures in Isomeric *cis*-$PtCl_2\{Ph_2PCH_2N(Ar)CH_2PPh_2\}$ Complexes Using Regioselective-N(Aryl)-Group Manipulation
Reprinted from: 2021, 26, 6809, doi:10.3390/molecules26226809 111

Russell M. Main, David B. Cordes, Aamod V. Desai, Alexandra M. Z. Slawin, Paul Wheatley and A. Robert Armstrong et al.
Solvothermal Synthesis of a Novel Calcium Metal-Organic Framework: High Temperature and Electrochemical Behaviour
Reprinted from: 2021, 26, 7048, doi:10.3390/molecules26227048 133

Samuel R. Lawrence, Matthew de Vere-Tucker, Alexandra M. Z. Slawin and Andreas Stasch
Synthesis of a Hexameric Magnesium 4-pyridyl Complex with Cyclohexane-like Ring Structure via Reductive C-N Activation
Reprinted from: 2021, 26, 7214, doi:10.3390/molecules26237214 147

Brian A. Chalmers, D. M. Upulani K. Somisara, Brian A. Surgenor, Kasun S. Athukorala Arachchige, J. Derek Woollins and Michael Bühl et al.
Synthetic and Structural Study of *peri*-Substituted Phosphine-Arsines
Reprinted from: 2021, 26, 7222, doi:10.3390/molecules26237222 157

Katrin Ackermann, Alexandra Chapman and Bela E. Bode
A Comparison of Cysteine-Conjugated Nitroxide Spin Labels for Pulse Dipolar EPR Spectroscopy
Reprinted from: 2021, 26, 7534, doi:10.3390/molecules26247534 173

R. Alan Aitken, Andrew D. Harper and Alexandra M. Z. Slawin
Rationalisation of Patterns of Competing Reactivity by X-ray Structure Determination: Reaction of Isomeric (Benzyloxythienyl)oxazolines with a Base
Reprinted from: 2021, 26, 7690, doi:10.3390/molecules26247690 189

Article

[Cr$^{III}_8$Ni$^{II}_6$]$^{n+}$ Heterometallic Coordination Cubes

Helen M. O'Connor [1], Sergio Sanz [1], Aaron J. Scott [1], Mateusz B. Pitak [2], Wim T. Klooster [2], Simon J. Coles [2], Nicholas F. Chilton [3], Eric J. L. McInnes [3], Paul J. Lusby [1], Høgni Weihe [4], Stergios Piligkos [4,*] and Euan K. Brechin [1,*]

[1] EaStCHEM School of Chemistry, The University of Edinburgh, David Brewster Road, Edinburgh EH3 5JF, UK; oconnoh7@tcd.ie (H.M.O.); s.calvo@fz-juelich.de (S.S.); Aaron.Scott@ed.ac.uk (A.J.S.); Paul.Lusby@ed.ac.uk (P.J.L.)

[2] UK National Crystallography Service, Chemistry, Highfield Campus, University of Southampton, Southampton SO17 1BJ, UK; Mateusz.Pitak@matthey.com (M.B.P.); W.T.Klooster@soton.ac.uk (W.T.K.); S.J.Coles@soton.ac.uk (S.J.C.)

[3] Department of Chemistry, The University of Manchester, Oxford Road, Manchester M13 9PL, UK; nicholas.chilton@manchester.ac.uk (N.F.C.); eric.mcinnes@manchester.ac.uk (E.J.L.M.)

[4] Department of Chemistry, University of Copenhagen, Universitetsparken 5, Copenhagen DK-2100, Denmark; weihe@chem.ku.dk

* Correspondence: piligkos@kiku.dk (S.P.); ebrechin@ed.ac.uk (E.K.B.)

Abstract: Three new heterometallic [Cr$^{III}_8$Ni$^{II}_6$] coordination cubes of formulae [Cr$^{III}_8$Ni$^{II}_6$L$_{24}$(H$_2$O)$_{12}$](NO$_3$)$_{12}$ (**1**), [Cr$^{III}_8$Ni$^{II}_6$L$_{24}$(MeCN)$_7$(H$_2$O)$_5$](ClO$_4$)$_{12}$ (**2**), and [Cr$^{III}_8$Ni$^{II}_6$L$_{24}$Cl$_{12}$] (**3**) (where HL = 1-(4-pyridyl)butane-1,3-dione), were synthesised using the paramagnetic metalloligand [Cr^{III}L$_3$] and the corresponding Ni^{II} salt. The magnetic skeleton of each capsule describes a face-centred cube in which the eight Cr^{III} and six Ni^{II} ions occupy the eight vertices and six faces of the structure, respectively. Direct current magnetic susceptibility measurements on (**1**) reveal weak ferromagnetic interactions between the Cr^{III} and Ni^{II} ions, with J_{Cr-Ni} = + 0.045 cm^{-1}. EPR spectra are consistent with weak exchange, being dominated by the zero-field splitting of the Cr^{III} ions. Excluding wheel-like structures, examples of large heterometallic clusters containing both Cr^{III} and Ni^{II} ions are rather rare, and we demonstrate that the use of metalloligands with predictable bonding modes allows for a modular approach to building families of related polymetallic complexes. Compounds (**1**)–(**3**) join the previously published, structurally related family of [M$^{III}_8$M$^{II}_6$] cubes, where M^{III} = Cr, Fe and M^{II} = Cu, Co, Mn, Pd.

Keywords: molecular magnetism; supramolecular chemistry; heterometallic clusters; magnetometry; EPR spectroscopy

1. Introduction

Heterometallic coordination complexes have seen application in areas as diverse as metalloprotein chemistry [1,2], catalysis [3], porous materials [4,5], and magnetism [6]. The latter includes three-dimensional (3D) networks [7], two-dimensional (2D) sheets [8], one-dimensional (1D) chains [9], and zero-dimensional (0D) (molecular) polygons and polyhedra [10,11], investigating controllable exchange interactions [12], enhanced magnetocaloric effects [13], spin frustration [14], slow relaxation of the magnetisation [15,16], and quantum coherence timescales [17]. A search of the Cambridge Structural Database (CSD) reveals that heterometallic wheels of varying size and metal ratios dominate the chemistry of polymetallic clusters containing both Cr^{III} and Ni^{II} ions with a nuclearity of four or more. Examples include [Cr$_7$Ni], [Cr$_9$Ni], [Cr$_8$Ni$_2$], [Cr$_7$Ni$_2$], [Cr$_6$Ni$_2$], [Cr$_2$Ni$_5$], [CrNi$_6$] wheels and discs (centred/Anderson wheels), and [Cr$_{14}$Ni$_2$] and [Cr$_{28}$Ni$_4$] 'linked rings' [18–23]. Surprisingly, the search reveals only two other unique structural motifs, a rather unusual [Cr$_3$Ni$_2$] linear complex [24], and an 'S-shaped' [Cr$_{12}$Ni$_3$] chain [25]. We have previously reported a metalloligand approach that enabled us to synthesise high-nuclearity heterometallic coordination capsules of paramagnetic transition metal ions in a modular and predictable fashion [26–29]. This metalloligand, based on the tritopic [M^{III}L$_3$] moiety shown in Figure 1 (HL = 1-(4-pyridyl)butane-1,3-dione), features a tris(acac) co-ordinated octahedral transition metal ion, in which the ligand is functionalised with a p-pyridyl donor group. In the *fac*-isomer of this metalloligand, the three N-donor groups are orientated in such a way that combination with a square-planar metal ion leads to the entropically favoured self-assembly of a cubic structure [30]. Herein, we report the syntheses, structures, and magnetic properties of three novel tetradecanuclear [Cr^{III}$_8$Ni^{II}$_6$]$^{n+}$ cubes, namely [Cr^{III}$_8$Ni^{II}$_6$L$_{24}$(H$_2$O)$_{12}$](NO$_3$)$_{12}$ (1), [Cr^{III}$_8$Ni^{II}$_6$L$_{24}$(MeCN)$_7$(H$_2$O)$_5$](ClO$_4$)$_{12}$ (2), and [Cr^{III}$_8$Ni^{II}$_6$L$_{24}$Cl$_{12}$] (3), which join the growing family of [M^{III}$_8$M^{II}$_6$] cubes constructed from [M^{III}L$_3$] and a variety of M^{II} salts (M^{III} = Cr, Fe; M^{II} = Cu, Co, Mn, Pd) [26–29].

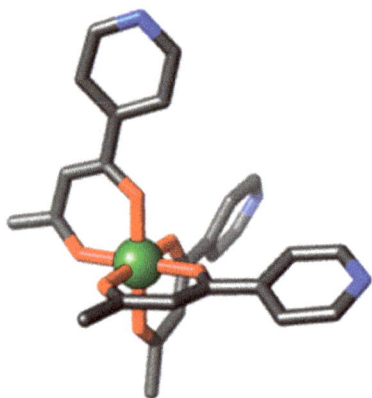

Figure 1. Molecular structure of [M^{III}L$_3$], where HL = 1-(4-pyridyl)butane-1,3-dione. Colour code: M^{III} = green, O = red, N = blue, C = grey. H-atoms have been omitted for clarity.

2. Materials and Methods

2.1. Synthesis

1-(4-pyridyl)butane-1,3-dione (HL) and the metalloligand [Cr^{III}L$_3$] were prepared by previously published procedures [26,31,32]. All reactions were carried out under aerobic conditions. Solvents and reagents were used as received from commercial suppliers. Caution: perchlorate salts of metal complexes with organic ligands are potentially explosive.

Synthesis of [Cr^{III}$_8$Ni^{II}$_6$L$_{24}$(H$_2$O)$_{12}$](NO$_3$)$_{12}$ (1). To a solution of [Cr^{III}L$_3$] (54 mg, 0.1 mmol) in 10 mL of dichloromethane, a solution of Ni(NO$_3$)$_2$·6H$_2$O (30 mg, 0.1 mmol) was added in 10 mL of methanol. The solution was stirred for 18 h before being filtered and allowed to stand. Dark orange X-ray quality crystals were obtained from the diffusion of diethyl ether into the mother liquor. Yield of (1) = 69%. Elemental analysis (%) calculated (found): C 46.16 (46.01) H 3.87 (3.78) N 8.97 (8.63).

Synthesis of [Cr^{III}$_8$Ni^{II}$_6$L$_{24}$(MeCN)$_7$(H$_2$O)$_5$](ClO$_4$)$_{12}$ (2). To a solution of [Cr^{III}L$_3$] (108 mg, 0.2 mmol) in 10 mL of dichloromethane, a solution of Ni(ClO$_4$)$_2$·6H$_2$O (73 mg, 0.2 mmol) was added in 10 mL of acetonitrile. The solution was stirred for 18 h before being filtered and allowed to stand. Brown X-ray quality crystals were obtained after 5 days from the diffusion of pentane into the mother liquor. Yield of (2) = 81%. Elemental analysis (%) calculated (found): C 44.34 (44.06) H 3.61 (3.59) N 6.97 (7.11).

Synthesis of [Cr^{III}$_8$Ni^{II}$_6$L$_{24}$Cl$_{12}$] (3). To a solution of [Cr^{III}L$_3$] (108 mg, 0.2 mmol) in 10 mL of dichloromethane, a solution of NiCl$_2$ (20 mg, 0.15 mmol) was added in 10 mL of tetrahydrofuran. The solution was stirred for 18 h before being filtered and allowed to stand. Brown X-ray quality crystals were obtained after room temperature evaporation of the mother liquor for 5 days. Yield of (3) = 58%. Elemental analysis (%) calculated (found): C 51.01 (50.79) H 3.81 (3.71) N 6.61 (6.68).

2.2. Crystallographic Details

Single-crystal X-ray diffraction data were collected for (1)–(3) at $T = 100$ K on a Rigaku AFC12 goniometer equipped with an enhanced sensitivity (HG) Saturn 724+ detector mounted at the window of an FR-E+ Superbright MoKα rotating anode generator with HF Varimax optics (70 µm focus) [33]. The CrysalisPro software package was used for instrument control, unit cell determination, and data reduction [34]. Due to very weak scattering power, single-crystal X-ray diffraction data for (1) and (2) were collected at $T = 30.15$ K using a synchrotron source ($\lambda = 0.6889$ Å) on the I19 beam line at Diamond Light Source on an undulator insertion device with a combination of double crystal monochromator, vertical and horizontal focussing mirrors, and a series of beam slits. The same software as above was used for data refinement. Crystals of all samples were sensitive to solvent loss, which resulted in crystal delamination and poor-quality X-ray diffraction data. To slow down crystal degradation, crystals of (1)–(3) were "cold-mounted" on MiTeGen Micromounts™ at $T = 203$ K using Sigma-Aldrich Fomblin Y® LVAC (3300 mol. wt.), with the X-Temp 2 crystal cooling system attached to the microscope [35]. This procedure protected crystal quality and permitted collection of usable X-ray data. Unit cell parameters in all cases were refined against all data. Crystal structures were solved using Intristic Phasing as implemented in SHELXT [36]. All non-hydrogen atoms were refined with anisotropic displacement parameters, and all hydrogen atoms were added at calculated positions and refined using a riding model with isotropic displacement parameters based on the equivalent isotropic displacement parameter (U_{eq}) of the parent atom. All three crystal structures contain large accessible voids and channels that are filled with diffuse electron density belonging to uncoordinated solvent, whose electron contribution was accounted for by the PLATON/SQUEEZE routine ((1) and (2)) [37], or by the SMTBX solvent masking routine, as implemented in OLEX2 software (3). To maintain reasonable molecular geometry, DFIX restraints were used in all three complexes.

Crystal Data for $[Cr^{III}_8Ni^{II}_6L_{24}(H_2O)_{12}](NO_3)_{12}$ (1). $C_{216}H_{216}Cr_8N_{24}Ni_6O_{60}$, $M_r = 4876.38$, monoclinic, $a = 25.754(3)$ Å, $b = 41.336(5)$ Å, $c = 43.217(5)$ Å, $\alpha = 90°$, $\beta = 90.6450(10)°$, $\gamma = 90°$, $V = 46,004(9)$ Å^3, $Z = 4$, $P2_1/n$, $D_c = 0.704$ g cm^{-3}, $\mu = 9.18$ mm^{-1}, $T = 100.15(10)$ K, 370,995 reflections measured, 81,102 unique ($R_{int} = 0.1902$), which were used in all calculations, wR_2 (all data) = 0.3687, and R_1 [$I > 2(I)$] = 0.1242. CCDC 1977309.

Crystal Data for $[Cr^{III}_8Ni^{II}_6L_{24}(MeCN)_7(H_2O)_5](ClO_4)_{12}$ (2). $C_{230}H_{218}Cr_8N_{31}Ni_6O_{54}$, $M_r = 5048.60$, monoclinic, $a = 25.788(6)$ Å, $b = 41.606(9)$ Å, $c = 45.869(11)$ Å, $\alpha = 90°$, $\beta = 90.785(2)°$, $\gamma = 90°$, $V = 49,210(20)$ Å^3, $Z = 4$, $P2_1/n$, $D_c = 0.681$ g cm^{-3}, $\mu = 0.412$ mm^{-1}, $T = 100.15(10)$ K, 391,278 reflections measured, 85,150 unique ($R_{int} = 0.2371$), which were used in all calculations, wR_2 (all data) = 0.4444, and R_1 [$I > 2(I)$] = 0.1521. CCDC 1977311.

Crystal Data for $[Cr^{III}_8Ni^{II}_6L_{24}Cl_{12}]$ (3). $C_{216}H_{192}Cl_{12}Cr_8N_{24}Ni_6O_{48}$, $M_r = 5085.58$, triclinic, $a = 28.171(16)$ Å, $b = 30.225(16)$ Å, $c = 32.40(2)$ Å, $\alpha = 72.27(6)°$, $\beta = 72.08(6)°$, $\gamma = 64.04(6)°$, $V = 22,417(27)$ Å^3, $Z = 2$, P-1, $D_c = 0.753$ g cm^{-3}, $\mu = 0.543$ mm^{-1}, $T = 100.0(1)$ K, 134,344 reflections measured, 66,105 unique ($R_{int} = 0.1446$), which were used in all calculations, wR_2 (all data) = 0.5544, and R_1 [$I > 2(I)$] = 0.1938. CCDC 1977312.

2.3. Magnetic and Spectroscopic Measurements

Direct current (dc) susceptibility and magnetisation data were measured on powdered, polycrystalline samples of (1) using a Quantum Design SQUID MPMS-XL magnetometer, operating between 1.8 and 300 K for dc applied magnetic fields ranging from 0 to 5 T. X-band EPR spectra were collected on powdered microcrystalline samples of (1) using a Bruker EMX spectrometer at the EPSRC UK National EPR Facility at The University of Manchester.

3. Results and Discussion

3.1. Structural Description

The heterometallic cubes $[Cr^{III}_8Ni^{II}_6L_{24}(H_2O)_{12}](NO_3)_{12}$ (1), $[Cr^{III}_8Ni^{II}_6L_{24}(MeCN)_7(H_2O)_5](ClO_4)_{12}$ (2), and $[Cr^{III}_8Ni^{II}_6L_{24}Cl_{12}]$ (3) were formed from the reaction of $[Cr^{III}L_3]$

with the corresponding Ni^{II} salt in CH_2Cl_2/MeOH, CH_2Cl_2/MeCN, and CH_2Cl_2/THF, respectively. All three structures (Figure 2) reveal a similar $[Cr^{III}{}_8Ni^{II}{}_6]$ cube-like metallic skeleton, with the eight Cr^{III} ions located at the corners and the six Ni^{II} ions located on the faces, approximately 1.4–2.3 Å above the Cr···Cr···Cr···Cr plane. The internal cavity volume of the cube is approximately 1400 Å^3.

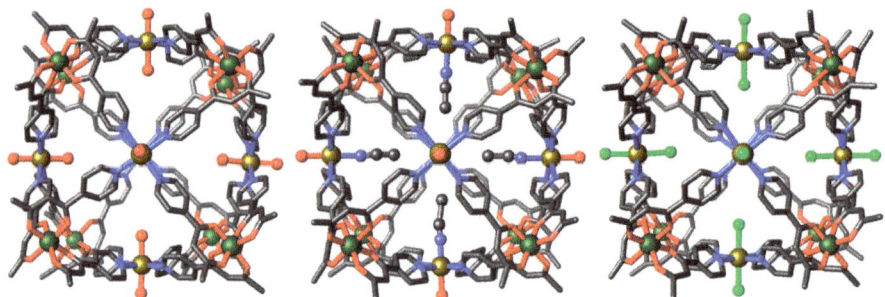

Figure 2. From left to right, molecular structures of (**1**), (**2**), and (**3**). Colour code: CrIII = green, NiII = yellow, O = red, N = blue, Cl = light green, C = grey. H-atoms have been omitted for clarity.

The Cr^{III} ions are all octahedral, possessing {CrO_6} coordination spheres with Cr^{III}-O distances between 1.9 and 2.2 Å, and *cis/trans* angles in the range 82.7–97°/171.3–179.4°, respectively. The equatorial positions of the octahedral Ni^{II} ions are occupied by four pyridyl donors from four distinct [$Cr^{III}L_3$] subunits, with Ni^{II}–N distances in the range 1.9–3.0 Å. For (**1**) and (**2**), the axial positions are occupied by twelve water and twelve acetonitrile/water molecules (Ni^{II}-O ≈2.1 Å; Ni^{II}-N ≈2.1 Å), respectively. The cubes are therefore cationic (12+). The charge balancing nitrate or perchlorate anions for (**1**) and (**2**) respectively, are located both within the central cavity of the cube and in the void spaces between cubes. In contrast to (**1**) and (**2**), complex (**3**) is neutral, with the axial positions of the Ni^{II} ions occupied by chloride anions (Ni^{II}-Cl ≈2.7 Å).

There are several close intermolecular contacts (Figure 3) between the cages in the extended structures of (**1**)–(**3**). In (**1**), the closest inter-cluster contact is between the aromatic protons of the pyridyl group and the O-atom (2.3 Å) of a neighbouring L$^-$ ligand. In (**2**) and (**3**), the closest contact is between the protons of the metalloligand methyl group, and the O-atom of a neighbouring L$^-$ ligand (2.3 Å) and the protons of a neighbouring methyl group (2.3 Å), respectively. Several other close inter-cluster contacts between neighbouring cubes exist, for example: Ar-H···O ≈2.5 Å and C-H···O ≈2.7 Å for (**1**), H_2C-H···O ≈2.5 Å and H_2O···H-CH_2 ≈2.7 Å for (**2**), and Ar-H···Cl ≈2.7 Å and C-H···Cl ≈2.8 Å for (**3**).

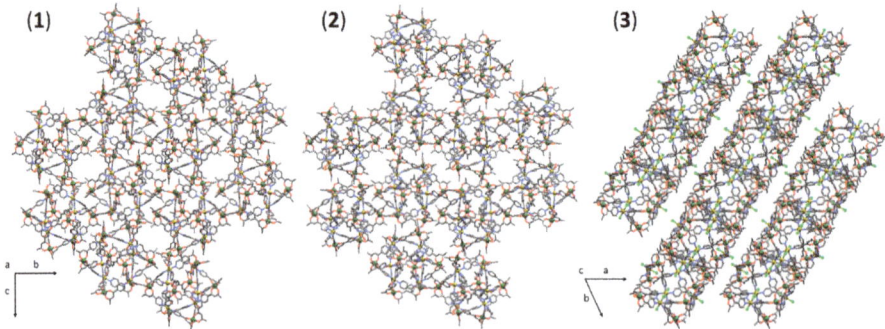

Figure 3. Packing diagrams of (**1**)–(**3**) viewed down the *a*-, *a*-, and *c*-axis, respectively. Colour code as in Figure 1.

3.2. Magnetic Properties

As complexes (1)–(3) are structurally analogous, and for the sake of brevity, we discuss only the behaviour of a representative example, complex (1). The dc molar magnetic susceptibility, χ_M, of a polycrystalline sample of (1) was measured in an applied magnetic field, B, of 0.1 T, over the 2–300 K temperature, T, range. The experimental results are shown in Figure 4 in the form of the $\chi_M T$ product versus temperature, where $\chi_M = M/B$, and M is the magnetization of the sample. Due to the loss of lattice solvent during the evacuation of the sample chamber of the SQUID magnetometer, leading to an uncertainty in the molar mass of the measured sample, the $T = 300$ K $\chi_M T$ product of (1) was scaled to 21.00 cm^3 mol^{-1} K, the expected value from the sum of Curie constants for a [Cr$^{III}_8$Ni$^{II}_6$] unit, with $g_{Cr} = g_{Ni} = 2.0$, where g_{Cr} and g_{Ni} are the g-factors of Cr^{III} and Ni^{II}, respectively. Note that this rescaled value has a maximum deviation of 15% from the unscaled data.

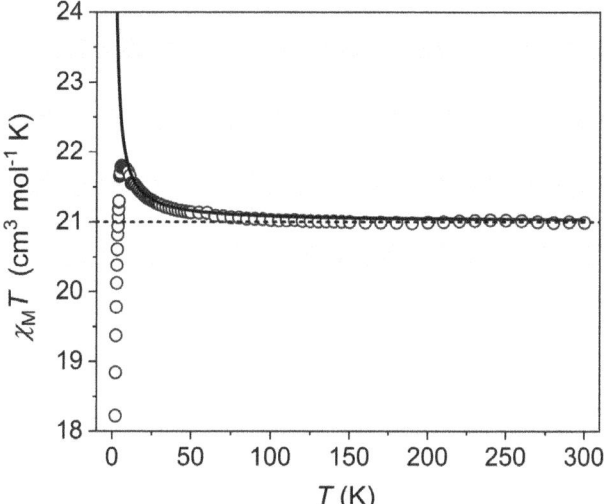

Figure 4. Plot of $\chi_M T$ (open circles) versus T for complex (1), with the sum of the Curie constants of the uncorrelated ions and the best-fit data represented by the dashed and solid lines, respectively.

Upon cooling, the value of $\chi_M T$ remains essentially constant to approximately $T = 75$ K, where it begins to increase, reaching a maximum of 21.8 cm^3 mol^{-1} K at $T = 6$ K. Below this temperature, $\chi_M T$ falls rapidly to a minimum value of 18.5 cm^3 mol^{-1} K at $T = 2.0$ K. The behaviour is suggestive of weak ferromagnetic exchange between the Cr^{III} and Ni^{II} ions, with the decrease in $\chi_M T$ below 6 K attributed to intermolecular antiferromagnetic exchange interactions, and/or zero-field splitting (zfs) effects primarily associated with the Ni^{II} ions. Quantitative analysis of the susceptibility data via standard matrix diagonalization techniques is non-trivial due to the large nuclearity of the cluster and the associated enormous dimensions of the spin-Hamiltonian matrices. Even the total spin (S) block matrices used in approaches based on Irreducible Tensor Operator algebra are of larger dimensions than what is realistic for exact numerical matrix diagonalization. Previously, we reported the use of computational techniques, known in theoretical nuclear physics as statistical spectroscopy [38], to analyse the structurally similar [M$^{III}_8$M$^{II}_6$]$^{n+}$ (M^{III} = Cr, Fe; M^{II} = Co, Cu, Ni; n = 0–12) cubes [26–28]. We now extend this methodology to quantify the exchange interactions present in (1). Due to the fact that the influence of the zfs of the Ni^{II} ions will mainly affect the measured properties at low temperatures, the use of the isotropic spin-Hamiltonian (1) is sufficient to model the exchange interactions between Cr^{III} and Ni^{II} ions in the T = 300–6 K region:

$$\hat{H}_{iso} = -2J_{Cr-Ni} \sum_{allCr-Nipairs} \hat{S}_{Cr} \cdot \hat{S}_{Ni} + \mu_B B g \sum_i \hat{S}_i^z \quad (1)$$

with i running over all constitutive metal centres, g is the isotropic g-factor, $\hat{S}$ is a spin-operator, J_{Cr-M} is the isotropic exchange parameter between Cr^{III} and M^{II} centres, and μ_B is the Bohr magneton. We assume common g-factors for both Cr^{III} and Ni^{II} ($g_{Cr} = g_{Ni} = 2.0$) since the 300 K $\chi_M T$ product of (1) was scaled to the sum of its Curie constants, as explained above. We neglect any J_{Cr-Cr} and J_{Ni-Ni} terms as these centres are not connected as first neighbours. Using Hamiltonian (1), J_{Cr-Ni} was determined to be +0.045 cm^{-1}. Variable-temperature and variable-field (VTVB) magnetization studies of (1) collected in the T = 2–10 K and B = 0.5–5 T temperature and field ranges (Figure 5) are consistent with this picture. M reaches a value of 32.8 μ_B at B = 5 T and T = 2 K, approaching the saturation value of 36 μ_B, consistent with relatively small exchange-induced splittings that lead to the m_S = 18 projection of the S = 18 total spin state, being the ground state at the highest measured magnetic field. The weak ferromagnetic exchange between the d^3 Cr^{III} ions and the d^8 Ni^{II} ions is as one would expect, mediated via the 1-(4-pyridyl)butane-1,3-dione ligand [26–28].

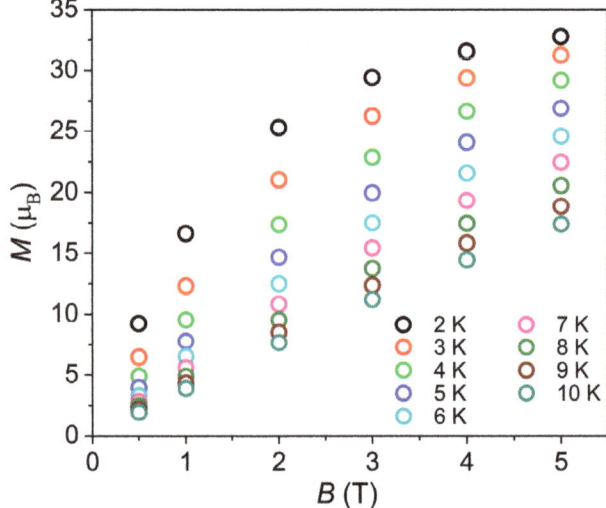

Figure 5. VTVB magnetisation data for (1) in the temperature and field ranges T = 2–10 K and B = 0.5–5 T.

3.3. EPR Spectroscopy

X-band EPR spectra of a powdered sample of (1) at 5 and 10 K are dominated by a feature at ca. 2 kG (Figure 6). This is similar to spectra from the isolated [$Cr^{III}L_3$] complex, and related [$Cr^{III}_8M^{II}_6$] and [$Cr^{III}_2M^{II}_3$] species [26,29], and arises from the Cr^{III} (S = 3/2) ions with a near-axial zero-field splitting of $|D_{Cr}|$ ca. 0.5–0.6 cm^{-1}. This is only consistent with a weak exchange interaction $|J_{Cr-Ni}|$ with respect to $|D_{Cr}|$, and hence consistent with the magnetic data. There are no clear features arising from the Ni^{II} (S = 1) ions, which implies that $|D_{Ni}|$ must be much larger than the microwave energy. We also observed this for a related [$Fe^{III}_8M^{II}_6$] cube, which only showed EPR features due to Fe^{III} [28]. This is consistent with $|D_{Ni}|$ values of 5–10 cm^{-1} determined from magnetization studies of isolated [Ni^{II}(pyridine)$_4$X$_2$] complexes [39], and with high-field EPR studies of Ni^{II} complexes with mixed N,O-donor sets [40].

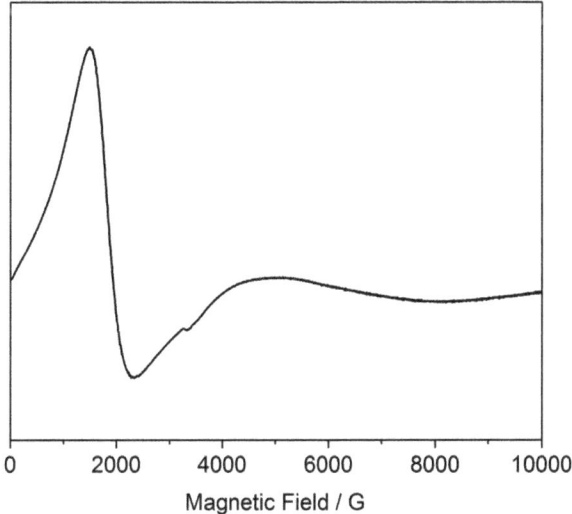

Figure 6. X-band (*ca.* 9.4 GHz) EPR spectrum of a powdered sample of (**1**) at 5 K.

4. Conclusions

We have shown that the modular self-assembly of [M^{III}L$_3$] metalloligands with simple M^{II} salts can be exploited to construct large heterometallic coordination compounds of Cr^{III} and Ni^{II}. Compounds (**1**)–(**3**) join a growing family of [M$^{III}_8$M$^{II}_6$] cubes, where M^{III} = Cr and Fe and M^{II} = Cu, Co, Mn, Pd, and Ni. The ability to build families of isostructural complexes containing different combinations of paramagnetic (and diamagnetic) metal centres aids the qualitative and quantitative understanding of magnetic properties and the underlying structural parameters that govern behaviour. Examples of large, heterometallic cages in which the 3*d* metal ions can be exchanged with other 3*d* metal ions are extremely rare.

Magnetic susceptibility and magnetization data show the presence of weak ferromagnetic exchange between the Cr^{III} and Ni^{II} ions, with J_{Cr-Ni} = +0.045 cm^{-1}. EPR spectroscopy is consistent with the exchange interactions being much weaker than the zero-field splittings of both the Cr^{III} and Ni^{II} ions.

Author Contributions: E.K.B., P.J.L., H.M.O. and S.S. designed the study. H.M.O., S.S. and A.J.S. performed the synthetic work. M.B.P., W.T.K. and S.J.C. performed the crystallographic work. H.W. and S.P. performed the magnetometry measurements and associated analysis. N.F.C. and E.J.L.M. performed the EPR spectroscopic experiments and associated analysis. H.M.O. and E.K.B. wrote the paper with input from all authors. All authors have read and agreed to the published version of the manuscript.

Funding: This research was funded by the EPSRC (UK), grant numbers EP/N01331X/1 and EP/P025986/1, and by the VILLUM FONDEN (Denmark), grant 13376. We also thank the EPSRC for funding the UK National EPR Facility.

Institutional Review Board Statement: Not applicable.

Informed Consent Statement: Not applicable.

Data Availability Statement: The data presented in this study are available on request from the corresponding authors.

Conflicts of Interest: The authors declare that there is no conflict of interest.

References

1. Barton, B.E.; Whaley, C.M.; Rauchfuss, T.B.; Gray, D.L. Nickel−Iron Dithiolato Hydrides Relevant to the [NiFe]-Hydrogenase Active Site. *J. Am. Chem. Soc.* **2009**, *131*, 6942–6943. [CrossRef]
2. Canaguier, S.; Field, M.; Oudart, Y.; Pécaut, J.; Fontecave, M.; Artero, V. A structural and functional mimic of the active site of NiFe hydrogenases. *Chem. Commun.* **2010**, *46*, 5876–5878. [CrossRef]
3. Buchwalter, P.; Rosé, J.; Braunstein, P. Multimetallic Catalysis Based on Heterometallic Complexes and Clusters. *Chem. Rev.* **2015**, *115*, 28–126. [CrossRef]
4. Rice, A.M.; Leith, G.A.; Ejegbavwo, O.A.; Dolgopolova, E.A.; Shustova, N.B. Heterometallic Metal–Organic Frameworks (MOFs): The Advent of Improving the Energy Landscape. *ACS Energy Lett.* **2019**, *4*, 1938–1946. [CrossRef]
5. Zhang, Y.Y.; Gao, W.X.; Lin, L.; Jin, G.X. Recent advances in the construction and applications of heterometallic macrocycles and cages. *Coord. Chem. Rev.* **2017**, *344*, 323–344. [CrossRef]
6. Vigato, P.A.; Tamburini, S. Advances in acyclic compartmental ligands and related complexes. *Coord. Chem. Rev.* **2008**, *252*, 1871–1995. [CrossRef]
7. Ferlay, S.; Mallah, T.; Ouahès, R.; Veillet, P.; Verdaguer, M. A room-temperature organometallic magnet based on Prussian blue. *Nature* **1995**, *378*, 701–703. [CrossRef]
8. Alexandru, M.-G.; Visinescu, D.; Shova, S.; Lloret, F.; Julve, M.; Andruh, M. Two-Dimensional Coordination Polymers Constructed by [$Ni^{II}Ln^{III}$] Nodes and [$W^{IV}(bpy)(CN)_6$]$^{2-}$ Spacers: A Network of [$Ni^{II}Dy^{III}$] Single Molecule Magnets. *Inorg. Chem.* **2013**, *52*, 11627–11637. [CrossRef]
9. Yao, M.-X.; Wei, Z.-Y.; Gu, Z.-G.; Zheng, Q.; Xu, Y.; Zuo, J.-L. Syntheses, Structures, and Magnetic Properties of Low-Dimensional Heterometallic Complexes Based on the Versatile Building Block [(Tp)Cr(CN)$_3$]$^-$. *Inorg. Chem.* **2011**, *50*, 8636–8644. [CrossRef]
10. Liu, W.; Wang, C.; Li, Y.; Zuo, J.; You, X. Structural and Magnetic Studies on Cyano-Bridged Rectangular Fe$_2$M$_2$ (M. = Cu, Ni) Clusters. *Inorg. Chem.* **2006**, *45*, 10058–10065. [CrossRef]
11. Beltran, L.M.C.; Long, J.R. Directed Assembly of Metal−Cyanide Cluster Magnets. *Acc. Chem. Res.* **2005**, *38*, 325–334. [CrossRef] [PubMed]
12. Rebilly, J.N.; Mallah, T. Synthesis of Single-Molecule Magnets Using Metallocyanates. *Single-Mol. Magn. Relat. Phenom.* **2006**, 103–131. [CrossRef]
13. Evangelisti, M.; Brechin, E.K. Recipes for enhanced molecular cooling. *Dalton Trans.* **2010**, *39*, 4672–4676. [CrossRef] [PubMed]
14. Schnack, J. Effects of frustration on magnetic molecules: A survey from Olivier Kahn until today. *Dalton Trans.* **2010**, *39*, 4677–4686. [CrossRef] [PubMed]
15. Milios, C.J.; Winpenny, R.E.P. Cluster-Based Single-Molecule Magnets. *Mol. Nanomagnets Relat. Phenom.* **2015**, 1–109. [CrossRef]
16. Coulon, C.; Pianet, V.; Urdampilleta, M.; Clérac, R. Single-Chain Magnets and Related Systems. *Mol. Nanomagnets Relat. Phenom.* **2015**, 143–184. [CrossRef]
17. Gaita-Ariño, A.; Luis, F.; Hill, S.; Coronado, E. Molecular spins for quantum computation. *Nat. Chem.* **2019**, *11*, 301–309. [CrossRef]
18. Larsen, F.K.; McInnes, E.J.L.; Mkami, H.E.; Overgaard, J.; Piligkos, S.; Rajaraman, G.; Rentschler, E.; Smith, A.A.; Boote, V.; Jennings, M.; et al. Synthesis and Characterization of Heterometallic {Cr$_7$M} Wheels. *Angew. Chem. Int. Ed.* **2003**, *115*, 105–109. [CrossRef]
19. Garlatti, E.; Guidi, T.; Ansbro, S.; Santini, P.; Amoretti, G.; Ollivier, J.; Mutka, H.; Timco, G.; Vitorica-Yrezabal, I.J.; Whitehead, G.F.S.; et al. Portraying entanglement between molecular qubits with four-dimensional inelastic neutron scattering. *Nat. Commun.* **2017**, *8*, 14543. [CrossRef]
20. Timco, G.A.; McInnes, E.J.L.; Pritchard, R.G.; Tuna, F.; Winpenny, R.E.P. Heterometallic Rings Made From Chromium Stick Together Easily. *Angew. Chem. Int. Ed.* **2008**, *47*, 9681–9684. [CrossRef]
21. Timco, G.A.; Batsanov, A.S.; Larsen, F.K.; Muryn, C.A.; Overgaard, J.; Teat, S.J.; Winpenny, R.E.P. Influencing the nuclearity and constitution of heterometallic rings via templates. *Chem. Commun.* **2005**, 3649–3651. [CrossRef] [PubMed]
22. Fraser, H.W.L.; Nichol, G.S.; Uhrín, D.; Nielsen, U.G.; Evangelisti, M.; Schnack, J.; Brechin, E. Order in disorder: Solution and solid-state studies of [$M^{III}_2M^{II}_5$] wheels (M^{III} = Cr, Al; M^{II} = Ni, Zn). *Dalton Trans.* **2018**, *47*, 11834–11842. [CrossRef] [PubMed]
23. Kakaroni, F.E.; Collet, A.; Sakellari, E.; Tzimopoulos, D.I.; Siczek, M.; Lis, T.; Murri, M.; Milios, C.J. Constructing Cr^{III}-centered heterometallic complexes: [$Ni^{II}_6Cr^{III}$] and [$Co^{II}_6Cr^{III}$] wheels. *Dalton Trans.* **2018**, *47*, 58–61. [CrossRef] [PubMed]
24. Manole, O.S.; Batsanov, A.S.; Struchkov, Y.T.; Timko, G.A.; Synzheryan, L.D.; Gerbeleu, N.V. Synthesis and crystalline-structure of pentanuclear heterometallic pivaloylacetylacetonate complexes. *Koord. Khim.* **1994**, *20*, 231–237.
25. Heath, S.L.; Laye, R.H.; Muryn, C.A.; Lima, N.; Sessoli, R.; Shaw, R.; Teat, S.J.; Timco, G.A.; Winpenny, R.E.P. Templating Open- and Closed-Chain Structures around Metal Complexes of Macrocycles. *Angew. Chem. Int. Ed.* **2004**, *43*, 6132–6135. [CrossRef]
26. Sanz, S.; O'Connor, H.M.; Pineda, E.M.; Pedersen, K.S.; Nichol, G.S.; Mønsted, O.; Weihe, H.; Piligkos, S.; McInnes, E.J.L.; Lusby, P.J.; et al. [$Cr^{III}_8M^{II}_6$]$^{12+}$ Coordination Cubes (M^{II}=Cu, Co). *Angew. Chem. Int. Ed.* **2015**, *54*, 6761–6764. [CrossRef]
27. Sanz, S.; O'Connor, H.M.; Comar, P.; Baldansuren, F.T.A.; Pitak, M.B.; Coles, S.J.; Weihe, H.; Chilton, N.F.; McInnes, E.J.L.; Lusby, P.J.; et al. Modular [$Fe^{III}_8M^{II}_6$]$^{n+}$ (MII = Pd, Co, Ni, Cu) Coordination Cages. *Inorg. Chem.* **2018**, *57*, 3500–3506. [CrossRef]
28. O'Connor, H.M.; Sanz, S.; Pitak, M.B.; Coles, S.J.; Nichol, G.S.; Piligkos, S.; Lusby, P.J.; Brechin, E.K. [$Cr^{III}_8M^{II}_6$]$^{n+}$ (M^{II} = Cu, Co) face-centred, metallosupramolecular cubes. *CrystEngComm* **2016**, *18*, 4914–4920. [CrossRef]

29. Sanz, S.; O'Connor, H.M.; Martí-Centelles, V.; Comar, P.; Pitak, M.B.; Coles, S.J.; Lorusso, G.; Palacios, E.; Evangelisti, M.; Baldansuren, F.T.A.; et al. [M$^{III}_2$M$^{II}_3$]$^{n+}$ trigonal bipyramidal cages based on diamagnetic and paramagnetic metalloligands. *Chem. Sci.* **2017**, *8*, 5526–5535. [CrossRef]
30. Chakrabarty, R.; Mukherjee, P.S.; Stang, P.J. Supramolecular Coordination: Self-Assembly of Finite Two- and Three-Dimensional Ensembles. *Chem. Rev.* **2011**, *111*, 6810–6918. [CrossRef]
31. Wu, H.-B.; Wang, Q.-M. Construction of Heterometallic Cages with Tripodal Metalloligands. *Angew. Chem. Int. Ed.* **2009**, *48*, 7343–7345. [CrossRef] [PubMed]
32. Singh, B.; Lesher, G.Y.; Pluncket, K.C.; Pagani, E.D.; Bode, D.C.; Bentley, R.G.; Connell, M.J.; Hamel, L.T.; Silver, P.J. Novel cAMP PDE III inhibitors: 1,6-naphthyridin-2(1H)-ones. *J. Med. Chem.* **1992**, *35*, 4858–4865. [CrossRef] [PubMed]
33. Coles, S.J.; Gale, P.A. Changing and challenging times for service crystallography. *Chem. Sci.* **2011**, *3*, 683–689. [CrossRef]
34. Rigaku, O.; CrysAlis, P.R.O.; Rigaku Oxford Diffraction. CrysAlisPro. 2016. Available online: https://www.rigaku.com/zh-hans/products/smc/crysalis (accessed on 18 January 2021).
35. Kottke, T.; Stalke, D. Crystal handling at low temperatures. *J. Appl. Crystallogr.* **1993**, *26*, 615–619. [CrossRef]
36. Sheldrick, G.M. SHELXT—Integrated space-group and crystal-structure determination. *Acta Cryst.* **2015**, *71*, 3–8. [CrossRef]
37. Dolomanov, O.V.; Blake, A.J.; Champness, N.R.; Schröder, M. OLEX: New software for visualization and analysis of extended crystal structures. *J. Appl. Crystallogr.* **2003**, *36*, 1283–1284. [CrossRef]
38. Wong, S.S.M. *Nuclear Statistical Spectroscopy*; Oxford University Press: Oxford, UK; Clarendon Press: Oxford, UK, 1986.
39. Otieno, T.; Thompson, R.C. Antiferromagnetism and metamagnetism in 1,4-diazine and pyridine complexes of nickel(II). *Can. J. Chem.* **1995**, *73*, 275–283. [CrossRef]
40. Krzystek, J.; Ozarowski, A.; Telser, J. Multi-frequency, high-field EPR as a powerful tool to accurately determine zero-field splitting in high-spin transition metal coordination complexes. *Coord. Chem. Rev.* **2006**, *250*, 2308–2324. [CrossRef]

Article

Synthesis and Single Crystal Structures of *N*-Substituted Benzamides and Their Chemoselective Selenation/Reduction Derivatives

Guoxiong Hua [1], Cameron L. Carpenter-Warren [1], David B. Cordes [1], Alexandra M. Z. Slawin [1,†] and J. Derek Woollins [1,2,*]

[1] EaStCHEM School of Chemistry, University of St Andrews, St Andrews Fife KY16 9ST, UK; gh15@st-andrews.ac.uk (G.H.); clcw@st-andrews.ac.uk (C.L.C.-W.); dbc21@st-andrews.ac.uk (D.B.C.); amzs@st-andrews.ac.uk (A.M.Z.S.)
[2] Department of Chemistry, Khalifa University, Abu Dhabi 127788, United Arab Emirates
* Correspondence: jdw3@st-andrews.ac.uk; Tel./Fax: +44-1334-463861
† Dedication: The co-authors dedicate this paper to Prof. Slawin on the occasion of her 60th birthday.

Abstract: A series of *N*-aryl-*N*-(2-oxo-2-arylethyl) benzamides and cinnamides has been prepared. The reaction of the benzamides with Woollins' reagent, a highly efficient chemoselective selenation/reduction reagent, gave the corresponding *N*-aryl-*N*-(arylenethyl) benzoselenoamides in good yields. Five representative single crystal X-ray structures are discussed.

Keywords: *N*-Substituted Benzamides; Woollins' reagent; selenation reagent; reduction reagent; single crystal X-ray structures

1. Introduction

2,4-Bis (phenyl)-1,3-diselenadiphosphetane-2,4-diselenide (Woollins' reagent, WR) has played a role in synthetic chemistry in the past two decades [1–13]. It has been successfully applied as an efficient building unit to synthesize a series of eight-, nine-, and ten-membered selenophosphorus heterocycles with P-Se-Se-P linkage [14], as well as unique octaselenocyclododecane with four carbon atoms and eight selenium atoms in this twelve-membered cycle [15]. Another attractive application has been that it acts as a highly chemoselective reagent, e.g., the reduction of a wide range of 1,4-enediones and 1,4-ynediones in methanol led to saturated 1,4-diketones [16] and the selective reduction of the double bond of 2-α,β-unsaturated thiazo- and selenazolidinones gave the corresponding saturated heterocycles [17]. Woollins' reagent has also been used as a reducing agent to transfer porpholactone into dihydroporpholactone or into adjacent-tetrahydroporpholactone [18].

Organoselenium compounds have received growing attention during the last decades due to their importance as useful precursors in synthetic chemistry [19,20], as new synthetic materials [20], and their biological and medicinal significance [21]. To continue our interest in the chemistry of Woollins' reagent towards various organic substrates, we report an investigation on the use of WR as a selenation/reduction reagent for transferring *N*-aryl-*N*-(2-oxo-2-arylethyl) benzamides into the corresponding *N*-aryl-*N*-(arylenethyl) benzoselenoamides.

2. Results and Discussion

2.1. Synthesis and Characterization

The synthesis of anilinoacetophenones **1–3**, *N*-aryl-*N*-arylamidoacetophenones **4–6** and *N*-aryl-*N*-cinnamidoacetophenones **7–9** was carried out using a modified literature method [22]. The reaction of anilines and an equivalent of the appropriate bromoacetophenones in dry acetonitrile at room temperature gave anilinoacetophenones **1–3** in 81–87% yields, respectively. Anilinoacetophenones **1** and **2** are new compounds, while **3** is a known compound, prepared previously by a similar method [23,24]; however, its single-crystal X-ray structure has not been reported previously. Acylation of anilinoacetophenones **1–3** with the appropriate acid chlorides in 1,2-dichloroethane at reflux led to the N-aryl-N-arylamidoacetophenones **4–6** and N-aryl-N-cinnamidoacetophenones **7–9** in 76–91% yields, as shown in Scheme 1. All the new compounds were characterized by standard analytical and spectroscopic techniques. **1–9** show the anticipated [M]$^+$ or [M + H]$^+$ peak in their mass spectra, satisfactory accurate mass measurements, and appropriate isotopic distributions; the ^{1}H NMR spectra display all the characteristic peaks of the phenyl backbones in compounds and the characteristic peaks of the NH group in compound **1–3**. The ^{13}C NMR spectra of compounds **1–9** display the characteristic signals of the C=O groups.

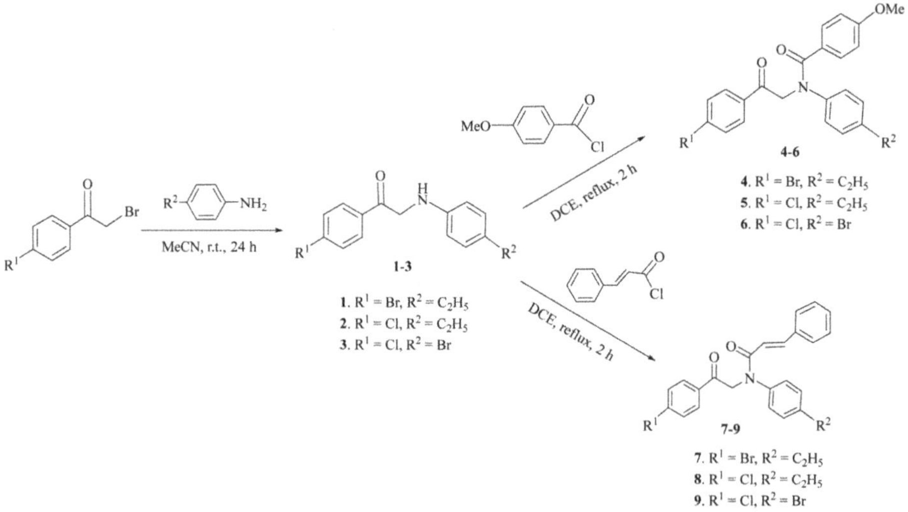

Scheme 1. Synthesis of anilinoacetophenones **1–3**, N-aryl-N-arylamidoacetophenones **4–6** and N-aryl-N-cinnamidoacetophenones **7–9**.

Selenation of N-aryl-N-arylamidoacetophenones **4–6** by WR gave rise to N-aryl-N-arylethylbenzoselenoamides **10–12** in 50%, 46% and 40% yields, respectively, rather than the expected 1,3-selenazole products (Scheme 2). One C=O group has been converted to C=Se and the other reduced to CH$_2$ to give the final product N-Aryl-N-arylethylbenzoselenoamides **10–12**. It is well known that WR is an efficient chemoselective reduction agent for diketones [16], α,β-unsaturated thioazo and selenoazolidinones [6]. Based on the literature research and our findings, a possible mechanism for the selective reduction of N-substituted-N-phenylamidoacetophenones **4–6** is broadly similar to that of NaSeH and LiSeH as selective reducing agents of α,β-unsaturated carbonyl compounds [13] and of PhSe-SePh as a reducing agent for electron deficient olefins [25], and it is probable that the reduction proceed through a Micheal reaction [26–28].

Reacting N-aryl-N-cinnamidoacetophenones **7–9** with WR under similar reaction conditions did not lead to a reaction, with the starting materials being recovered (Scheme 3). We speculate that the extra C=C bond in N-substituted-N-phenylamidoacetophenones **7–9** which gives a conjugated structure, may be more stable and robust than N-substituted-N-phenylamidoacetophenones **4–6** towards WR.

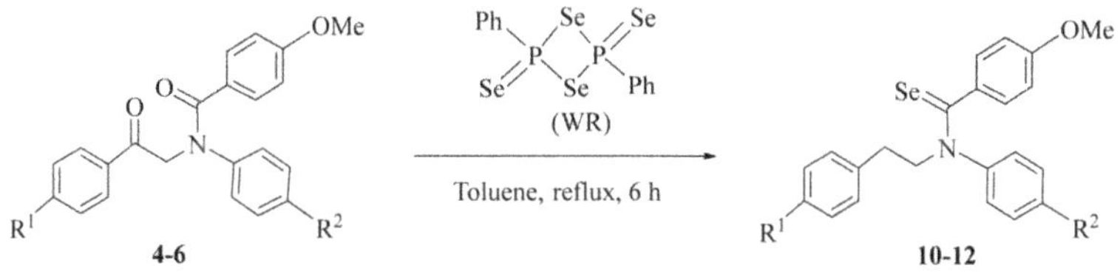

Scheme 2. Selenation of *N*-aryl-*N*-phenylamidoacetophenones **4–6**.

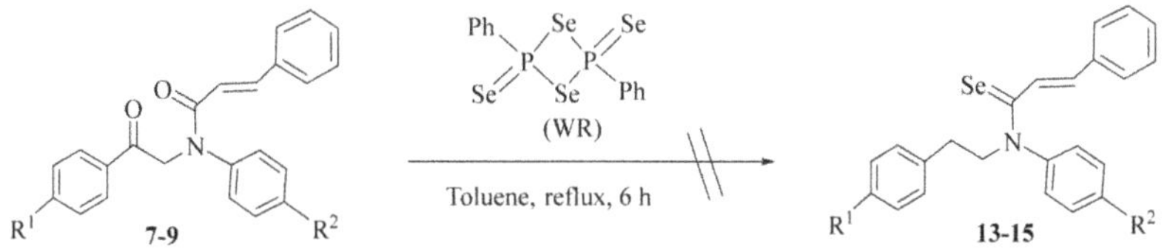

4. R^1 = Br, R^2 = C$_2$H$_5$
5 R^1 = Cl, R^2 = C$_2$H$_5$
6. R^1 = Cl, R^2 = Br

10, R^1 = Br, R^2 = C$_2$H$_5$
11, R^1 = Cl, R^2 = C$_2$H$_5$
12, R^1 = Cl, R^2 = Br

7. R^1 = Br, R^2 = C$_2$H$_5$
8. R^1 = Cl, R^2 = C$_2$H$_5$
9. R^1 = Cl, R^2 = Br

13. R^1 = Br, R^2 = C$_2$H$_5$
14. R^1 = Cl, R^2 = C$_2$H$_5$
15. R^1 = Cl, R^2 = Br

Scheme 3. Attempted selenation of *N*-aryl-*N*-cinnamidoacetophenones **7–9**.

The new selenium derivatives **10–12** are quite stable both as solids and in solution, in air and moist atmospheres, and are soluble in common organic solvents. They show the anticipated molecular ion peaks [M + H]$^+$ in their CI spectra and [M]$^+$ in their EI spectra, and satisfactory accurate mass measurements (EI). All the characteristic peaks of the phenyl backbones were found, and the characteristic peaks of the NH group disappeared in their ^{1}H NMR spectra. Their ^{13}C NMR spectra all show the normal signals for the C=Se groups (δ_C, 204.8–206.8 ppm). Their ^{77}Se NMR spectra exhibit singlet signals at δ_{Se} = 598.4, 601.5 and 601.4 ppm for **10–12**, respectively.

2.2. Single Crystal Structure Analysis

Single crystals of **3, 7, 10–12** suitable for X-ray crystallographic analysis were grown by slow evaporation of dichloromethane solutions of the compound in air at room temperature. Selected crystallographic data are given in Table 1 and the resulting molecular structures are illustrated in Figures 1 and 2.

The molecular structure of anilinoacetophenone **3** (Figure 1) shows a planar arrangement, with a mean deviation of non-hydrogen atoms from the plane of 0.047 Å. Adjacent molecules of **3** interact to form hydrogen-bonded dimers via a pair of NH···O hydrogen bonds at a H···O distance of 2.59(4) Å, and N···O separation of 3.360(5) Å. The structure of the *N*-aryl-*N*-cinnamidoacetophenone **7** shows the compound in a *twisted-T* conformation (Figure 1). As expected, the acetophenone group retains its planarity (mean deviation of non-hydrogen atoms from the plane of 0.004 Å). Meanwhile, the other two phenyl ring planes are twisted out of the acetophenone plane, with angles between planes of 71.36 and 50.74° for C(10)-C(15) and C(21)-C(26), respectively.

Table 1. Details of the X-ray data collections and refinements for compounds **3, 7, 10–12**.

Compound	3	7	10	11	12
Formula	$C_{14}H_{11}BrClNO$	$C_{25}H_{22}ClNO_2$	$C_{24}H_{24}BrNOSe$	$C_{24}H_{24}ClNOSe$	$C_{22}H_{19}BrClNOSe$
M	324.60	403.91	501.32	456.87	507.72
Temperature/K	93	93	173	173	173
Crystal system	triclinic	monoclinic	triclinic	triclinic	triclinic
Space group	$P\bar{1}$	$P2_1/c$	$P\bar{1}$	$P\bar{1}$	$P\bar{1}$
a/Å	5.718 (3)	6.00023 (18)	9.2004 (17)	9.1375 (12)	9.98830 (10)
b/Å	7.296 (4)	22.8404 (6)	10.541 (3)	10.6864 (14)	10.0160 (2)
c/Å	15.461 (8)	14.7270 (4)	11.146 (3)	11.1485 (15)	12.3778 (18)
α	95.472 (13)		101.842 (4)	103.062 (4)	99.855 (14)
β	90.3200 (10)	93.111 (3)	94.351 (6)	95.439 (3)	110.822 (8)
γ	906.126 (12)		90.478 (6)	91.412 (4)	108.988 (12)
U/Å^3	638.3 (6)	2015.33 (10)	1054.6 (5)	1054.5 (2)	1035.2 (2)
Z	2	4	2	2	2
μ/mm^{-1}	3.424	0.211	3.695	1.922	3.890
Reflections collected	8491	34,715	14,106	21,665	21,452
Independent reflections	2296	4515	3785	3832	3776
R_{int}	0.0769	0.0669	0.0448	0.0283	0.0314
R_1 [$I > 2\sigma(I)$]	0.0529	0.0341	0.0240	0.0304	0.0207
wR_2	0.1479	0.0883	0.0768	0.0973	0.0569

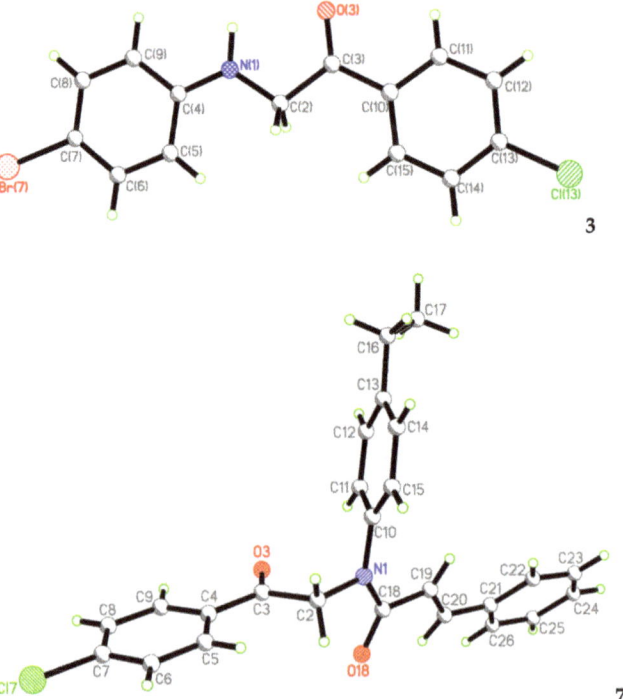

Figure 1. Single crystal X-ray structures of anilinoacetophenone **3** and N-aryl-N-cinnamidoacetophenone **7**.

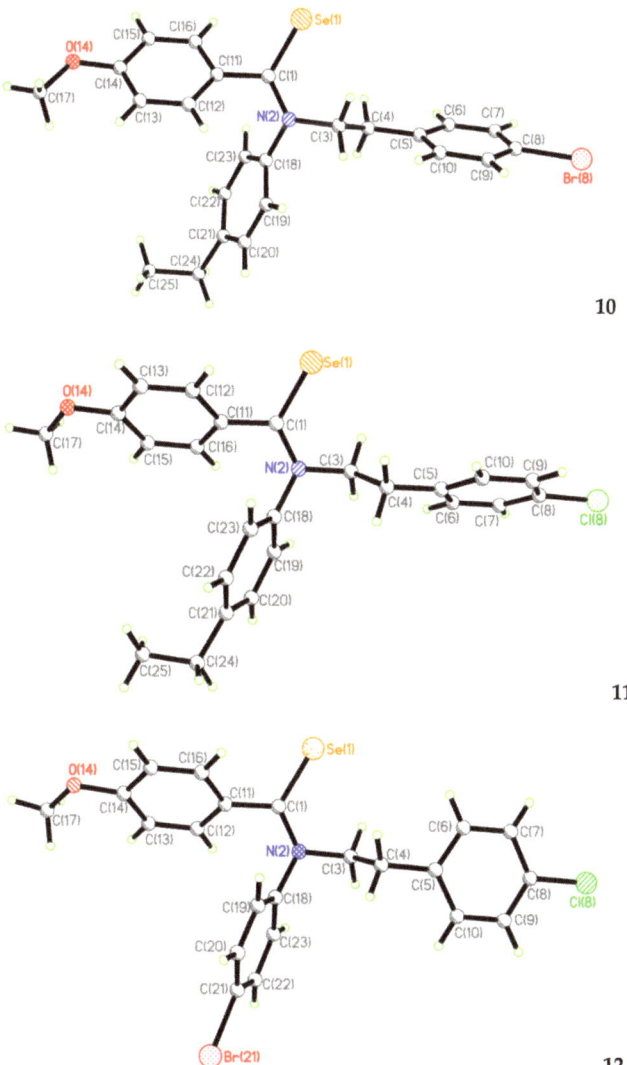

Figure 2. Single crystal X-ray structures of selenoamides **10**, **11** and **12**.

The X-ray structures of **10**, **11** and **12** are depicted in Figure 2, each displaying similar *twisted-T* conformations. The three aryl rings [C(5)-C(10) (ring 1), C(11)-C(16) (ring 2) and C(18)-C(23) (ring 3) in all three compounds] are all twisted with respect to each other, the angles between ring planes being 28.92, 69.02 and 62.00° in **10**, 18.92, 62.23 and 64.51° in **11** and 55.76, 61.91 and 68.23° in **12** for rings 1 and 2, 1 and 3, and 2 and 3, respectively. This pattern of ring twists puts all the rings out of the plane of the selenoamide, except for ring 1 in compound **12**, which is near parallel to the selenoamide, inclined at 9.24°. The C=Se double bond lengths [1.832(3) Å in **10**, 1.833(2) Å in **11** and 1.8264(19) Å in **12**] are very similar, falling at the middle of the range of C=Se bonds in known selenoamides [1.81(5) to 1.856(4) Å]. [29] The conformations of **10**–**12** are all very similar, that between **10** and **11**, being almost identical. The difference between this conformation and that adopted by **12** is in the orientation of ring 1, which differs by ~67° between the two conformations.

3. Materials and Methods

3.1. General

Unless otherwise stated, all reactions were carried out under an oxygen-free nitrogen atmosphere using pre-dried solvents and standard Schlenk techniques, subsequent chromatographic and work up procedures were performed in air. ^{1}H (400.1 MHz), ^{13}C (100.6 MHz) and ^{77}Se-{^{1}H} (51.5 MHz, referenced to external Me$_2$Se) NMR spectra were recorded at 25 °C (NMR Jeol GSX270). IR spectra were recorded as KBr pellets in the range of 4000–250 cm^{-1} on a Perkin-Elmer 2000 FTIR/Raman spectrometer Mass spectrometry was performed by the EPSRC National Mass Spectrometry Service Centre, Swansea.

Crystallography

X-ray diffraction data for compounds **3**, **7** and **10–12** were collected at either 93 K or 173 K using a Rigaku FR-X Ultrahigh Brilliance Microfocus RA generator/confocal optics with XtaLAB P200 diffractometer [Mo Kα radiation (λ = 0.71075 Å)]. Intensity data were collected using ω steps accumulating area detector images spanning at least a hemisphere of reciprocal space. Data for all compounds were collected using CrystalClear 2.1 [30] and processed (including correction for Lorentz, polarization and absorption) using either CrystalClear [30] or CrysAlisPro 1.171.38.43. [31] Structures were solved by dual-space (SHELXT-2014/4 [32]) or Patterson (PATTY [33]) methods and refined by full-matrix least-squares against F^2 (SHELXL-2018/3 [34]). Non-hydrogen atoms were refined anisotropically, and hydrogen atoms were refined using a riding model, except for the amine hydrogen in **3** which was located from the difference Fourier map and refined isotropically subject to a distance restraint. All calculations were performed using the CrystalStructure 4.3 interface [35]. Selected crystallographic data are presented in Table 1. Deposition numbers 2071413–2071417 contains the supplementary crystallographic data for this paper. These data are provided free of charge by the joint Cambridge Crystallographic Data Centre and Fachinformationszentrum Karlsruhe Access Structures service www.ccdc.cam.ac.uk/structures.

3.2. Synthesis

3.2.1. General Procedure for Synthesis of Compounds 1–3

The appropriate aniline (20 mmol) and the phenacyl bromide (10 mmol) were combined in MeCN (40 mL) and allowed to stir at room temperature for 24 h. The amine salt was filtered off and the filtrate was concentrated under vacuum. The residue was dissolved in EtOAc (40 mL) and washed sequentially with H$_2$O (50 mL), 5% citric acid (50 mL) and brine (25 mL). The organic layer was dried over Na$_2$SO$_4$, filtered through a pad of silica gel and the solvent was evaporated to give the product aminoacetophenones **1–3** in good yield.

1-(4-Bromophenyl)-2-((4-ethylphenyl)amino)ethan-1-one (**1**). Brown solid (83% yield). M.p. 151–153 °C. Selected IR (KBr, cm^{-1}), 1679(vs), 1617(s), 1583(s), 1522(vs), 1585(m), 1388(m), 1351(m), 1311(m), 1216(m), 1178(m), 1140(m), 1068(s), 992(vs), 812(vs), 576(m), 501(m). ^{1}H NMR (CDCl$_3$, δ), 7.80 (d, J(H,H) = 8.6 Hz, 2H), 7.69 (d, J(H,H) = 8.4 Hz, 2H), 7.32 (s, 1H), 7.09 (d, J(H,H) = 8.6 Hz, 2H), 6.68 (d, J(H,H) = 8.6 Hz, 2H), 4.60 (s, 2H), 2.58 (d, J(H,H) = 7.6 Hz, 2H), 1.22 (d, J(H,H) = 7.6 Hz, 3H) ppm. ^{13}C NMR (CDCl$_3$, δ), 194.4, 145.0, 133.9, 133.7, 132.2, 129.3, 129.0, 128.7, 113.2, 50.7, 28.0, 16.0 ppm. Accurate mass measurement [ESI$^+$, m/z]: found 318.0482 [M + H]$^+$, calculated mass for C$_{16}$H$_{16}$BrNOH: 318.0486.

1-(4-Chlorophenyl)-2-((4-ethylphenyl)amino)ethan-1-one (**2**). Dark yellow solid (81% yield). M.p. 148–149 °C. Selected IR (KBr, cm^{-1}), 1678(vs), 1618(m), 1598(s), 1522(s), 1489(m), 1440(m), 1395(m), 1352(s), 1312(m), 1218(s), 1090(s), 993(s), 815(vs), 577(m), 529(m). ^{1}H NMR (CDCl$_3$, δ), 7.99 (d, J(H,H) = 8.6 Hz, 2H), 7.52 (d, J(H,H) = 8.6 Hz, 2H), 7.32 (s, 1H), 7.09 (d, J(H,H) = 8.4 Hz, 2H), 6.68 (d, J(H,H) = 8.4 Hz, 2H), 4.41 (s, 2H), 2.60 (q, J(H,H) = 7.6 Hz, 2H), 1.23 (t, J(H,H) = 7.6 Hz, 3H) ppm. ^{13}C NMR (CDCl$_3$, δ), 194.2, 145.0, 140.3, 133.9, 133.3, 129.3, 129.2, 129.1, 128.7, 113.2, 50.8, 28.0, 16.0 ppm. Accurate mass measurement [ESI$^+$, m/z]: found 274.0992 [M + H]$^+$, calculated mass for C$_{16}$H$_{16}$ClNOH: 274.0994.

2-((4-Bromophenyl)amino)-1-(4-chlorophenyl)ethan-1-one (**3**). Greenish yellow solid (87% yield). M.p. 165–166 °C. Selected IR (KBr, cm^{-1}), 1678(vs), 1595(s), 1510(s), 1491(s), 1400(m), 1357(s), 1256(m), 1218(m), 1094(s), 991(s), 814(s), 797(m), 574(m), 499(m). ^{1}H NMR (CDCl$_3$, δ), 7.98 (d, J(H,H) = 8.6 Hz, 2H), 7.52 (d, J(H,H) = 8.6 Hz, 2H), 7.32 (d, J(H,H) = 8.4 Hz, 2H), 7.28 (s, 1H), 6.60 (d, J(H,H) = 8.4 Hz, 2H), 4.56 (s, 2H) ppm. ^{13}C NMR (CDCl$_3$, δ), 194.2, 145.0, 140.3, 133.9, 133.3, 129.3, 129.2, 129.2, 128.7, 113.2, 50.8, 28.0, 16.0 ppm. Accurate mass measurement [ESI$^+$, m/z]: found 323.9789 [M + H]$^+$, calculated mass for C$_{14}$H$_{11}$BrClNOH: 323.9791.

3.2.2. General Procedure for Synthesis of Compounds **4–9**

The appropriate aminoacetophenone (5.0 mmol) was dissolved in dichloroethane (25 mL) and refluxed for 2 h with the appropriate acid chloride (5.0 equiv). Volatiles were evaporated in vacuo, and the residue was recrystallized from ethyl acetate to give the expected products **4–9**.

N-(2-(4-Bromophenyl)-2-oxoethyl)-N-(4-ethylphenyl)-4-methoxybenzamide (**4**). Yellow solid (91% yield). M.p. 127–129 °C. Selected IR (KBr, cm^{-1}), 1678(vs), 1617(m), 1590(m), 1521(s), 1352(m), 1306(m), 1262(m), 1219(m), 1091(m), 994(s), 844(m), 815(s), 772(m), 696(m), 613(m), 546(m), 503(m). ^{1}H NMR (CDCl$_3$, δ), 8.09 (d, J(H,H) = 8.9 Hz, 2H), 7.90 (d, J(H,H) = 8.7 Hz, 2H), 7.68 (d, J(H,H) = 8.6 Hz, 2H), 7.09 (d, J(H,H) = 8.6 Hz, 2H), 6.97 (d, J(H,H) = 8.9 Hz, 2H), 6.68 (d, J(H,H) = 8.5 Hz, 2H), 4.59 (s, 2H), 3.91 (s, 3H), 2.59 (q, J(H,H) = 7.6 Hz, 2H), 1.22 (t, J(H,H) = 7.6 Hz, 3H) ppm. ^{13}C NMR (CDCl$_3$, δ), 194.5, 171.3, 164.2, 145.0, 133.9, 133.7, 132.4, 133.2, 132.0, 129.3, 129.0, 128.7, 121.7, 113.8, 113.3, 55.5, 50.8, 28.0, 16.0 ppm. Accurate mass measurement [CI$^+$, m/z]: found 452.0859 [M + H]$^+$, calculated mass for C$_{24}$H$_{22}$BrNO$_3$H: 452.0861.

N-(2-(4-Chlorophenyl)-2-oxoethyl)-N-(4-ethylphenyl)-4-methoxybenzamide (**5**). Bright yellow solid (81% yield). M.p. 126–128 °C. Selected IR (KBr, cm^{-1}), 1678(vs), 1617(s), 1590(s), 1512(vs), 1396(m), 1352(m), 1307(m), 1262(m), 1218(m), 1179(m), 1090(s), 994(s), 815(s), 772(m), 613(m), 578(m), 546(m), 503(m). ^{1}H NMR (CDCl$_3$, δ), 8.10 (d, J(H,H) = 8.9 Hz, 2H), 7.99 (d, J(H,H) = 8.6 Hz, 2H), 7.51 (d, J(H,H) = 8.6 Hz, 2H), 7.09 (d, J(H,H) = 8.4 Hz, 2H), 6.97 (d, J(H,H) = 8.9 Hz, 2H), 6.68 (d, J(H,H) = 8.4 Hz, 2H), 4.61 (s, 2H), 3.91 (s, 3H), 2.59 (q, J(H,H) = 7.6 Hz, 2H), 1.23 (t, J(H,H) = 7.6 Hz, 3H) ppm. ^{13}C NMR (CDCl$_3$, δ), 194.2, 171.5, 164.0, 145.0, 140.3, 133.9, 133.3, 132.4, 129.3, 129.2, 129.0, 128.8, 121.7, 113.7, 113.2, 55.5, 50.8, 28.0, 16.0 ppm. Accurate mass measurement [CI$^+$, m/z]: found 408.1366 [M + H]$^+$, calculated mass for C$_{24}$H$_{22}$ClNO$_3$H: 408.1367.

N-(4-Bromophenyl)-N-(2-(4-chlorophenyl)-2-oxoethyl)-4-methoxybenzamide (**6**). Gray solid (83% yield). M.p. 150–152 °C. Selected IR (KBr, cm^{-1}), 1680(s), 1601(s), 1574(m), 1513(m), 1487(m), 1427(s), 1301(s), 1260(s), 1166(s), 1025(m), 926(m), 844(s), 816(m), 772(s), 696(m), 613(s), 547(s), 503(m), 484(m). ^{1}H NMR (CDCl$_3$, δ), 8.09 (d, J(H,H) = 8.9 Hz, 2H), 7.98 (d, J(H,H) = 8.6 Hz, 2H), 7.52 (d, J(H,H) = 8.6 Hz, 2H), 7.32 (d, J(H,H) = 8.9 Hz, 2H), 6.97 (d, J(H,H) = 8.8 Hz, 2H), 6.61 (d, J(H,H) = 8.8 Hz, 2H), 4.57 (s, 2H), 3.91 (s, 3H) ppm. ^{13}C NMR (CDCl$_3$, δ), 193.5, 171.3, 164.0, 145.8, 140.6, 133.0, 132.4, 132.1, 129.3, 129.2, 121.6, 114.7, 113.8, 110.8, 109.7, 55.5, 50.2 ppm. Accurate mass measurement [CI$^+$, m/z]: found 458.0157 [M + H]$^+$, calculated mass for C$_{22}$H$_{17}$BrClNO$_3$H: 458.0159.

N-(2-(4-Bromophenyl)-2-oxoethyl)-N-(4-ethylphenyl)cinnamamide (**7**). Yellow solid (79% yield). M.p. 126–127 °C. Selected IR (KBr, cm^{-1}), 1697(s), 1656(vs), 1620(s), 1584(s), 1510(s), 1401(m), 1376(m), 1327(m), 1209(s), 1069(s), 1005(s), 982(s), 840(m), 812(s), 703(s), 698(s), 569(m), 549(s). ^{1}H NMR (CDCl$_3$, δ), 7.87 (d, J(H,H) = 7.6 Hz, 2H), 7.73 (d, J(H,H) = 15.6 Hz, 1H), 7.63 (d, J(H,H) = 7.8 Hz, 2H), 7.38–7.26 (m, 8H), 6.49 (d, J(H,H) = 15.6 Hz, 1H), 5.18 (s, 2H), 2.72 (q, J(H,H) = 7.6 Hz, 2H), 1.29 (t, J(H,H) = 7.6 Hz, 3H) ppm. ^{13}C NMR (CDCl$_3$, δ), 192.9, 166.4, 144.3, 142.7, 140.0, 135.1, 134.1, 132.1, 129.9, 129.6, 129.5, 129.1, 128.7, 128.1, 128.0, 118.0, 56.5, 28.5, 15.4 ppm. Accurate mass measurement [CI$^+$, m/z]: found 448.0916 [M + H]$^+$, calculated mass for C$_{25}$H$_{22}$BrNO$_2$H: 448.0912.

N-(2-(4-Chlorophenyl)-2-oxoethyl)-N-(4-ethylphenyl)cinnamamide (**8**). Dark yellow solid (80% yield). M.p. 155–157 °C. Selected IR (KBr, cm^{-1}), 1697(s), 1653(s), 1617(s), 1587(s),

1510(s), 1402(m), 1380(s), 1329(m), 1214(s), 1089(s), 1000(m), 981(s), 819(s), 766(m), 703(m), 548(m), 525(m). ^{1}H NMR (CDCl$_3$, δ), 7.92 (d, J(H,H) = 7.6 Hz, 2H), 7.70 (d, J(H,H) = 15.1 Hz, 1H), 7.43 (d, J(H,H) = 7.8 Hz, 2H), 7.28–7.21 (m, 8H), 6.45 (d, J(H,H) = 16.1 Hz, 1H), 5.16 (s, 2H), 2.68 (q, J(H,H) = 7.6 Hz, 2H), 1.25 (t, J(H,H) = 7.6 Hz, 3H) ppm. ^{13}C NMR (CDCl$_3$, δ), 192.8, 166.5, 144.4, 142.8, 140.1, 140.0, 135.2, 133.7, 129.7, 129.5, 129.1, 129.0, 128.8, 128.1, 128.0, 118.0, 56.6, 28.6, 15.5 ppm. Accurate mass measurement [CI$^+$, m/z]: found 404.1408 [M + H]$^+$, calculated mass for C$_{25}$H$_{22}$ClNO$_2$H: 404.1412.

N-(4-Bromophenyl)-N-(2-(4-chlorophenyl)-2-oxoethyl)cinnamamide (**9**). Off-white solid (76% yield). M.p. 174–176 °C. Selected IR (KBr, cm^{-1}), 1696(s), 1657(s), 1617(s), 1587(m), 1486(vs), 1408(m), 1373(s), 1207(s), 1092(m), 1070(m), 1013(s), 816(s), 763(m), 696(m), 484(s). ^{1}H NMR (CDCl$_3$, δ), 7.95 (d, J(H,H) = 8.6 Hz, 2H), 7.74 (d, J(H,H) = 15.5 Hz, 1H), 7.58 (d, J(H,H) = 8.6 Hz, 2H), 7.48 (d, J(H,H) = 8.6 Hz, 2H), 7.39–7.32 (m, 6H), 6.46 (d, J(H,H) = 15.5 Hz, 2H), 5.18 (s, 2H) ppm. ^{13}C NMR (CDCl$_3$, δ), 192.4, 166.1, 143.5, 141.5, 140.2, 134.8, 133.4, 132.9, 130.0, 129.9, 129.5, 129.2, 128.8, 128.1, 122.0, 117.3, 56.3 ppm. Accurate mass measurement [CI$^+$, m/z]: found 456.0184 [M + H]$^+$, calculated mass for C$_{23}$H$_{17}$BrClNO$_2$H: 456.0189.

3.2.3. General Procedure for Synthesis of Compounds 10–12

A mixture of the appropriate benzamide or cinnamide with an equivalent of **WR** in dry toluene was refluxed for 6 h. Following cooling to room temperature and filtration to remove unreacted solid, the filtrate was evaporated in vacuo, the residue was dissolved in 2 mL of dichloromethane and purified by silica gel column chromatography (1:1 hexane/dichloromethane as eluant) to give the products **10–12**. Cinnamides **7–9** did not show any reaction with **WR**, returning the starting materials.

N-(4-Bromophenethyl)-N-(4-ethylphenyl)-4-methoxybenzoselenoamide (**10**). Greyish yellow paste (0.25g, 50%). Selected IR (KBr, cm^{-1}), 1601(s), 1508(s), 1488(m), 1448(s), 1399(s), 1300(m), 1250(vs), 1170(s), 1073(m), 1029(m), 830(m), 806(m). ^{1}H NMR (CD$_2$Cl$_2$, δ), 8.03 (d, J(H,H) = 8.0 Hz, 1H), 7.45 (d, J(H,H) = 8.4 Hz, 2H), 7.26–7.23 (m, 4H), 7.10–7.02 (m, 3H), 6.94 (d, J(H,H) = 8.0 Hz, 2H), 6.10 (d, J(H,H) = 8.4 Hz, 2H), 4.72 (t, J(H,H) = 8.1 Hz, 2H), 3.74(s, 3H), 3.24 (t, J(H,H) = 8.1 Hz, 2H), 2.60 (q, J(H,H) = 7.6 Hz, 2H), 1.21 (t, J(H,H) = 7.6 Hz, 3H) ppm. ^{13}C NMR (CD$_2$Cl$_2$, δ), 206.6, 159.5, 143.7, 140.0, 137.6, 131.5, 130.7, 129.0, 128.6, 126.2, 120.2, 114.4, 112.4, 62.4, 55.2, 31.4, 28.3, 15.0 ppm. ^{77}Se NMR (CDCl$_3$, δ), 598.4 ppm. Accurate mass measurement [CI$^+$, m/z]: found 502.0283 [M + H]$^+$, calculated mass for C$_{24}$H$_{24}$BrNOSeH: 502.0285.

N-(4-Chlorophenethyl)-N-(4-ethylphenyl)-4-methoxybenzoselenoamide (**11**). Reddish yellow paste (0.21 g, 46%). Selected IR (KBr, cm^{-1}), 1602(s), 1508(s), 1488(m), 1448(s), 1398(s), 1300(m), 1251(vs), 1170(s), 1030(m), 832(m), 809(m). ^{1}H NMR (CDCl$_3$, δ), 7.25–7.23 (m, 4H), 7.19 (d, J(H,H) = 8.6 Hz, 2H), 7.02 (d, J(H,H) = 8.0 Hz, 2H), 6.85 (d, J(H,H) = 8.0 Hz, 2H), 6.55 (d, J(H,H) = 8.0 Hz, 2H), 6.55 (d, J(H,H) = 8.0 Hz, 2H), 4.67 (d, J(H,H) = 8.0 Hz, 2H), 3.69 (s, 3H), 3.21 (d, J(H,H) = 8.0 Hz, 2H), 2.54 (q, J(H,H) = 8.1 Hz, 2H), 1.18 (t, J(H,H) = 8.0 Hz, 3H) ppm. ^{13}C NMR (CDCl$_3$, δ), 206.8, 159.7, 143.7, 139.8, 136.8, 132.5, 130.4, 129.2, 128.8, 128.7, 126.2, 112.7, 62.8, 55.3, 31.5, 28.4, 15.2 ppm. ^{77}Se NMR (CDCl$_3$, δ), 601.5 ppm. Accurate mass measurement [CI$^+$, m/z]: found 458.0789 [M + H]$^+$, calculated mass for C$_{24}$H$_{24}$ClNOSeH: 458.0790.

N-(4-Bromophenyl)-N-(4-chlorophenethyl)-4-methoxybenzoselenoamide (**12**). Pale orange paste (0.20 g, 40%). Selected IR (KBr, cm^{-1}), 1601(s), 1489(m), 1485(s), 1446(m), 1392(m), 1302(m), 1251(vs), 1170(vs), 1068(m), 1011(m), 832(m), 801(m). ^{1}H NMR (CDCl$_3$, δ), 7.33 (d, J(H,H) = 8.2 Hz, 2H), 7.27–7.16 (m, 6H), 6.82 (d, J(H,H) = 8.2 Hz, 2H), 6.58 (d, J(H,H) = 8.3 Hz, 2H), 4.63 (q, J(H,H) = 8.4 Hz, 2H), 3.72 (s, 3H), 3.16 (d, J(H,H) = 8.4 Hz, 3H) ppm. ^{13}C NMR (CDCl$_3$, δ), 204.8, 160.0, 145.0, 139.6, 136.5, 132.7, 130.3, 129.6, 129.2, 128.8, 128.0, 121.1, 113.0, 62.3, 55.4, 29.8 ppm. ^{77}Se NMR (CDCl$_3$, δ), 601.4 ppm. Accurate mass measurement [CI$^+$, m/z]: found 507.9580 [M + H]$^+$, calculated mass for C$_{22}$H$_{19}$BrClNOSeH: 507.9582.

4. Conclusions

In summary, we have disclosed Woollins' reagent used as a highly efficient chemoselective selenation/reduction reagent for benzamide leading to N-aryl-N-(arylenethyl)benzoselenoamides. The reported results enhance the application of Woollins' reagent further, providing an efficient route to the preparation of the unusual substituted selenoamides.

Author Contributions: G.H. and J.D.W. conceived and designed the project; G.H. performed the experiments; D.B.C., C.L.C.-W. and A.M.Z.S. did the crystallography; All authors have read and agreed to the published version of the manuscript.

Funding: This research received no external funding.

Institutional Review Board Statement: Not applicable.

Informed Consent Statement: Not applicable.

Data Availability Statement: Not applicable.

Acknowledgments: We are grateful to the University of St Andrews for financial support and the EPSRC National Mass Spectrometry Service Centre (Swansea) for mass spectral measurements.

Conflicts of Interest: The authors declare no conflict of interest.

Sample Availability: Samples of all compounds are not available from the authors.

References

1. Hua, G.; Woollins, J.D. Formation and reactivity of phosphorus-selenium rings. *Angew. Chem. Int. Ed.* **2009**, *48*, 1368–1377. [CrossRef] [PubMed]
2. Hua, G.; Zhang, Q.; Li, Y.; Slawin, A.M.Z.; Woollins, J.D. Novel heterocyclic selenazadiphospholaminediselenides, zwitterionic carbamimidoyl(phenyl)phosphinodiselenoic acids and selenoureas derived from cyanamides. *Tetrahedron* **2009**, *65*, 6074–6082. [CrossRef]
3. Hua, G.; Li, Y.; Fuller, A.L.; Slawin, A.M.Z.; Woollins, J.D. Facile synthesis and structure of novel 2,5-disubstituted 1,3,4-selenadiazoles. *Eur. J. Org. Chem.* **2009**, *2009*, 1612–1618. [CrossRef]
4. Gómez Castaño, J.A.; Romano, R.M.; Beckers, H.; Willner, H.; Della Védova, C.O. Trifluoroselenoacetic acid, $CF_3C(O)SeH$: Preparation and properties. *Inorg. Chem.* **2010**, *49*, 9972–9977. [CrossRef]
5. Hua, G.; Li, Y.; Fuller, A.L.; Slawin, A.M.Z.; Woollins, J.D. Synthesis and X-ray structures of new phosphorus-selenium heterocycles with an E-P(Se)-E' (E, E' = N, S, Se) linkage. *New J. Chem.* **2010**, *34*, 1565–1571. [CrossRef]
6. Huang, Y.; Jahreis, G.; Lücke, C.; Wildemann, D.; Fischer, G. Modulation of the Peptide Backbone Conformation by the Selenoxo Photoswitch. *J. Am. Chem. Soc.* **2010**, *132*, 7578–7579. [CrossRef] [PubMed]
7. Hua, G.; Fuller, A.; Slawin, A.M.; Woollins, J.D. Novel five- to ten-membered organoselenium heterocycles from the selenation of aryl-diols. *Eur. J. Org. Chem.* **2010**, *2010*, 2607–2615. [CrossRef]
8. Hua, G.; Henry, J.B.; Li, Y.; Mount, A.R.; Slawin, A.M.Z.; Woollins, J.D. Synthesis of novel 2,5-diarylselenophenes from selenation of 1,4-diarylbutane-1,4-diones or methanol/arylacetylenes. *Org. Biomol. Chem.* **2010**, *8*, 1655–1660. [CrossRef]
9. Hua, G.; Fuller, A.L.; Slawin, A.M.; Woollins, J.D. Formation of new organoselenium heterocycles and ring reduction of ten-membered heterocycles into seven-memered heterocycle. *Polyhedron* **2011**, *30*, 805–808. [CrossRef]
10. Hua, G.; Fuller, A.L.; Buehl, M.; Slawin, A.M.; Woollins, J.D. Selenation/Thionation of α-Amino Acids: Formation and X-ray Structures of Diselenopiperazine and Dithiopiperazine and Related Compounds. *Eur. J. Org. Chem.* **2011**, *2011*, 3067–3073. [CrossRef]
11. Hua, G.; Cordes, D.B.; Li, Y.; Slawin, A.M.; Woollins, J.D. Symmetrical organophosphorus spiroheterocycles from selenation of carbohydrazides. *Tetrahedron Lett.* **2011**, *52*, 3311–3314. [CrossRef]
12. Wong, R.C.S.; Ooi, M.L. A new approach to coordination chemistry involving phosphorus-selenium based ligands: Ring opening, deselenation and phosphorus—phosphorus coupling of Woollins' Reagent. *Inorg. Anica Chim. Acta* **2011**, *366*, 350–356. [CrossRef]
13. Gray, I.P.; Bhattacharyya, P.; Slawin, A.M.Z.; Woollins, J.D. A new synthesis of $(PhPSe_2)_2$ (Woollins' Reagnet) and its use in the synthesis of novel P-Se heterocycles. *Chem. Eur. J.* **2005**, *11*, 6221–6227. [CrossRef]
14. Hua, G.; Li, Y.; Slawin, A.M.; Woollins, J.D. Synthesis and structure of eight-, nine- and ten-membered rings with P-Se-Se-P linkages. *Angew. Chem.* **2008**, *120*, 2899–2901.
15. Hua, G.; Griffin, J.M.; Ashbrook, S.E.; Slawin, A.M.Z.; Woollins, J.D. Octaselenocycododecane. *Angew. Chem. Int. Ed.* **2011**, *123*, 4209–4212. [CrossRef]
16. Mandal, M.; Chatterjee, S.; Jaisankar, O. Woollins' reagent: A chemoselective reducing agent for 1,4-enediones and 1,4-ynediones to saturated 1,4-diones. *Synlett* **2012**, *23*, 2615. [CrossRef]
17. Pizzo, C.; Graciela, M. Woollins' reagent promotes selective reduction of α, β-unsaturated thiazo and selenazolidinones. *Tetrahedron Lett.* **2017**, *58*, 1445–1447. [CrossRef]

18. Yu, Y.; Furuyama, T.; Tang, J.; Wu, Z.Y.; Chen, J.Z.; Kobayashi, N.; Zhang, J.L. Stable iso-bacteriochlorin mimics from porpholactone: Effect of α,β-oxazolone moiety on the frontier π-molecular orbitals. *Inorg. Chem. Front.* **2015**, *2*, 671–677. [CrossRef]
19. Wirth, T. Organoselenium chemistry in stereoselective reactions. *Angew. Chem. Int. Ed.* **2000**, *39*, 3740–3749. [CrossRef]
20. Rhoden, C.R.; Zeni, G. New development of synthesis and reactivity of seleno- and tellurophenes. *Org. Biomol. Chem.* **2011**, *9*, 1301–1303. [CrossRef] [PubMed]
21. Uemoto, T.; Emmanuel, M. S- Se-, Te-(perfluoroalkyl)dibenzothiophenium, -selenophenium, and -tellurophenium salts. *Adv. Heterocycl. Chem.* **1995**, *64*, 323–339. [CrossRef]
22. Nogueira, C.W.; Zeni, G.; Rocha, J.B.T. Organoselenium and Organotellurium Compounds: Toxicology and Pharmacology. *Chemin Rev.* **2005**, *36*, 6255–6286. [CrossRef]
23. Lakner, F.J.; Parker, M.A.; Rogovoy, B.; Khvat, A.; Ivachtchenko, A. Synthesis of novel trisubstituted imidazolines. *Synthesis* **2009**, *12*, 1987–1990.
24. Porretta, G.C.; Biava, M.; Fioravanti, R.; Fischetti, M.; Melino, C.; Venza, F.; Bolle, P.; Tita, B. Research on antibacterial and antifungal agents. VIII. Synthesis and antimicrobial activity of 1,4-diarylpyrroles. *Eur. J. Med. Chem.* **1992**, *27*, 717–722. [CrossRef]
25. Nishiyama, Y.; Yoshida, M.; Ohkawa, S.; Hamanaka, S. New agents for the selective reduction of the carbon-carbon double bond of alpha/betal unsaturated carbonyl compounds. *J. Org. Chem.* **1991**, *56*, 6720–6722. [CrossRef]
26. Mesquita, K.D.; Waskow, B.; Schumacher, R.F.; Perin, G.; Jacob, R.G.; Alves, D. Glycerol/hypophosphorus acid and PhSeSePh: An efficient and selective system for reactions in the carbon-carbon double bond of (E)-chalcones. *J. Braz. Chem. Soc.* **2014**, *25*, 1261–1269.
27. Lalezari, I.; Ghanbarpour, F.; Niazi, M.; Jafari-Namin, R. Selenium heterocycles XIV. 2,6-Diaryltetrahydroselenopyran-4-ones. *J. Heterocycl. Chem.* **1974**, *11*, 469–470. [CrossRef]
28. Miyashita, M.; Yoshikoehi, A. Facile and highly efficient conjugate addition of benzeneselenol to α,β-unsaturated carbonyl compounds. *Synthesis* **1980**, *8*, 664–666. [CrossRef]
29. Li, Y.; Hua, G.-X.; Slawin, A.M.Z.; Woollins, J.D. The X-ray Crystal Structures of Primary Aryl Substituted Selenoamides. *Molecules* **2009**, *14*, 884–892. [CrossRef] [PubMed]
30. *CrystalClear-SM Expert v2.1*; Rigaku Americas: The Woodlands, TX, USA; Rigaku Corporation: Tokyo, Japan, 2015.
31. *CrysAlisPro v1.171.38.43.*; Rigaku Oxford Diffraction; Rigaku Corporation: Oxford, UK, 2015.
32. Sheldrick, G.M. SHELXT—Integrated space-group and crystal structure determination. *Acta Crystallogr. Sect. A* **2015**, *71*, 3–8. [CrossRef]
33. Beurskens, P.T.; Beurskens, G.; de Gelder, R.; Garcia-Granda, S.; Gould, R.O.; Israel, R.; Smits, J.M.M. *DIRDIF-99*; Crystallography Laboratory, University of Nijmegen: Nijmegen, The Netherlands, 1999.
34. Sheldrick, G.M. Crystal structure refinement with SHELXL. *Acta Crystallogr. Sect. C* **2015**, *71*, 3–8. [CrossRef] [PubMed]
35. *CrystalStructure v4.3.0*; Rigaku Americas: The Woodlands, TX, USA; Rigaku Corporation: Tokyo, Japan, 2018.

Hydricity of 3d Transition Metal Complexes from Density Functional Theory: A Benchmarking Study

Alister S. Goodfellow and Michael Bühl *

School of Chemistry, University of St Andrews, St Andrews KY16 9ST, UK; ag266@st-andrews.ac.uk
* Correspondence: buehl@st-andrews.ac.uk

Abstract: A range of modern density functional theory (DFT) functionals have been benchmarked against experimentally determined metal hydride bond strengths for three first-row TM hydride complexes. Geometries were found to be produced sufficiently accurately with RI-BP86-D3(PCM)/def2-SVP and further single-point calculations with PBE0-D3(PCM)/def2-TZVP were found to reproduce the experimental hydricity accurately, with a mean absolute deviation of 1.4 kcal/mol for the complexes studied.

Keywords: DFT; 3d metal complex; benchmark; hydricity

1. Introduction

At the forefront of modern chemistry is sustainability, with a drive towards 'greener' processes and development. Catalysis has always been a tool used to reduce energetic costs and to promote specific reactions, improving selectivity and the efficacy of various transformations. Traditionally, homogeneous catalysts have been based upon expensive and unsustainable metals such as platinum, palladium and rhodium [1] and while these 4d and 5d transition metals (TMs) have been very successful in this application, their future use is limited by concerns over sustainability and price. The scarcity of the metals has resulted in a high economic and social cost in their extraction. The metals themselves are also toxic, which may lead to issues of contamination in extraction, chemical transformations, or in the application of these catalysts in industrial processes. To alleviate these issues, development has moved towards the use of 3d TMs, which are largely more abundant, less toxic and more sustainable [2–5].

First-row TM catalysts have undergone huge development in the past few years and are being optimised to ultimately become competitive against their unsustainable heavy metal congeners. Iron and manganese are attractive due to their low toxicity and high abundance, and work on catalytic hydrogenation reactions using these metals is very topical (see Scheme 1). For a review on the development of manganese based pincer catalysts, see Garbe et al. [6].

Density functional theory is a powerful tool for the elucidation of reaction mechanisms and for the rational design of these catalytic systems. To accurately study such systems, the DFT functional used must be suitable for the system, accurately predicting bond energies that are crucial to the functionality of the molecule. 3d TMs are notoriously tricky to study with DFT, especially bare TMs where the electronic structure is not always predicted correctly [7,8]. For investigation in the field of homogeneous catalysis, especially in hydrogenation reactions as shown in Scheme 1, the metal hydride bond will be of prime importance for the catalytic activity, and the functional used must be accurate in the description of these bond strengths.

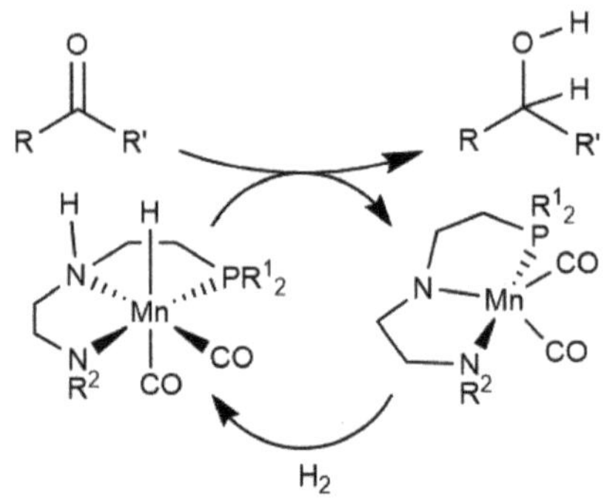

Scheme 1. A representation of the catalytic application of Mn-based systems from the group of Beller [9].

A number of papers have used DFT computations to examine the reactivity of these catalytic systems and a variety of methodology has been employed (Table 1). The choice of methodology may be based upon previous work in related catalysis, perhaps 4d metal systems, or simply by using methods that have already been applied in the literature. For example, early computations from the Beller group used B3PW91, based upon the performance of this hybrid functional in previous work (including benchmarking studies) across 3d, 4d and 5d transition metal complexes though more recent work from this group has used the hybrid PBE0 functional [9–15].

An alternate approach was adopted by Ge et al. [16], where the methodology was selected based upon the testing of a range of functionals for a multi-step transformation. M06 was selected for their work as the computed barrier heights were found to be closest to the average value across the functionals tested. A benchmarking study was performed by Gamez et al. [17] on a number of high-spin Mn-based complexes using a range of different functionals. It was found that pure functionals (i.e., non-hybrid) favoured low-spin forms of these Mn complexes, with weak-field nitrogen and oxygen based ligands, experimentally known to be high-spin complexes. B3LYP was found to produce poor geometries and PBE0 was recommended for use on these systems. Dispersion corrections were also found to be of importance in optimisation. The accurate description of spin states is important in the study of 3d metal complexes, where the energetic splitting of metal-based 3d orbitals may be small and multi-state reactivity may be possible.

Table 1. A summary of the literature methodology used for DFT calculations on 3d metal catalytic systems.

System	Methodology [a]	Dispersion [b,c]	Solvent Model [c]	References
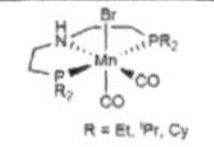	B3PW91/TZVP			[9,10,15]
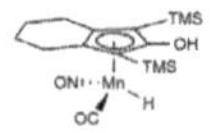	M06/6-31++G(d,p)	M	PCM	[16]
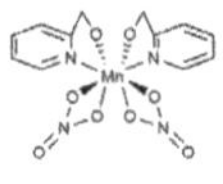	PBE0/def2-TZVP			[17]
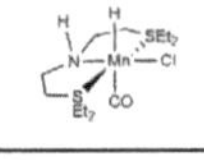	ωB97X-D/ECP (Mn)/6-31++G(d,p)	GD2	PCM	[18]

Table 1. Cont.

System	Methodology [a]	Dispersion [b,c]	Solvent Model [c]	References
	M06/ECP (Mn,Co)/6-31(+)G(d,p)	M	PCM	[19]
	ωB97X-D/ECP (Mn)/cc-PVTZ	GD2	PCM	[20]
	M06/TZVP// ωB97X-D/TZVP (Mn)/SVP	M sp GD2 opt	PCM	[21]
	B3PW91/TZVP			[22]
	ωB97X-D/ECP (Mn)/cc-pVDZ	GD2	PCM	[23]
	M06L/def2-TZVP// B3LYP/def2-SVP	M sp	SMD sp	[24]
	PBE0/ECP (Mn)/6-31G(d,p)		PCM	[25]
	PBE0/6-311+G(d,p)// PBE0/6-31G(d,p)			[26]
	DSD-PBE96-NL/def2-QZVP// PBE0-D3/def2-TZVP	GD3BJ opt		[27]
	M06/6-311++G(d,p)// M06/ECP (Mn)/6-31G(d)	M	SMD sp	[28]
	PBE0/6-311+G(d)// PBE0/6-31G(d)		PCM sp	[29]
	PBE0/def2-QZVPP// PBE0-D3/def2-TZVP	GD3BJ opt	SMD opt	[30]
	PBE0/ECP (Mn)/6-31G(d,p)			[31]
	B3LYP/ECP (Mn)/6-311++G(d,p)		CPCM	[32]
	ωB97X/6-31++G(d,p)		PCM	[33]

Table 1. Cont.

System	Methodology [a]	Dispersion [b,c]	Solvent Model [c]	References
	BP86/6-311+G(d)// BP86/6-311G(d) (Co)/6-31G(d)		PCM sp	[34]
	B3PW91/ECP (Co)/6-31++G(d,p)// B3PW91/ECP (Co)/6-31++G(d,p) (B,N,P)/6-31G(d)	GD sp GD opt	CPCM sp	[35]
	M06/TZVP// BP86/ECP (Fe)/SVP	M sp	PCM sp	[36]
	B3PW91/def2-TZVP// B3PW91/TZVP	GD3BJ sp	SMD sp	[37]
	B3LYP*/ECP (Fe)/6-311+G(2df,2p)// B3LYP/ECP (Fe)/6-31G(d,p)	GD3BJ sp GD3BJ opt	SMD sp	[38]

[a] For simplicity, the use of any pseudopotential has been indicated with ECP. [b] M denotes that there is implicit dispersion included, due to the parametrisation of the Minnesota family of functionals. GD is used to indicate the inclusion of Grimme's empirical dispersion correction, with BJ denoting the use of Becke-Johnson dampening. [c] 'sp' and 'opt' denote the use of dispersion corrections or solvation models in either the single-point or geometry optimisation calculation, respectively.

We now present a validation study on the performance of these and other functionals for the calculation of hydricity for a few select complexes. Hydricity is the heterolytic bond dissociation energy of a metal hydride to a metal cation and hydride (Equation (1)), predicted to be an important property for the study of hydrogenation catalysis.

$$MH \rightleftharpoons M^+ + H^- \qquad \Delta G°_{H^-} \qquad (1)$$

A sizeable amount of $\Delta G°_{H^-}$ data is available from experiment [39], the ultimate benchmark. Arguably, this quantity is key in determining the ease of elementary steps such as transfer of the hydride to an electrophilic substrate (cf. upper arrow in Scheme 1). As it turns out, there is a notable variability in the performance of different functionals in their ability to reproduce these hydricity values, and some clear recommendations for functionals to choose can be made.

2. Methodology

2.1. Computational Details

Geometry optimisations were performed primarily at the GGA level of DFT using the BP86 functional, employing a variety of basis sets from the redefinitions of the Ahlrichs family of bases [40–43]. Structures obtained by X-ray crystallography for closely related systems were used as starting points in geometry optimisations [44–46]. This is akin to the methodology of Neale [47] used in the investigation of Fe(II) and Co(III) catalysis and of Bühl and Kabrede [48] in the benchmarking of geometries for a wide range of first-row transition metal complexes. Def2-TZVP was used for the investigations of proton and metal cation solvation and throughout further benchmarking, a variety of basis from this family were used; def2-SVP, with def2-TZVP describing the metal atom of Fe or Co, and the addition of a diffuse function on the hydridic hydrogen, taking 1/3 of the exponent from the diffuse 's' function in the def2-TZVP basis for hydrogen. Dispersion corrections were included throughout, predominantly with the D3 empirical correction from Grimme [49], including Becke-Johnson dampening [50]. Implicit solvation was also included using the Polarisable Continuum Model with the integral equation formalism (IEF-PCM, herein referred to as PCM) employing the parameters of acetonitrile ($\epsilon = 35.688$). The explicit treatment of a small number of individual solvent molecules acting as ligands was also considered (Section 3.1). The nature of the minima obtained were verified by computation of harmonic frequencies at the same level of theory. Single-point energies were obtained using a number of functionals from a range of levels on Jacob's ladder [51] using the def2-TZVP basis, with an ultrafine integration grid (99 radial shells with 590 angular points per shell). The impact of using a larger, quadruple-ζ basis, def2-QZVP, was also explored. The functionals considered were; B3LYP [52–54], B3PW91 [52,55–57], BLYP [53,54,58], BP86 [58,59], CAM-B3LYP [60], M06 [61], M06-2X [61], M06-HF [62,63], M06-L [64], M11 [65], M11-L [66], MN12-L [67], MN15 [68], MN15-L [69], PBE [55,70], PBE0 [71], PBE0-1/3 [72], revPBE [73], revPBE0 [73], TPSS [74], TPSSh [74,75], ωB97X-D [76]. Dispersion was included for all single-point calculations as mentioned earlier, except for the Minnesota family of functionals, which are parametrised to account for dispersion, and the ωB97X-D functional includes the D2 correctional term from Grimme [77]. Thermochemistry was evaluated at 468 atm (to mimic bulk acetonitrile, see Section 1 in the ESI for details) and 298.15 K using thermodynamic corrections obtained at the level of the geometry optimisation, combined with single-point energetics. Density fitting, invoking the RI approximation with the Gaussian keyword 'auto', was used where possible for pure functionals. Calculations were performed primarily using Gaussian 09, D.01 [78], with some single-points evaluated using Gaussian 16, C.01 [79] where the functional was unavailable in Gaussian 09. Some geometries were also evaluated using the more recent optimiser of Gaussian 16 to remove spurious imaginary frequencies that could not be removed with Gaussian 09. Where Gaussian 16 was used, defaults from the older Gaussian 09 were employed to ensure that there was no change in implementation across different versions (see Table S5 in the ESI).

2.2. Calculation of Hydricity

To determine a functional suitable for the study of 3d TM catalysts (for example, see Scheme 1), we have chosen to benchmark three 3d TM hydride complexes, for which there are experimentally determined hydricity values (ΔG_{H^-}, Table 2) [39]. Direct evaluation of hydricity according to Equation (1) is challenging computationally. In this work we use a thermodynamic cycle with a protonation as key step, where the ionic stoichiometry is maintained across the reaction (i.e. charge separation is avoided, Equation (2)). This has been done to ensure that there is no artefact arising from an imbalance of ionic species, as their large stabilisation in a polar solvent is difficult to compute accurately using methods employing implicit solvation, such as PCM (as would be the case by directly modelling Equation (1)). By combination with H_2 heterolysis obtained from experiment [39] (Equation (3)), the thermochemical cycle may be completed to obtain the thermodynamic hydricity.

$$MH + H^+ \rightleftharpoons M^+ + H_2 \qquad \Delta G_1^\circ \qquad (2)$$

$$H_2 \rightleftharpoons H^+ + H^- \qquad \Delta G_{H_2}^\circ = 76.0 \text{ kcal/mol [39]} \qquad (3)$$

Table 2. Three 3d transition metal complexes with experimentally determined thermodynamic hydricity values (ΔG_{H^-}).

Complex	FeCp(CO)$_2$H	Co(dppe)$_2$H	Co(P$_4$N$_2$)H
ΔG_{H^-} (kcal/mol)	61.7 [39,80]	49.9 [81,82]	31.8 [82]
Positively charged cationic complex fragment after hydride dissociation [a]			

[a] Hydrogen atoms have been removed for clarity. The viewing angle is looking towards the vacant site on the metal centre where the hydridic hydrogen atom was residing previously, emphasising the accessibility of the now vacant site by a solvent molecule. Electron density (isodensity value of 0.001 a.u.) mapped with electrostatic potential [colour-coded between 0.00 a.u. (red) and +0.35 a.u. (blue)].

With limited availability of Mn based data, neutral Co and Fe hydride complexes have been chosen from data by Wiedner et al. [39] as representatives of 3d metal hydride systems. Benchmarking studies have previously been performed on 3d TM diatomics [83], but the applicability of these studies to such larger complexes remains unclear.

2.3. Acetonitrile Proton Clusters

The proton involved in Equation (2) will be modelled through a complex with discrete solvent molecules. Experimentally, acetonitrile had been used as a solvent and it was observed by Himmel et al. [84] that, in acetonitrile, proton-solvent clusters were present in the form of extended hydrogen bonded arrays, [H(NCMe)$_n$]$^+$, with values of n up to 8. For the purpose of this study, we determined that the proton-solvent cluster was suitably described with two explicit solvent molecules ($n = 2$) forming an inner solvent shell and approximation of the outer solvent shell with implicit solvation using a polarisable continuum model (further details are available in Section 2 of the ESI).

3. Results and Discussion

3.1. Explicit Treatment of Solvent Molecules

Implicit solvation is used as a way to approximate the solvation of gas phase molecules, with continuum models used to describe a solvation shell around a molecule as a dielectric; however, this does not describe specific interactions, where a solvent molecule may act as a ligand to stabilise an ionic species (Equation (4)). In such cases, the explicit treatment of a solvent molecule is required to describe the role as a ligand in the coordination sphere and as such, the reaction shown in Equation (2) may be better represented by the reaction in Equation (5), where both cations are solvated by explicit solvent molecules.

$$[M(NCMe)]^+ \rightleftharpoons M^+ + MeCN \qquad \Delta G°_{BDE} \qquad (4)$$

$$MH + [H(NCMe)_2]^+ \rightleftharpoons [M(NCMe)]^+ + MeCN + H_2 \qquad \Delta G°_2 \qquad (5)$$

If a solvent molecule is coordinated to the metal centre after hydride dissociation, according to Equation (1), this binding energy is implicitly included in the experimental determination of thermodynamic hydricity by Wiedner et al. [39] (Equation (2)). Across the three complexes used (Table 2), both the sterics and accessibility of the solvent to the coordination shell are significantly different. In cationic metal complexes, where a vacant coordination site is exposed and accessible, a solvent molecule may strongly bind to the metal centre, stabilising the complex to a higher degree than would be predicted with only the inclusion of a dielectric model. Here, the bond dissociation energy of the ligand (Equation (4)) becomes important and a solvent molecule should be treated explicitly if there is a strong interaction between the cationic metal complex and solvent molecule.

The Fe cationic complex has a low steric bulk and vacant and accessible coordination sphere, $[FeCp(CO)_2]^+$. In contrast, the two Co complexes have large ligands attached to each metal centre and are far more sterically crowded (see 3D surface plots in Table 2). We found that, for the most accurate computed hydricity values, the Fe cationic species must be treated with an explicitly bound solvent molecule, whereas the Co complexes do not require this (Equations (6)–(8)). While there is an element of functional dependence, this methodology has been tested and was found to produce the lowest mean absolute errors (MAEs) in the calculation of hydricity when compared to the values obtained by Wiedner et al. [39] (further details are available in Section 3 of the ESI).

$$FeCp(CO)_2H + [H(NCMe)_2]^+ \rightleftharpoons [FeCp(CO)_2(NCMe)]^+ + MeCN + H_2 \qquad (6)$$

$$Co(P_4N_2)H + [H(NCMe)_2]^+ \rightleftharpoons [Co(P_4N_2)]^+ + 2\,MeCN + H_2 \qquad (7)$$

$$Co(dppe)_2H + [H(NCMe)_2]^+ \rightleftharpoons [Co(dppe)_2]^+ + 2\,MeCN + H_2 \qquad (8)$$

3.2. Choice of Single-Point Functional

To quantitatively predict the thermodynamic hydricity of a complex, the ability of the functional used to produce accurate energies is of prime importance. Single-point energy calculations have been performed with a range of functionals, on geometry optimised structures at the level of RI-BP86-D3(PCM$_{MeCN}$), with a large triple-ζ basis, def2-TZVP, from the group of Ahlrichs [40–43].

The functional used in the single-point calculation was found to have a significant impact upon the ability to quantitatively reproduce thermodynamic hydricity values obtained from experiment. Errors in the form of mean absolute error, have been calculated for each functional across the three complexes (Equation (9) and Figure 1).

$$MAE = \frac{1}{n}\Sigma|E_{calc} - E_{exp}| \qquad (9)$$

Clearly, the Minnesota functionals of M06-2X and M06-HF, both highly parameterised for main group chemistry, are poor in their description of these 3d TM complexes. The lowest MAEs were obtained from RI-BP86, PBE0 and revPBE0 (1.7, 1.3 and 0.9 kcal/mol). The percentage of HF exchange has been shown to be important by Moltved et al. [83] and this effect can be seen in the comparison of PBE0 with PBE0-1/3, with 25% and 33% HF exchange, respectively. Both forms of the hybrid PBE functional performed well in terms of overall performance, yet PBE0-1/3 had over twice the MAE of PBE0 (1.3 and 2.9 kcal/mol, respectively).

Based purely upon the MAEs obtained here, revPBE0 would be the functional of choice; however, we recommend to use PBE0 (based upon an improved performance over revPBE0 on geometries produced with the smaller def2-SVP basis, MAEs of 1.4 and

1.8 kcal/mol, see ESI Figure S7 and Section 3.3.2), which has been shown to perform well in many previous studies of TM systems [17,85] and in the differentiation of spin state energetics [17].

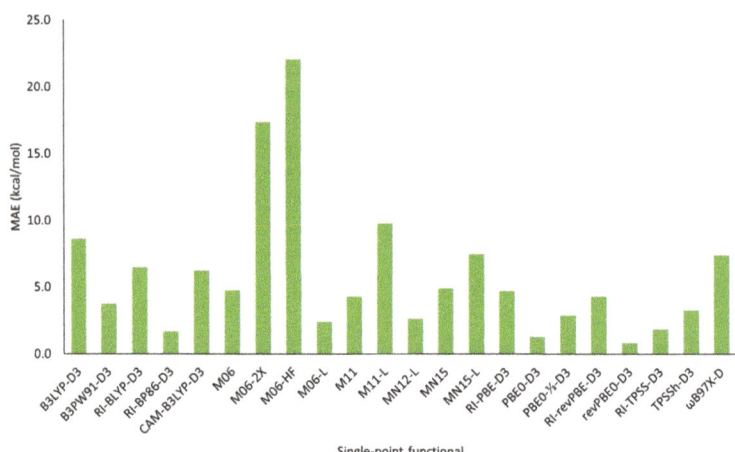

Figure 1. MAE in the calculation of hydricity across each of the three complexes with a range of single-point functionals. Performed with implicit solvation (PCM_{MeCN}) and a def2-TZVP basis on optimised geometries from RI-BP86-D3(PCM_{MeCN})/def2-TZVP.

3.3. Validation of Methodology

We have carefully tested the robustness of the benchmark data summarised in Figure 1 against other choices of methodological details. These comprise:

3.3.1. Choice of Basis Set for Single-Point Energies

Single-point calculations in this work were performed using the def2-TZVP basis from Ahlrichs. The impact of a larger, quadruple-ζ basis, def2-QZVP, was examined on the smallest Fe system for a selection of single-points (Figure S4 in the ESI).

For the additional expense of using a quadruple-ζ basis in the description of the system, there was no significant improvement to be found. Def2-TZVP has been shown to be of sufficient accuracy for single-point calculations and a greater emphasis is placed upon the choice of functional for these 3d TM hydride complexes.

3.3.2. Level of Geometry Optimisation

RI-BP86 was initially chosen for use in geometry optimisation as a 'tried and tested' method [48]. From comparison of BP86 with PBE0, M06 and M06-L, it was found that there was no benefit to the use of the more expensive hybrid, meta-hybrid or meta-GGA functional over the pure functional in the geometry optimisation (MAEs of 1.4, 4.8, 4.5 and 4.3 kcal/mol, respectively, Figure S5 in the ESI).

Furthermore, under consideration was the use of a different basis in the optimisation. In general, use of the largest, triple-ζ, def2-TZVP basis set was found to perform best, with a reduction in MAE compared to the smaller double-ζ, def2-SVP basis. Though, we did find that the description of the hydridic hydrogen with an additional diffuse function, or of the metal centre with a larger basis, was not always found to improve the MAE (see Figure S7 in the ESI). While the triple-ζ basis was found to offer a slight improvement, the associated expense of using this larger basis means that this is impractical for the fractional improvement and the smaller, double-ζ basis is deemed sufficient for geometry optimisation of these systems (MAEs of 1.3 and 1.4 kcal/mol, respectively, from single-points of PBE0-D3(PCM_{MeCN})/def2-TZVP, but with CPU times for the frequency calculation, at the level of geometry optimisation, for Co(dppe)$_2$H of approximately 12.5 days compared to 2 days for def2-TZVP and def2-SVP basis sets, respectively). Additionally, there was little change in the optimised geometry between the use of def2-SVP and def2-TZVP, with central metal ligand bond distances reproduced to within 0.006 Å (see Table S3 in the ESI for a comparison to X-Ray crystal structures from Ciancanelli et al. [81] and Ariyaratne et al. [86]).

Theoretically, for the most accurate description of the system, both dispersion and solvation should be considered and indeed we found that the inclusion of both minimised the MAEs in the calculation of hydricity across the three complexes (Figure S8 in the ESI).

3.3.3. Choice of Solvation Model

The choice of solvent model has been examined at the level of single-point calculations and there was no significant variation of the hydricity values obtained by using three different solvation models; IEF-PCM, C-PCM and SMD (Table S4 in the ESI).

4. Conclusions

We have accurately reproduced experimentally measured values of hydricity for three 3d TM complexes. While a mixture of functionals have been used in the literature for studies on 3d metal homogeneous catalysis, we propose a methodology that has been shown to accurately reproduce a key M−H bond strength, central to the reactivity of these compounds.

While low on the Jacob's ladder of functionals, the pure GGA, BP86, has been shown to produce accurate energetics for the hydricity of 3d TM hydrides. The hybrid functional, PBE0 has also been shown to perform well and is recommended for energy calculations over BP86 due to the improved ability to more reliably differentiate between spin states of 3d TM complexes [17]. The lowest mean absolute errors were found with the inclusion of both dispersion corrections and implicit solvation.

A double-ζ basis, def2-SVP, was used in geometry optimisation with the RI-BP86-D3(PCM) and led to a MAE of 1.4 kcal/mol after evaluation of subsequent single-point at the level of PBE0-D3(PCM)/def2-TZVP. A larger triple-ζ basis, def2-TZVP, used at the stage of the geometry optimisation led to a lower MAE of 1.3 kcal/mol which, was not shown to offer any significant improvement for the additional cost.

For a balance between accuracy and expense, we recommend the methodology of PBE0-D3(PCM)/def2-TZVP//RI-BP86-D3(PCM)/def2-SVP for use on systems involving 3d TM hydride complexes.

Supplementary Materials: The following are available online at, additional computational details and graphical and tabular material.

Author Contributions: Conceptualisation, M.B.; methodology, A.S.G. and M.B.; investigation, A.S.G.; writing—original draft, A.S.G.; Supervision, M.B. All authors have read and agreed to the published version of the manuscript.

Funding: This research received no external funding. The publication of this work received support from the St Andrews Institutional Open Access Fund.

Institutional Review Board Statement: Not applicable.

Informed Consent Statement: Not applicable.

Data Availability Statement: The research data supporting this publication can be accessed at https://doi.org/10.17630/7cf7b7e9-3edc-4ad4-b016-3b9b5548f9ac (accessed on 2 July 2021).

Acknowledgments: We thank EastCHEM and the School of Chemistry for support through the EaSI-CAT program. Calculations were performed on a local compute cluster maintained by H. Früchtl.

Conflicts of Interest: The authors declare no conflict of interest.

References

1. Herrmann, W.A.; Cornils, B. Organometallic Homogeneous Catalysis - Quo vadis? *Angew. Chem. (Int. Ed. Engl.)* **1997**, *36*, 1048–1067. [CrossRef]
2. Reed-Berendt, B.G.; Polidano, K.; Morrill, L.C. Recent advances in homogeneous borrowing hydrogen catalysis using earth-abundant first row transition metals. *Org. Biomol. Chem.* **2019**, *17*, 1595–1607. [CrossRef]
3. Carney, J.R.; Dillon, B.R.; Thomas, S.P. Recent Advances of Manganese Catalysis for Organic Synthesis. *Eur. J. Org. Chem.* **2016**, *2016*, 3912–3929. [CrossRef]
4. Mukherjee, A.; Milstein, D. Homogeneous Catalysis by Cobalt and Manganese Pincer Complexes. *ACS Catal.* **2018**, *8*, 11435–11469. [CrossRef]
5. Bender, T.A.; Dabrowski, J.A.; Gagné, M.R. Homogeneous catalysis for the production of low-volume, high-value chemicals from biomass. *Nat. Rev. Chem.* **2018**, *2*, 35–46. [CrossRef]
6. Garbe, M.; Junge, K.; Beller, M. Homogeneous Catalysis by Manganese-Based Pincer Complexes. *Eur. J. Org. Chem.* **2017**, *2017*, 4344–4362. [CrossRef]
7. Cramer, C.J.; Truhlar, D.G. Density functional theory for transition metals and transition metal chemistry. *Phys. Chem. Chem. Phys.* **2009**, *11*, 10757–10816. [CrossRef]
8. Vogiatzis, K.D.; Polynski, M.V.; Kirkland, J.K.; Townsend, J.; Hashemi, A.; Liu, C.; Pidko, E.A. Computational Approach to Molecular Catalysis by 3d Transition Metals: Challenges and Opportunities. *Chem. Rev.* **2019**, *119*, 2453–2523. [CrossRef] [PubMed]
9. Elangovan, S.; Topf, C.; Fischer, S.; Jiao, H.; Spannenberg, A.; Baumann, W.; Ludwig, R.; Junge, K.; Beller, M. Selective Catalytic Hydrogenations of Nitriles, Ketones, and Aldehydes by Well-Defined Manganese Pincer Complexes. *J. Am. Chem. Soc.* **2016**, *138*, 8809–8814. [CrossRef]
10. Wei, Z.; De Aguirre, A.; Junge, K.; Beller, M.; Jiao, H. Exploring the mechanisms of aqueous methanol dehydrogenation catalyzed by defined PNP Mn and Re pincer complexes under base-free as well as strong base conditions. *Catal. Sci. Technol.* **2018**, *8*, 3649–3665. [CrossRef]
11. Junge, K.; Wendt, B.; Jiao, H.; Beller, M. Iridium-Catalyzed Hydrogenation of Carboxylic Acid Esters. *ChemCatChem* **2014**, *6*, 2810–2814. [CrossRef]
12. Wei, Z.; Junge, K.; Beller, M.; Jiao, H. Hydrogenation of phenyl-substituted CN, CN, CC, CC and CO functional groups by Cr, Mo and W PNP pincer complexes–A DFT study. *Catal. Sci. Technol.* **2017**, *7*, 2298–2307. [CrossRef]
13. Alberico, E.; Lennox, A.J.; Vogt, L.K.; Jiao, H.; Baumann, W.; Drexler, H.J.; Nielsen, M.; Spannenberg, A.; Checinski, M.P.; Junge, H.; et al. Unravelling the Mechanism of Basic Aqueous Methanol Dehydrogenation Catalyzed by Ru-PNP Pincer Complexes. *J. Am. Chem. Soc.* **2016**, *138*, 14890–14904. [CrossRef]
14. Jiao, H.; Junge, K.; Alberico, E.; Beller, M. A Comparative Computationally Study About the Defined M(II) Pincer Hydrogenation Catalysts (M = Fe, Ru, Os). *J. Comput. Chem.* **2016**, *37*, 168–176. [CrossRef] [PubMed]
15. Elangovan, S.; Garbe, M.; Jiao, H.; Spannenberg, A.; Junge, K.; Beller, M. Hydrogenation of Esters to Alcohols Catalyzed by Defined Manganese Pincer Complexes. *Angew. Chem.* **2016**, *128*, 15590–15594. [CrossRef]
16. Ge, H.; Chen, X.; Yang, X. A mechanistic study and computational prediction of iron, cobalt and manganese cyclopentadienone complexes for hydrogenation of carbon dioxide. *Chem. Commun.* **2016**, *52*, 12422–12425. [CrossRef]
17. Gámez, J.A.; Hölscher, M.; Leitner, W. On the applicability of density functional theory to manganese-based complexes with catalytic activity toward water oxidation. *J. Comput. Chem.* **2017**, *38*, 1747–1751. [CrossRef] [PubMed]
18. Qian, F.; Chen, X.; Yang, X. DFT and AIMD prediction of a SNS manganese pincer complex for hydrogenation of acetophenone. *Chem. Phys. Lett.* **2019**, *714*, 37–44. [CrossRef]
19. Chen, X.; Ge, H.; Yang, X. Newly designed manganese and cobalt complexes with pendant amines for the hydrogenation of CO_2 to methanol: A DFT study. *Catal. Sci. Technol.* **2017**, *7*, 348–355. [CrossRef]
20. Passera, A.; Mezzetti, A. Mn(I) and Fe(II)/PN(H)P Catalysts for the Hydrogenation of Ketones: A Comparison by Experiment and Calculation. *Adv. Synth. Catal.* **2019**, *361*, 4691–4706. [CrossRef]
21. Zubar, V.; Lebedev, Y.; Azofra, L.M.; Cavallo, L.; El-Sepelgy, O.; Rueping, M. Hydrogenation of CO_2-Derived Carbonates and Polycarbonates to Methanol and Diols by Metal–Ligand Cooperative Manganese Catalysis. *Angew. Chem. Int. Ed.* **2018**, *57*, 13439–13443. [CrossRef]
22. Garbe, M.; Junge, K.; Walker, S.; Wei, Z.; Jiao, H.; Spannenberg, A.; Bachmann, S.; Scalone, M.; Beller, M. Manganese(I)-Catalyzed Enantioselective Hydrogenation of Ketones Using a Defined Chiral PNP Pincer Ligand. *Angew. Chem.* **2017**, *129*, 11389–11393. [CrossRef]
23. Zhang, L.; Wang, Z.; Han, Z.; Ding, K. Manganese-Catalyzed anti-Selective Asymmetric Hydrogenation of α-Substituted β-Ketoamides. *Angew. Chem.* **2020**, *132*, 15695–15699. [CrossRef]
24. Wang, Y.; Zhu, L.; Shao, Z.; Li, G.; Lan, Y.; Liu, Q. Unmasking the Ligand Effect in Manganese-Catalyzed Hydrogenation: Mechanistic Insight and Catalytic Application. *J. Am. Chem. Soc.* **2019**, *141*, 17337–17349. [CrossRef]
25. Glatz, M.; Stöger, B.; Himmelbauer, D.; Veiros, L.F.; Kirchner, K. Chemoselective Hydrogenation of Aldehydes under Mild, Base-Free Conditions: Manganese Outperforms Rhenium. *ACS Catal.* **2018**, *8*, 4009–4016. [CrossRef]

26. Van Putten, R.; Uslamin, E.A.; Garbe, M.; Liu, C.; Gonzalez-de Castro, A.; Lutz, M.; Junge, K.; Hensen, E.J.; Beller, M.; Lefort, L.; et al. Non-Pincer-Type Manganese Complexes as Efficient Catalysts for the Hydrogenation of Esters. *Angew. Chem. Int. Ed.* **2017**, *56*, 7531–7534. [CrossRef]
27. Ryabchuk, P.; Stier, K.; Junge, K.; Checinski, M.P.; Beller, M. Molecularly Defined Manganese Catalyst for Low-Temperature Hydrogenation of Carbon Monoxide to Methanol. *J. Am. Chem. Soc.* **2019**, *141*, 16923–16929. [CrossRef] [PubMed]
28. Zeng, L.; Yang, H.; Zhao, M.; Wen, J.; Tucker, J.H.; Zhang, X. C1-Symmetric PNP Ligands for Manganese-Catalyzed Enantioselective Hydrogenation of Ketones: Reaction Scope and Enantioinduction Model. *ACS Catal.* **2020**, *10*, 13794–13799. [CrossRef]
29. Liu, C.; van Putten, R.; Kulyaev, P.O.; Filonenko, G.A.; Pidko, E.A. Computational insights into the catalytic role of the base promoters in ester hydrogenation with homogeneous non-pincer-based Mn-P,N catalyst. *J. Catal.* **2018**, *363*, 136–143. [CrossRef]
30. Van Putten, R.; Filonenko, G.A.; Gonzalez De Castro, A.; Liu, C.; Weber, M.; Müller, C.; Lefort, L.; Pidko, E. Mechanistic Complexity of Asymmetric Transfer Hydrogenation with Simple Mn-Diamine Catalysts. *Organometallics* **2019**, *38*, 3187–3196. [CrossRef]
31. Zirakzadeh, A.; de Aguiar, S.R.; Stöger, B.; Widhalm, M.; Kirchner, K. Enantioselective Transfer Hydrogenation of Ketones Catalyzed by a Manganese Complex Containing an Unsymmetrical Chiral PNP' Tridentate Ligand. *ChemCatChem* **2017**, *9*, 1744–1748. [CrossRef]
32. Rawat, K.S.; Mahata, A.; Choudhuri, I.; Pathak, B. Catalytic hydrogenation of CO_2 by manganese complexes: Role of π-acceptor ligands. *J. Phys. Chem. C* **2016**, *120*, 16478–16488. [CrossRef]
33. Yang, X. Hydrogenation of carbon dioxide catalyzed by PNP pincer iridium, iron, and cobalt complexes: A computational design of base metal catalysts. *ACS Catal.* **2011**, *1*, 849–854. [CrossRef]
34. Mondal, B.; Sengupta, K.; Rana, A.; Mahammed, A.; Botoshansky, M.; Dey, S.G.; Gross, Z.; Dey, A. Cobalt corrole catalyst for efficient hydrogen evolution reaction from H_2O under ambient conditions: Reactivity, spectroscopy, and density functional theory calculations. *Inorg. Chem.* **2013**, *52*, 3381–3387. [CrossRef] [PubMed]
35. Ganguly, G.; Malakar, T.; Paul, A. Theoretical studies on the mechanism of homogeneous catalytic olefin hydrogenation and amine-borane dehydrogenation by a versatile boryl-ligand-based cobalt catalyst. *ACS Catal.* **2015**, *5*, 2754–2769. [CrossRef]
36. Della Monica, F.; Vummaleti, S.V.; Buonerba, A.; Nisi, A.D.; Monari, M.; Milione, S.; Grassi, A.; Cavallo, L.; Capacchione, C. Coupling of Carbon Dioxide with Epoxides Efficiently Catalyzed by Thioether-Triphenolate Bimetallic Iron(III) Complexes: Catalyst Structure–Reactivity Relationship and Mechanistic DFT Study. *Adv. Synth. Catal.* **2016**, *358*, 3231–3243. [CrossRef]
37. Murugesan, K.; Wei, Z.; Chandrashekhar, V.G.; Neumann, H.; Spannenberg, A.; Jiao, H.; Beller, M.; Jagadeesh, R.V. Homogeneous cobalt-catalyzed reductive amination for synthesis of functionalized primary amines. *Nat. Commun.* **2019**, *10*, 1–9. [CrossRef] [PubMed]
38. Li, Y.Y.; Tong, L.P.; Liao, R.Z. Mechanism of Water Oxidation Catalyzed by a Mononuclear Iron Complex with a Square Polypyridine Ligand: A DFT Study. *Inorg. Chem.* **2018**, *57*, 4590–4601. [CrossRef] [PubMed]
39. Wiedner, E.S.; Chambers, M.B.; Pitman, C.L.; Bullock, R.M.; Miller, A.J.; Appel, A.M. Thermodynamic Hydricity of Transition Metal Hydrides. *Chem. Rev.* **2016**, *116*, 8655–8692. [CrossRef]
40. Schäfer, A.; Horn, H.; Ahlrichs, R. Fully optimized contracted Gaussian basis sets for atoms Li to Kr. *J. Chem. Phys.* **1992**, *97*, 2571–2577. [CrossRef]
41. Schäfer, A.; Huber, C.; Ahlrichs, R. Fully optimized contracted Gaussian basis sets of triple zeta valence quality for atoms Li to Kr. *J. Chem. Phys.* **1994**, *100*, 5829–5835. [CrossRef]
42. Weigend, F.; Ahlrichs, R. Balanced basis sets of split valence, triple zeta valence and quadruple zeta valence quality for H to Rn: Design and assessment of accuracy. *Phys. Chem. Chem. Phys.* **2005**, *7*, 3297–3305. [CrossRef] [PubMed]
43. Weigend, F. Accurate Coulomb-fitting basis sets for H to Rn. *Phys. Chem. Chem. Phys.* **2006**, *8*, 1057–1065. [CrossRef] [PubMed]
44. Chin, B.; Lough, A.J.; Morris, R.H.; Schweitzer, C.T.; D'Agostino, C. Influence of Chloride versus Hydride on H-H Bonding and Acidity of the Trans Dihydrogen Ligand in the Complexes trans-$[Ru(H_2)X(PR_2CH_2CH_2PR_2)_2]^+$, X = Cl, H, R = Ph, Et. Crystal Structure Determinations of $[RuCl(dppe)_2]PF_6$ and trans-$[Ru(H_2)Cl(dppe)_2]PF_6$. *Inorg. Chem.* **1994**, *33*, 6278–6288. [CrossRef]
45. Guerchais, V.; Astruc, D.; Nunn, C.M.; Cowley, A.H. Generation, Characterization, and Chemistry of the Methylene Complexes $[Fe(\eta^5\text{-}C_5Me_5)(CO)(L)(=CH_2)]^+$ (L = CO, PPh_3) and the X-ray Crystal Structure of $[Fe(\eta^5\text{-}C_5Me_5)(CO)_2(CH_2PPh_3)]^+BF_4^-$. *Organometallics* **1990**, *9*, 1036–1041. [CrossRef]
46. Klug, C.M.; Dougherty, W.G.; Kassel, W.S.; Wiedner, E.S. Electrocatalytic Hydrogen Production by a Nickel Complex Containing a Tetradentate Phosphine Ligand. *Organometallics* **2019**, *38*, 1269–1279. [CrossRef]
47. Neale, S.E. Computational Investigations of Homogeneous Catalysis and Spin-State Energetics in Fe (II) and Co (III) Complexes. Ph.D. Thesis, Heriot-Watt University, Edinburgh, UK, 2019.
48. Bühl, M.; Kabrede, H. Geometries of transition-metal complexes from density-functional theory. *J. Chem. Theory Comput.* **2006**, *2*, 1282–1290. [CrossRef]
49. Grimme, S.; Antony, J.; Ehrlich, S.; Krieg, H. A consistent and accurate ab initio parametrization of density functional dispersion correction (DFT-D) for the 94 elements H-Pu. *J. Chem. Phys.* **2010**, *132*, 154104. [CrossRef]
50. Grimme, S.; Ehrlich, S.; Goerigk, L. Effect of the Damping Function in Dispersion Corrected Density Functional Theory. *J. Comput. Chem.* **2011**, *32*, 1457–1465. [CrossRef]
51. Perdew, J.P.; Ruzsinszky, A.; Tao, J.; Staroverov, V.N.; Scuseria, G.E.; Csonka, G.I. Prescription for the design and selection of density functional approximations: More constraint satisfaction with fewer fits. *J. Chem. Phys.* **2005**, *123*, 062201. [CrossRef]

52. Becke, A.D. Density-functional thermochemistry. III. The role of exact exchange. *J. Chem. Phys.* **1993**, *98*, 5648–5652. [CrossRef]
53. Lee, C.; Yang, W.; Parr, R.G. Development of the Colle-Salvetti correlation-energy formula into a functional of the electron density. *Phys. Rev. B* **1988**, *37*, 785–789. [CrossRef] [PubMed]
54. Miehlich, B.; Savin, A.; Stoll, H.; Preuss, H. Results obtained with the correlation energy density functionals of Becke and Lee, Yang and Parr. *Chem. Phys. Lett.* **1989**, *157*, 200–206. [CrossRef]
55. Perdew, J.P.; Burke, K.; Wang, Y. Generalized gradient approximation for the exchange-correlation hole of a many-electron system. *Phys. Rev. B* **1996**, *54*, 16533–16539. [CrossRef]
56. Perdew, J.P.; Chevary, J.A.; Vosko, S.H.; Jackson, K.A.; Pederson, M.R.; Singh, D.J.; Fiolhais, C. Atoms, molecules, solids, and surfaces: Applications of the generalized gradient approximation for exchange and correlation. *Phys. Rev. B* **1992**, *46*, 6671–6687. [CrossRef] [PubMed]
57. Shi, J.M.; Peeters, F.M.; Hai, G.Q.; Devreese, J.T. Erratum: Donor transition energy in GaAs superlattices in a magnetic field along the growth axis. *Phys. Rev. B* **1993**, *48*, 4978–4978. [CrossRef] [PubMed]
58. Becke, A.D. Density-functional exchange-energy approximation with correct asymptotic behaviour. *Phys. Rev. A* **1988**, *38*, 3098–3100. [CrossRef] [PubMed]
59. Perdew, J.P. Density-functional approximation for the correlation energy of the inhomogeneous electron gas. *Phys. Rev. B* **1986**, *33*, 8822–8824. [CrossRef]
60. Yanai, T.; Tew, D.P.; Handy, N.C. A new hybrid exchange-correlation functional using the Coulomb-attenuating method (CAM-B3LYP). *Chem. Phys. Lett.* **2004**, *393*, 51–57. [CrossRef]
61. Zhao, Y.; Truhlar, D.G. The M06 suite of density functionals for main group thermochemistry, thermochemical kinetics, noncovalent interactions, excited states, and transition elements: Two new functionals and systematic testing of four M06-class functionals and 12 other function. *Theor. Chem. Acc.* **2008**, *120*, 215–241. [CrossRef]
62. Zhao, Y.; Truhlar, D.G. Comparative DFT study of van der Waals complexes: Rare-gas dimers, alkaline-earth dimers, zinc dimer and zinc-rare-gas dimers. *J. Phys. Chem. A* **2006**, *110*, 5121–5129. [CrossRef] [PubMed]
63. Zhao, Y.; Truhlar, D.G. Density functional for spectroscopy: No long-range self-interaction error, good performance for Rydberg and charge-transfer states, and better performance on average than B3LYP for ground states. *J. Phys. Chem. A* **2006**, *110*, 13126–13130. [CrossRef] [PubMed]
64. Zhao, Y.; Truhlar, D.G. A new local density functional for main-group thermochemistry, transition metal bonding, thermochemical kinetics, and noncovalent interactions. *J. Chem. Phys.* **2006**, *125*, 194101. [CrossRef] [PubMed]
65. Peverati, R.; Truhlar, D.G. Improving the accuracy of hybrid meta-GGA density functionals by range separation. *J. Phys. Chem. Lett.* **2011**, *2*, 2810–2817. [CrossRef]
66. Peverati, R.; Truhlar, D.G. M11-L: A local density functional that provides improved accuracy for electronic structure calculations in chemistry and physics. *J. Phys. Chem. Lett.* **2012**, *3*, 117–124. [CrossRef]
67. Peverati, R.; Truhlar, D.G. An improved and broadly accurate local approximation to the exchange-correlation density functional: The MN12-L functional for electronic structure calculations in chemistry and physics. *Phys. Chem. Chem. Phys.* **2012**, *14*, 13171–13174. [CrossRef]
68. Yu, H.S.; He, X.; Li, S.L.; Truhlar, D.G. MN15: A Kohn-Sham global-hybrid exchange-correlation density functional with broad accuracy for multi-reference and single-reference systems and noncovalent interactions. *Chem. Sci.* **2016**, *7*, 5032–5051. [CrossRef]
69. Yu, H.S.; He, X.; Truhlar, D.G. MN15-L: A New Local Exchange-Correlation Functional for Kohn-Sham Density Functional Theory with Broad Accuracy for Atoms, Molecules, and Solids. *J. Chem. Theory Comput.* **2016**, *12*, 1280–1293. [CrossRef] [PubMed]
70. Perdew, J.P.; Burke, K.; Ernzerhof, M. Errata: Generalized Gradient Approximation Made Simple [Phys. Rev. Lett. 77, 3865 (1996)]. *Phys. Rev. Lett.* **1997**, *78*, 1396. [CrossRef]
71. Adamo, C.; Barone, V. Toward reliable density functional methods without adjustable parameters: The PBE0 model. *J. Chem. Phys.* **1999**, *110*, 6158–6170. [CrossRef]
72. Guido, C.A.; Brémond, E.; Adamo, C.; Cortona, P. Communication: One third: A new recipe for the PBE0 paradigm. *J. Chem. Phys.* **2013**, *138*, 021104. [CrossRef]
73. Ernzerhof, M.; Perdew, J.P. Generalized gradient approximation to the angle- and system-averaged exchange hole. *J. Chem. Phys.* **1998**, *109*, 3313–3320. [CrossRef]
74. Tao, J.; Perdew, J.P.; Staroverov, V.N.; Scuseria, G.E. Climbing the density functional ladder: Nonempirical meta–generalized gradient approximation designed for molecules and solids. *Phys. Rev. Lett.* **2003**, *91*, 146401. [CrossRef]
75. Staroverov, V.N.; Scuseria, G.E.; Tao, J.; Perdew, J.P. Comparative assessment of a new nonempirical density functional: Molecules and hydrogen-bonded complexes. *J. Chem. Phys.* **2003**, *119*, 12129–12137; Erratum in *J. Chem. Phys.* **2004**, *121*, 11507. [CrossRef]
76. Chai, J.D.; Head-Gordon, M. Long-range corrected hybrid density functionals with damped atom-atom dispersion corrections. *Phys. Chem. Chem. Phys.* **2008**, *10*, 6615–6620. [CrossRef]
77. Grimme, S. Semiempirical GGA-Type Density Functional Constructed with a Long-Range Dispersion Correction. *J. Comput. Chem.* **2006**, *27*, 1787–1799. [CrossRef]
78. Frisch, M.J.; Trucks, G.W.; Schlegel, H.B.; Scuseria, G.E.; Robb, M.A.; Cheeseman, J.R.; Scalmani, G.; Barone, V.; Mennucci, B.; Petersson, G.A.; et al. *Gaussian 09, Revision D.01*; Gaussian Inc.: Wallingford, CT, USA, 2013.
79. Frisch, M.J.; Trucks, G.W.; Schlegel, H.B.; Scuseria, G.E.; Robb, M.A.; Cheeseman, J.R.; Scalmani, G.; Barone, V.; Petersson, G.A.; Nakatsuji, H.; et al. *Gaussian 16, Revision C.01*; Gaussian Inc.: Wallingford, CT, USA, 2019.

30. Estes, D.P.; Vannucci, A.K.; Hall, A.R.; Lichtenberger, D.L.; Norton, J.R. Thermodynamics of the metal-hydrogen bonds in (η^5-C$_5$H$_5$)M(CO)$_2$H (M = Fe, Ru, Os). *Organometallics* **2011**, *30*, 3444–3447. [CrossRef]
31. Ciancanelli, R.; Noll, B.C.; DuBois, D.L.; Rakowski DuBois, M. Comprehensive thermodynamic characterization of the metal-hydrogen bond in a series of cobalt-hydride complexes. *J. Am. Chem. Soc.* **2002**, *124*, 2984–2992. [CrossRef]
32. Wiedner, E.S.; Appel, A.M.; Dubois, D.L.; Bullock, R.M. Thermochemical and mechanistic studies of electrocatalytic hydrogen production by cobalt complexes containing pendant amines. *Inorg. Chem.* **2013**, *52*, 14391–14403. [CrossRef] [PubMed]
33. Moltved, K.A.; Kepp, K.P. Chemical Bond Energies of 3d Transition Metals Studied by Density Functional Theory. *J. Chem. Theory Comput.* **2018**, *14*, 3479–3492. [CrossRef] [PubMed]
34. Himmel, D.; Goll, S.K.; Leito, I.; Krossing, I. A unified pH scale for all phases. *Angew. Chem. Int. Ed.* **2010**, *49*, 6885–6888. [CrossRef] [PubMed]
35. Bühl, M.; Reimann, C.; Pantazis, D.A.; Bredow, T.; Neese, F. Geometries of third-row transition-metal complexes from density-functional theory. *J. Chem. Theory Comput.* **2008**, *4*, 1449–1459. [CrossRef] [PubMed]
36. Ariyaratne, J.K.P.; Bierrum, A.M.; Green, M.L.H.; Ishaq, M.; Prout, C.K.; Swanwick, M.G. Evidence for near-neighbour interactions in some substituted methyl derivatives of transition metals including the molecular crystal structure determinations of (π-C$_5$H$_5$)Fe(CO)$_2$CH$_2$CO$_2$H and (π-C$_5$H$_5$)Mo(CO)$_3$CH$_2$CO$_2$H. *J. Chem. Soc. A* **1969**, 1309–1321. [CrossRef]

Article

A Reusable Efficient Green Catalyst of 2D Cu-MOF for the Click and Knoevenagel Reaction

Kaushik Naskar [1], Suvendu Maity [1], Himadri Sekhar Maity [2] and Chittaranjan Sinha [1,*]

1. Department of Chemistry, Jadavpur University, Kolkata 700032, India; naskar.kaushik123@gmail.com (K.N.); suvendumaity99@gmail.com (S.M.)
2. Department of Chemistry, Indian Institute of Technology, Kharagpur 721302, India; himadri.maity84@gmail.com
* Correspondence: crsjuchem@gmail.com

Abstract: [Cu(CPA)(BDC)]$_n$ (CPA = 4-(Chloro-phenyl)-pyridin-4-ylmethylene-amine; BDC = 1,4-benzenedicarboxylate) has been synthesized and structurally characterized by single crystal X-Ray diffraction measurement. The structural studies establish the copper (II) containing 2D sheet with (4,4) square grid structure. The square grid lengths are 10.775 and 10.769 Å. Thermal stability is assessed by TGA, and subsequent PXRD data establish the crystallinity. The surface morphology is evaluated by FE-SEM. The N$_2$ adsorption−desorption analysis demonstrates the mesoporous feature (∼6.95 nm) of the Cu-MOF. This porous grid serves as heterogeneous green catalyst with superficial recyclability and thermal stability and facilitates organic transformations efficiently such as, Click and Knoevenagel reactions in the aqueous methanolic medium.

Keywords: 2D Cu-MOF; square grid structure; Click reaction; Knoevenagel reaction; green chemistry

1. Introduction

Metal-organic frameworks (MOFs) are burgeoning targets for their potential applications in gas separation and storage [1–3], energy research [4–10], sensing of ions and molecules [11–13], bio-imaging [14], drug delivery [15,16], reusable and recycling sustainable catalyses [17,18], magnetism [19,20], etc. In the last few decades, such materials have received more attention towards development of more flexible catalytic materials owing to their varying symmetric and large pore volume, high surface area, tunable pore size and versatile functionality together with the diversity of metal knots and their redox states, functional groups, and the retention of crystallinity after catalytic reactions.

Zeolite imidazolate frameworks (ZIFs) [21–23] have arisen as a novel type of highly porous materials along with the advantages in the field of conventional MOF catalyst. By using ZIFs as solid acid catalysts, several organic transformation reactions have been carried out, such as Knoevenagel condensation [24,25], aldol condensation [26,27], Suzuki cross-coupling [28,29], Friedel–Crafts alkylation [30,31], and epoxide ring-opening reaction [32,33]. Current research in this area has been mostly focused to develop low-cost reusable recycling green catalysts to support sustainable development. Designing of multifunctional CPs or MOFs as catalyst to investigate the catalytic efficiency of various organic transformation reactions is the main focus of present research [34,35].

The copper ions play key roles in various types of biological process, such as galactose, tyrosine radical, hemocyanin, etc., also having the ability to transform various organic reactions because of its variable types of redox behaviors. The copper ion containing catalysts can be incorporated in different redox forms which may be available relatively at lower potential such as, (a) Cu(I) salts (usually in the presence of a base and/or a ligand) [36], (b) in situ reduction of Cu(II) salts (e.g., copper sulphate with sodium ascorbate or ascorbic acid) [37], (c) in situ oxidation of Cu(0) metal [38], and (d) easy comproportionation of Cu(0) and Cu(II), generally limited to special applications (e.g., biological systems) [39].

Daturi et al. demonstrated mixed valence state containing MOFs of Cu^{II}/Cu^{I} ions in the HKUST-1 [40]. Moreover, Steiner and Zhang revealed that the mixed valence state Cu-MOF had dual pore size distribution and presented superior water stability compared to HKUST-1 [41]. These studies provide guidance of the catalytic activity of Cu-based MOFs. In recent times, Cu-MOFs (e.g., supported heterogeneous catalyst) have been upgraded to diminish the contamination of metal with the end product and to reuse it.

Generally, Cu(I)-catalyzed azide-alkyne cycloaddition (CuAAC) reaction, where copper present in (+1) oxidation state, is one of the best methods for the preparation of 1,4-disubstituted-1,2,3-triazoles in a regioselective manner, (as a sole regioisomer) either in aqueous or in organic medium. 1,2,3-Triazoles have received considerable interest because of their useful applications in the field of pharmaceutical agents, agrochemicals, photographic materials, etc. Moreover, in situ reduction of copper(II) complexes to copper(I) in the presence of reducing agent (e.g., sodium ascorbate, alcohols, etc.) is the most common and reliable Click reaction conditions during the assembly of diverse molecules [42]. Beside the Click chemistry, copper catalyzed Knoevenagel condensation is widely used reaction in manufacturing numerous chemical compounds which are useful for pharmaceuticals industries [43,44]. In general, this type of reaction is catalyzed by weak bases such as primary, secondary, and tertiary amines under homogeneous conditions, or it may go upward by using 40 mol% catalysts with having difficulties in catalyst recycling and recovery.

Herein, we have successfully fabricated a mesoporous 2D Cu-MOF, $[Cu(CPA)(BDC)]_n$ (CPA = 4-Chloro-phenyl)-pyridin-4-ylmethylene-amine; BDC = 1,4-benzenedicarboxylate) which can facilitate organic transformation reactions highly efficient for Click and Knoevenagel reactions in aqueous medium. The Cu-MOF was produced and well-characterized by PXRD, TGA, SXRD, SEM, BET, etc. and reusable as a heterogeneous green catalyst. Moreover, Cu-MOF offers several advantages like thermal stability and superficial recyclability in contrast with conventional catalysts (Scheme 1).

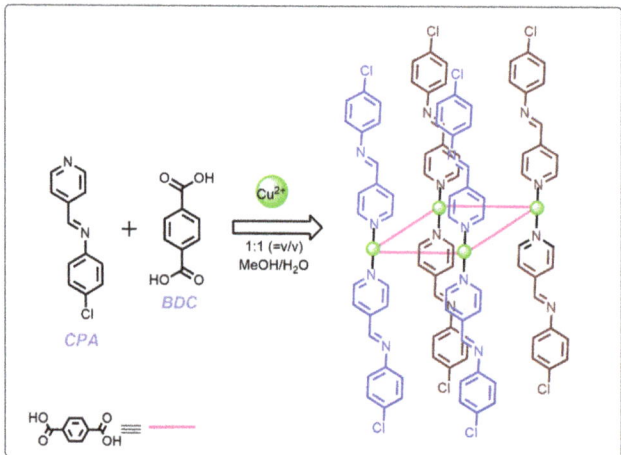

Scheme 1. Schematic representation of synthesis of Cu-MOF (BDC = 1,4-benzenedicarboxylate).

2. Experimental Section

2.1. Synthesis of (4-Chloro-phenyl)-pyridin-4-ylmethylene-amine (CPA)

In the literature, there are few reported procedures [45,46] for the CPA ligand synthesis, but herein, we have modified the synthetic procedure for better yield. A mixture of the appropriately Pyridine-4-carbaldehyde (5 mmol) and the appropriately substituted aniline (5 mmol) in dry toluene (50 mL) is refluxed for 5 h in the presence of molecular sieves (75 g; Davison, grade 514Å, effective pore size 4, 8–12 mesh beads). At the end of the reaction, the molecular sieves are filtered, washed with toluene, and the combined filtrates rotary evaporated to remove toluene. The syrupy residual material is crystallized from hexane or hexane-petroleum ether mixture to yield 97.2%, faint yellow crystalline product; IR spectrum (KBr pellet): v_{imine} C=N, 1629; aromatic C=N, 1617; C-Cl and 687 cm^{-1} (Figure S1); ^{1}H-NMR (DMSO-d$_6$, 400 MHz): δ = 7.37 (2H, d, J = 8.4 Hz), 7.52 (2H, d, J = 8.4 Hz), 7.85 (2H, d, J = 4.4 Hz), 8.17 (1H, s), 8.77 (2H, d, J = 4.4 Hz) (Figure S2); ESI-MS, m/z 217.1077 [M + H]$^+$ (Figure S3); elemental analysis (%) calcd for $C_{12}H_9ClN_2$: C, 66.52; H, 4.19; N, 12.93. Found: 66.67; H, 4.25; N, 12.98.

2.2. Synthesis of Cu-MOF, [Cu(CPA)(BDC)]$_n$

A solution of CPA (0.043 g, 0.2 mmol) in MeOH (5 mL) was slowly and cautiously layered to a solution of Cu(NO$_3$)$_2$·3H$_2$O (0.048 g, 0.2 mmol), in H$_2$O (5 mL) using 5 mL of 1:1 (=v/v) buffer solution of MeOH and H$_2$O, followed by layering of BDC (0.033 g, 0.2 mmol) neutralizing with Et$_3$N (0.021 g, 0.2 mmol) in 5 mL of EtOH. The greenish-blue-colored crystals were obtained after several days with yield 70% (0.031 g). IR spectrum (KBr pellet, cm^{-1}): v_{as}(COO$^-$), 1623; v_{sys}(COO$^-$), 1392 and v(C-Cl), 687 (Figure S1). Elemental analysis (%) calcd for $C_{20}H_{12.42}ClCuN_2O_4$: C 54.13, H 2.82, N 6.31; found: C 54.06, H 2.95, N 6.28.

2.3. Crystal Structure Determination of Cu-MOF

Single dark green coloured crystals suitable for data collection obtained during synthesis were chosen under an optical microscope and mounted on glass fibres and data collection using a Bruker SMART APEX II diffractometer equipped with graphite-monochromated MoKα radiation (λ = 0.71073 Å) at 293 K. Least-squares refinements of all reflections within the hkl range −12 ≤ h ≤ 12, −18 ≤ k ≤ 18, −16 ≤ l ≤ 16 (1) used to evaluate the crystal-orientation matrices and unit cell parameters. The collected data (I > 2σ(I)) were integrated using the SAINT program [47], and the absorption correction was made with SADABS [48]. The molecular structure was solved using the SHELXL-2016/6 [49] package. Non-hydrogen atoms were refined with anisotropic thermal parameters. Hydrogen atoms were placed in their geometrically idealized positions and constrained to ride on their parent atoms. All calculations were carried out using the SHELXL-2016/6 [49], SHELXT 2014/4 [50], and PLATON 99 [51] programs. The crystallographic data of Cu-MOF are depicted in Table 1.

Table 1. Crystal data and refinement parameters for Cu-MOF.

CCDC No.	2094389
formula	$C_{20}H_{12.4}ClCuN_2O_4$
formula weight	443.74
crystal system	monoclinic
space group	$P2_1/n$
a (Å)	10.2699 (4)
b (Å)	15.3188 (6)
c (Å)	14.1931 (6)
β (°)	105.404 (2)
V (Å^3)	2152.68 (15)
T (K)	293 (2)
Z	4
D_{calcd} (g/cm^3)	1.369
μ (mm^{-1})	1.164
λ (Å)	0.71073
θ range (°)	2.98–25.01
total reflections	3795
unique reflections	2336
refined parameters	296
R_1 [a] [I > 2σ(I)]	0.0533
wR_2 [b]	0.1569
Goodness-of-fit	1.008
difference between peak and hole (e·Å^{-3})	0.595–0.443

[a] $R_1 = \Sigma||F_o| - |F_c||/\Sigma|F_o|$, [b] $wR_2 = [\Sigma w(F_o^2 - F_c^2)^2/\Sigma w(F_o^2)^2]^{1/2}$, for 1, $w = 1/[\sigma^2(F_o^2) + (0.0771P)^2 + 3.2709P]$, where $P = (F_o^2 + 2F_c^2)/3$.

2.4. Characterization of MOF Catalyst

The Cu-MOF catalyst samples were measured using a Bruker D8 ADVANCE X-ray diffractometer system with Cu-Kα radiation (λ = 1.541 Å). Powder X-ray diffraction (PXRD) was executed at room temperature, and most of the PXRD patterns of as-synthesized were well-matched with the simulated patterns from single-crystal data. Thus, it signified that the bulk purity of the sample was retained. In addition, the PXRD was verified after the catalytic performance to clarifying the robustness of Cu-MOF shown in Figure S4. The thermal stability of **1** was performed within the temperature range of 32–650 °C under N_2 atmosphere via PerkinElmer–TGA 4000 thermogravimetric analyzer and TGA analysis revealed that Cu-MOF is moderately stable up to ~360 °C (Figure S5). To authenticate, morphology of catalyst samples (before and after) was observed by using the Field Emission Scanning Electron Microscopy (FESEM; JEOL, JSM-6700F) (Figure S6).

2.5. Gas Adsorption Measurements

The nitrogen adsorption–desorption isotherm was studied by using the dehydrated Cu-MOF sample in a Micromeritics ASAP 2420 surface area analyzer at liquid nitrogen temperature (77 K) (see Figure 1a). Prior to gas adsorption, in a tube, pretreated sample was placed and degassed for 4 h at 120 °C to remove the adsorbed solvent.

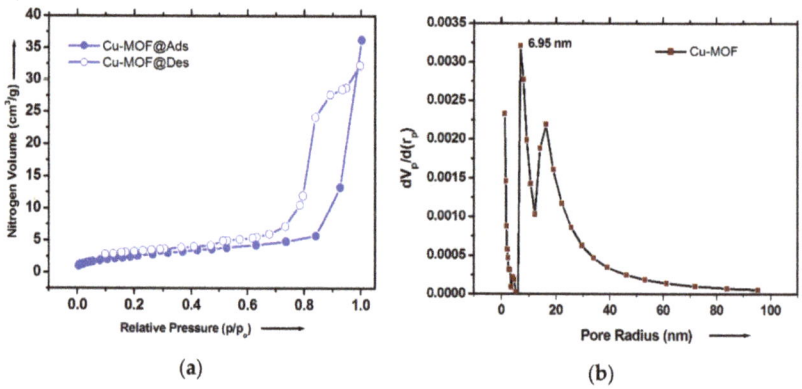

Figure 1. (a) N_2 adsorption isotherms for Cu-MOF at 77 K and (b) Howarth–Kawazoe (HK) pore-size distribution calculated from N_2 adsorption at 77 K for Cu-MOF.

2.6. Catalytic Response

Click Reaction: In a 50 mL round bottom flask, benzyl bromide (1 mmol), sodium azide (1.1 mmol), phenylacetylene derivatives (1 mmol), sodium ascorbate (5 mol%), K_2CO_3 (0.5 equivalent), and 10 mol% Cu-MOF were placed, followed by 5 mL of H_2O-MeOH (1:4) solvent mixture. The flask was connected through a reflux condenser and heated to ~70 °C. After the completion of the reaction, the product was separated by solvent extraction process. The products were purified by column chromatography. Finally, products were characterized via ^{1}H-NMR by using a 500 MHz Bruker NMR and ^{13}C-NMR spectroscopy (Figure S7).

Knoevenagel Reaction: A solution of different substituted benzaldehyde and active methylene compounds such as malononitrile and ethyl cyanoacetate (1:1.2, molar ratio), where the nucleophilic addition occurs by the active hydrogen compound to a carbonyl group followed by a dehydration reaction, and a molecule of water is eliminated. The product is often an α,β-unsaturated ketone in the presence of Cu-MOF (3 mol %) was stirred under room temperature conditions for 15 min (Figure S8). The solid yellow product was isolated with an excellent yield. Similarly, Knoevenagel products were also characterized through ^{1}H-NMR and ^{13}C-NMR spectroscopy by using a 400 MHz and 500 MHz Bruker NMR (See Figure S9). Moreover, the kinetic transformation of the reaction was analyzed through GC. The comparative catalytic activities of some MOFs for the Knoevenagel condensation reaction are summarized in Table S1.

3. Results and Discussion

3.1. Structural Descriptions of [Cu(CPA)(BDC)]$_n$,

The Single Crystal X-ray crystallographic analysis shows that Cu-MOF was crystallized in the monoclinic space group $P2_1/n$ with Z = 4. The asymmetric unit comprises half of the secondary building unit in Figure 2. The paddle-wheel unit of [Cu$_2$(O$_2$CC)$_4$] was repeated along with having a crystallographic inversion center in it. In Cu-MOF, the geometry around each Cu(II) is distortional pentagonal pyramid geometry where four carboxylate groups are bridged by the BDC^{2-}, and in the axial site of the Cu(II), atoms are coordinated by a CPA as a terminal ligand as shown in Figure 2.

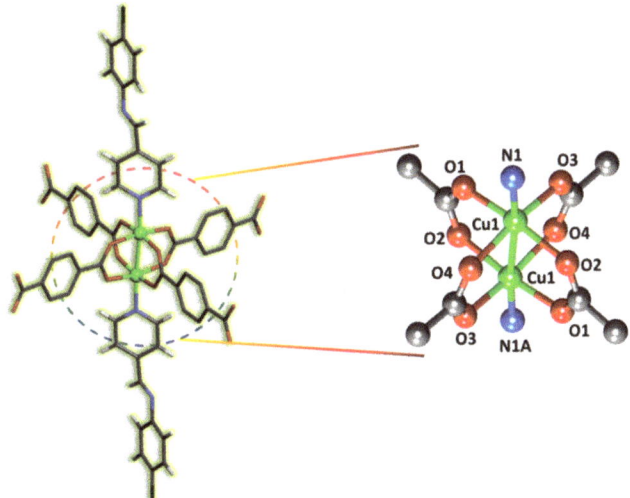

Figure 2. X-ray crystal structure illustrates the copper paddle-wheel secondary building unit in Cu-MOF.

Structural motif of Cu-MOF demonstrates the coordination of BDC^{2-} to Cu(II) metal nodes as bidentate chelating mode and acts as a carboxylato-O donor to Cu(II) (Cu(1)–O(1), 1.955(3) Å; Cu(1)-O(2), 1.958(3) Å; Cu(1)-O(3), 1.961(4) Å; and Cu(1)-O(4), 1.964(3) Å) and the remaining coordinations (axial sites) are originated from the Pyridyl-N donor of CPA ligand (Cu(1)-N(1), 2.137(9) Å; Cu(1)-N(1A), 2.177(8) Å) (Table S2). The paddle-wheel units are further extended by dicarboxylate (BDC^{2-}) bridged spacer ligands to construct the 2D sheet with (4, 4) square grid structure as shown in Figure 3a.

Moreover, these square grid lengths are slightly differ by 10.775 and 10.769 Å as shown in Figure 3a. The axial CPA ligands are projected on both sides of each layer. The empty voids in Cu-MOF are due to the square grids that are interlinked by π–π interactions of CPA ligands from the adjacent layers (Figure 3b). The π–π stacking distances are ~3.974 Å which are facilitated to decrease the interlayer distance between benzene rings, (CPA)···benzene rings (BDC) and thus leading to construct 3D self-assembled framework structure. The solvent-accessible void volume (246.7 Å^3) has been estimated to be 11.5% of the total unit cell volume (2152.7 Å^3).

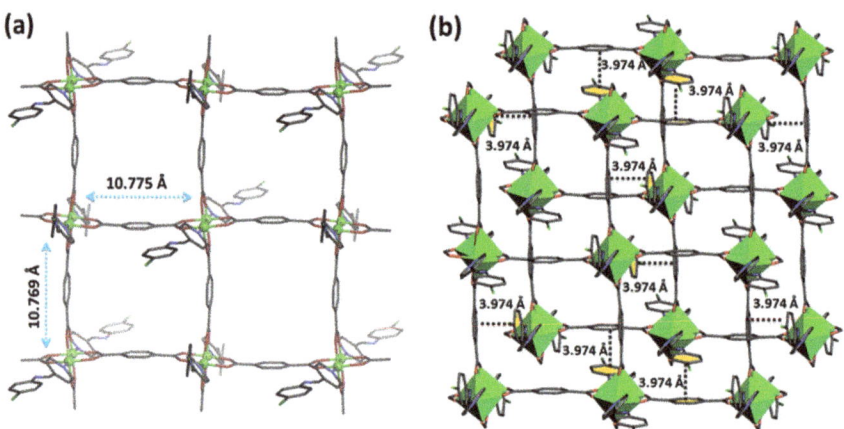

Figure 3. (a) a portion of the (4,4) net structure and (b) a perspective view of the 3D supramolecular aggregate of Cu-MOF by π–π interactions.

The N_2 adsorption–desorption isotherm for Cu-MOF was shown in Figure 1a at 77 K. The Cu-MOF exhibits BET specific surface area ~65.32 m^2 g^{-1}. The observed N_2 adsorption-desorption isotherm is of type-IV category attributing to typically mesoporous structure. A hysteresis loop was observed in the relative pressure (p/p$_0$) range from 0.73 to 0.99, which is of H_2-type typically observed for different size/shape of pores inside particles. Interestingly, such hysteresis loop originated in Cu-MOF may be because of swelling of the network structure by the presence of condensed nitrogen. Again, Nitrogen adsorption/desorption Howarth–Kawazoe (HK) pore size distributions indicated the mesopores with diameter ranges from 7 to 16 nm likely due to the interparticle voids. However, the resulting π–π stacking layered structure has strengthened significant pores in the Cu-MOFs, which may allow the guest molecules to achieve the catalytic performance.

3.2. Catalytic Activity of Cu-MOF

3.2.1. Optimization of Click Reaction

To further explore the catalytic activity of the Cu-MOF catalyst, we have chosen the synthesis of various 1,2,3 triazole derivatives catalyzed by Cu-MOF. In order to get an understanding into the optimum catalytic conditions, the reaction of sodium azide, phenylacetylene, benzyl bromide, and sodium ascorbate has been used as a model reaction. The outcome with variation of solvent, temperature, and amount of Cu-MOF catalyst on the reaction is studied in Table 2. In this reaction, the copper catalyst catalyzed azide-alkyne cycloaddition reaction via redox reaction, where Cu-MOF catalyst acts as heterogeneous catalyst. Generally, this click reaction oxidation state of copper ion changes from Cu(II) to Cu(I). [40] The similar kind of role of copper ion is observed in the catalytic reaction.

A variety of solvents are used for this reaction in the presence of 5 mol% of Cu-MOF catalyst (Table 2, entries 1–11). The yield of the reaction using H_2O-MeOH (1:4) is maximum at 70 °C in comparison to other solvents or mixtures of solvents. After choosing H_2O-MeOH as the solvent for the reaction, the amount of Cu-MOF catalyst and the effect of temperature are also investigated. The yield of 1-benzyl-4-phenyl-1H-1,2,3-triazole is 83% when the catalyst amount is 5 mol%. As evident from Table 2, on increasing the amount of catalyst to 10 mol%, the yield reaches 93%. With an increase in the amount of Cu-MOF catalyst from 10 to 15 mol%, the yield of 1-benzyl-4-phenyl-1H-1,2,3-triazole does not change significantly. Thus, we have performed the reaction in the presence of 10 mol% catalyst as the optimum amount. At room temperature, the yield of the product is 11% (entry 19). Increase in the reaction temperature increases the yield of the product which has been optimized at 90 °C. Even in the absence of Cu-MOF catalyst, no product is isolated (entry 13). Therefore, 10 mol% catalyst at 70 °C in H$_2$O-MeOH (1:4) has been chosen under optimum conditions for the synthesis of 1,2,3-triazole derivatives.

Table 2. Optimization of reaction conditions [a].

Entry	Catalyst	Amount of Catalyst	Solvent	Time (h)	T (°C)	Yield (%) [b]
1	Cu-MOF	5 mol%	Neat	7	70	trace
2	Cu-MOF	5 mol%	DMF	7	70	42
3	Cu-MOF	5 mol%	THF	7	70	43
4	Cu-MOF	5 mol%	CH$_3$CN	7	70	48
5	Cu-MOF	5 mol%	H$_2$O	7	70	68
6	Cu-MOF	5 mol%	Toluene	7	70	29
7	Cu-MOF	5 mol%	Ethanol	7	70	64
8	Cu-MOF	5 mol%	Methanol	7	70	59
9	Cu-MOF	5 mol%	H$_2$O-MeOH	7	70	83
10	Cu-MOF	5 mol%	H$_2$O-THF	7	70	49
11	Cu-MOF	5 mol%	H$_2$O-DMF	7	70	55
12	Cu-MOF	5 mol%	H$_2$O-MeOH	9	70	39
13	—	-	H$_2$O-MeOH	12	80	–
14	Cu-MOF	3 mol%	H$_2$O-MeOH	7	70	54
15	Cu-MOF	5 mol%	H$_2$O-MeOH	7	70 [c]	83
16	Cu-MOF	5 mol%	H$_2$O-MeOH	7	70 [d]	22
17	Cu-MOF	10 mol%	H$_2$O-MeOH	7	70	93
18	Cu-MOF	15 mol%	H$_2$O-MeOH	7	70	93
19	Cu-MOF	10 mol%	H$_2$O-MeOH	7	rt	11
20	Cu-MOF	10 mol%	H$_2$O-MeOH	7	90	93
21	Cu-MOF	10 mol%	H$_2$O-MeOH	7	70 [e]	93
22	Cu-MOF	10 mol%	H$_2$O-MeOH	7	70 [f]	57

[a] Reaction conditions: phenylacetylene (1 mmol), benzyl bromide (1 mmol), sodium azide (1.1 mmol), sodium ascorbate (5 mol%), K$_2$CO$_3$ (0.5 equivalent), solvent (5 mL). [b] Isolated yield. [c] 10 mol% sodium ascorbate. [d] Without sodium ascorbate. [e] K$_2$CO$_3$ (1 equivalent), [f] Without K$_2$CO$_3$.

3.2.2. Substrate Scope

To investigate the generality and versatility of this method, the optimized reaction conditions are subsequently applied to the reaction of a variety of organic halides with different aliphatic and aromatic terminal alkynes (Table 3). It is observed that all reactions underwent efficiently and provided the corresponding products in good to excellent yield within a reasonable time. Electron donating and electron withdrawing substituents in the aromatic ring of the terminal alkyne do not affect significantly the product yield (Table 3, entries 2 and 4). Even the terminal alkyne containing halo-substituted aromatic ring undergoes this reaction very efficiently. The hetero-aryl substituted acetylene, 3-ethynylthiophene shows clean reaction to produce the corresponding 1-benzyl-4-(thiophen-3-yl)-1,2,3-triazole (Table 3, entry 5). Variation of halide moiety in benzyl halides shows that the benzyl bromide is more reactive than benzyl chloride under the optimized reaction conditions. According to the results précised in Table 3, the regioselectivity of the reaction is high in all cases, with only 1,4-regioisomers being produced.

Table 3. Cu-MOF catalyzed one-pot azide preparation from organic halide followed by cycloaddition with terminal alkyne.

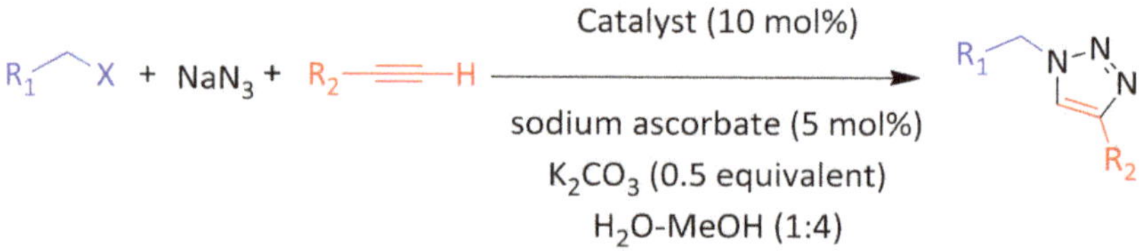

Entry	Aliphatic Halide	Alkyne	Triazole	Yield (%) [a]
1	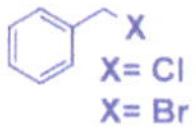 X= Cl X= Br		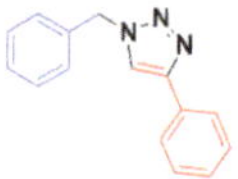	84 93
2	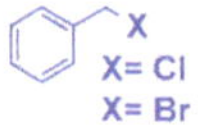 X= Cl X= Br	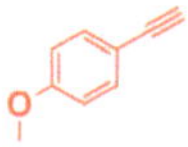	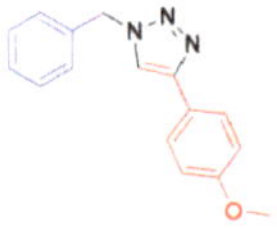	81 89
3	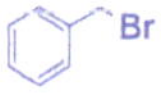 Br	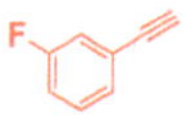	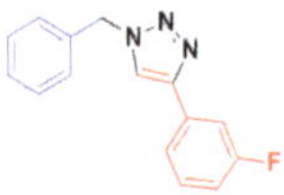	78
4	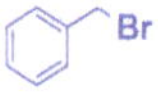 Br	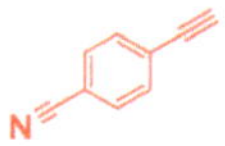	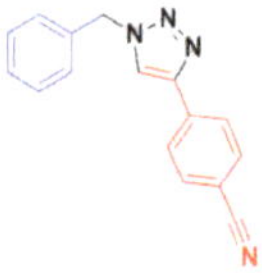	80
5	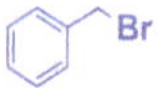 Br		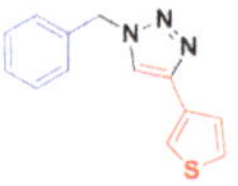	82

[a] **Reaction conditions:** phenylacetylene (1 mmol), benzyl bromide (1 mmol), sodium azide (1.1 mmol), sodium ascorbate (5 mol%), K_2CO_3 (0.5 equivalent), H_2O-MeOH (5 mL).

The catalyst leaching has been studied as follows. The reaction is thus terminated by removal of the catalyst from the reaction mixture at half the reaction time. Residual mixture is kept separated for reaction under the same conditions. Our results indicate that the yield of product does not increase. To determine the amount of copper leaching from the catalyst, ICP-AES analysis of the liquid phase is conducted. The result of ICP-AES analysis shows no detectable amount of copper in the liquid phase. This result indicates that copper ions are not leached from the catalyst surface during the reaction.

3.3. Knoevenagel Condensation Catalytic Experiments

The well-known Knoevenagel condensation is a reaction between an active methylene compound and a carbonyl compound (aldehydes or ketones). This important C−C coupling reaction which is widely used in the synthesis of fine chemicals has also been investigated. The Cu-MOF plays an important role during the reaction of different substituted benzaldehyde with malononitrile and ethyl cyanoacetate (1:1.2, molar ratio) in the presence of Cu-MOF (3 mol %). As inspired by the green chemistry principles, the above reaction is needed to carry out under room temperature conditions. The catalytic transformation is executed in H_2O:MeOH (3:1) (the mixed solvent solution) for 15 min in aerobic condition, and the solid yellow product is isolated with excellent yield (summarized in Table 4). Aliquots are withdrawn from the reaction mixture at different time intervals and thus are analyzed by GC giving kinetic data during the course of the reaction. The GC analysis also indicates that the condensation reaction has finished in 15 min. The Cu-MOF works as the heterogeneous catalysts during this organic transformation. Using this catalyst in the above reaction, the important observations are very little reaction time, room temperature, aqueous medium and high yield product. Such type of reaction can be considered as green chemistry.

Table 4. Knoevenagel condensation catalytic experiments.

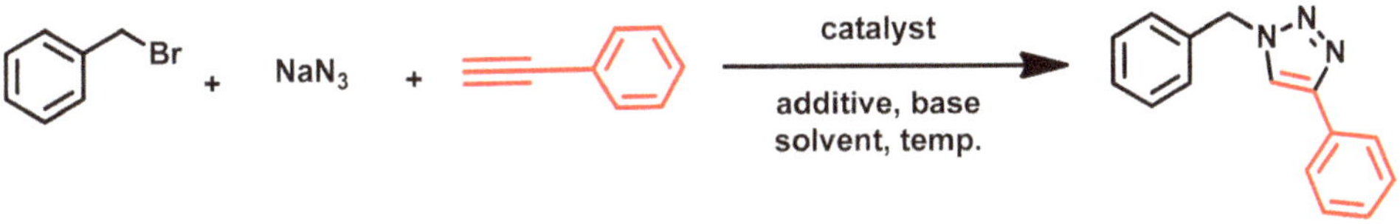

Entry	R_1	Time	Product	Conversion (%)
1.	-OH	15 min	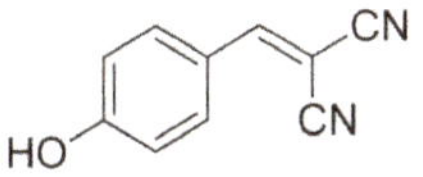 HO-C6H4-CH=C(CN)2	>99
2.	-OH	15 min	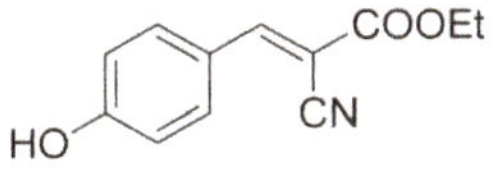 HO-C6H4-CH=C(CN)(COOEt)	>98
3.	-OMe	15 min	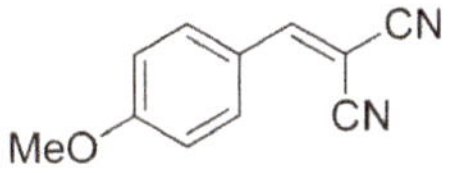 MeO-C6H4-CH=C(CN)2	>99
4.	-OMe	15 min	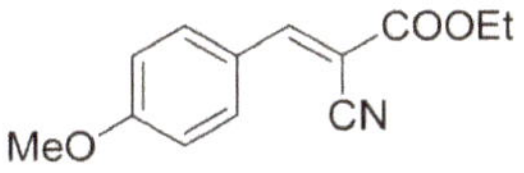 MeO-C6H4-CH=C(CN)(COOEt)	>97
5.	-OMe, -OH	15 min	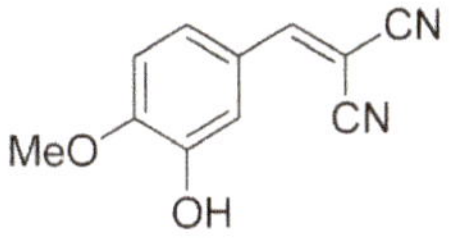 (MeO)(OH)-C6H3-CH=C(CN)2	>99
6.	-OMe, -OH	15 min	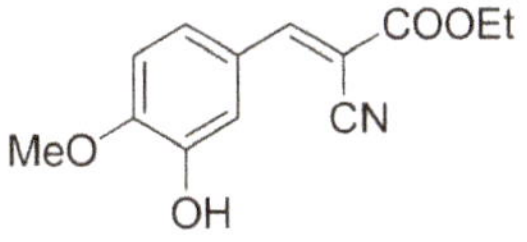 (MeO)(OH)-C6H3-CH=C(CN)(COOEt)	>99
7.	-OMe, -OH	15 min	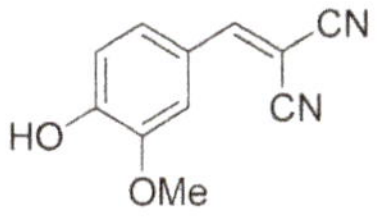 (HO)(OMe)-C6H3-CH=C(CN)2	>99
8.	-OMe, -OH	15 min	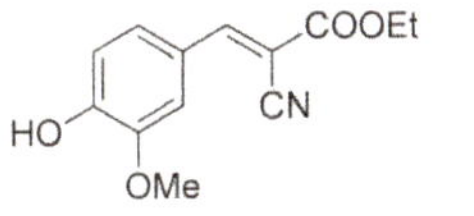 (HO)(OMe)-C6H3-CH=C(CN)(COOEt)	>98

3.4. Recyclability Cu-MOF Catalyst

A recovery and reusability study of the catalyst is performed for the production of 1-benzyl-4-phenyl-1H-1,2,3-triazole. The Cu-MOF catalyst is recovered and dried after washing with ethanol followed by acetone before reuse. We have found recyclability up to four runs although there is marginal linear loss of activity of the catalyst due to lower mass recovery in each step (Figure 4). The Au@Cu(II)-MOF effective heterogeneous catalyst for successive oxidation of alcohol and condensation reactions produce the good yield in toluene at 110 °C in air with more than 95% which is stable up to 5th cycle [52]. The other catalyst Pd(0)@UiO-68-AP which does not contain copper ion exhibits bifunctional heterogeneous catalyst for stepwise organic transformations. This promotes benzyl alcohol oxidation–Knoevenagel condensation in a stepwise way with very good yield (99%) in MeOH at room temperature [53].

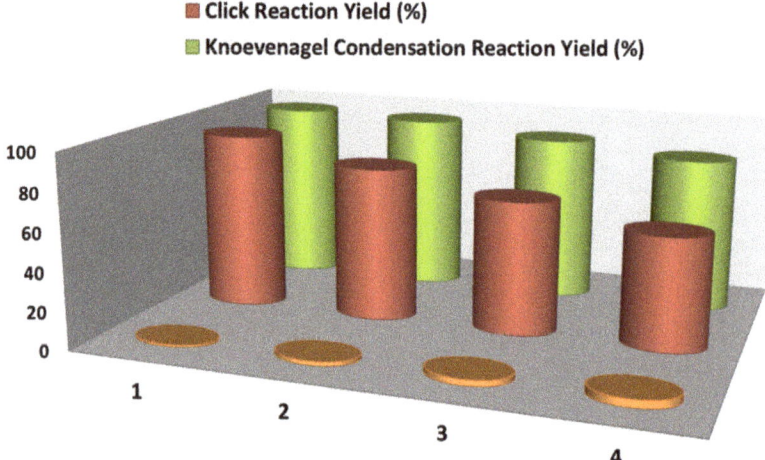

Figure 4. Recyclability chart of Cu-MOF upto 4th cycle.

The heterometallic 3D networks containing [Co^{3+}–Zn^{2+}] and [Co^{3+}–Cd^{2+}] exhibit interesting network topologies including an unprecedented one. However, these networks act as the heterogeneous and reusable catalysts for the Knoevenagel condensation reactions as well as cyanation reactions of assorted aldehydes with good yields [54]. Copper is a non noble and inexpensive abundant metal. Therefore, the use of copper ion in metal organic framework is highly encouraged to develop new catalysts for green reaction.

4. Conclusions

In summary, Cu-MOF catalyst acts as a highly efficient, eco-friendly, bifunctional, and heterogeneous catalytic nature. Therefore, this catalyst readily promotes Knoevenagel condensation reactions and also accelerates Click Reaction to form 1,2,3 triazole derivatives with good catalytic activity as well as excellent conversions even in water and MeOH at room temperature using a low amount of catalyst and a short reaction time. The porous square grid structure of Cu-MOF performs this bifunctional activity in solid catalysts for a broad scope of organic reactions.

Supplementary Materials: The following are available online. Figure S1: IR spectrum of CPA and Cu-MOF (1). Figure S2: ^{1}H-NMR of CPA. Figure S3: ESI-MS of CPA. Figure S4: PXRD plots of Cu-MOF. Figure S5: Thermogravimetric analysis plots of Cu-MOF. Figure S6: FE-SEM image of crystalline morphologies of Cu-MOF. Table S1: The catalytic activity of Some MOFs in the Knoevenagel condensation reaction. Table S2: List of selective bond lengths and bond angles of Cu-MOF, List of ^{1}H-NMR and ^{13}C-NMR spectroscopy of Click Reactions (Figure S7). Figure S8: Solvent dependent catalytic assay of Knoevenagel Reaction. Figure S9: List of ^{1}H-NMR and ^{13}C-NMR spectroscopy of Knoevenagel Reactions.

Author Contributions: K.N., S.M. and C.S. conceived and designed the project; K.N. and H.S.M. performed the experiments; K.N. and S.M. did the crystallography. All authors have read and agreed to the published version of the manuscript.

Funding: The authors are grateful for the financial support by the Council of Scientific and Industrial Research (CSIR, Sanction No. 01(2894)/17/EMR-II), New Delhi, India, and JU RUSA 2.0 Scheme (R-11/100/19).

Institutional Review Board Statement: Not applicable.

Informed Consent Statement: Not applicable.

Data Availability Statement: Not applicable.

Conflicts of Interest: The authors declare no conflict of interest.

Sample Availability: Samples of the compounds are not available from the authors.

References

1. Katz, M.J.; Howarth, A.J.; Moghadam, P.Z.; DeCoste, J.B.; Snurr, R.Q.; Huppa, J.T.; Farha, O.K. High volumetric uptake of ammonia using Cu-MOF-74/Cu-CPO-27. *Dalton Trans.* **2016**, *45*, 4150–4153. [CrossRef]
2. Li, H.; Wang, K.; Sun, Y.; Lollar, C.T.; Li, J.; Zhou, H.-C. Recent advances in gas storage and separation using metal–organic frameworks. *Mater. Today* **2018**, *21*, 108–121. [CrossRef]
3. Forgan, R.S.; Smaldone, R.A.; Gassensmith, J.J.; Furukawa, H.; Cordes, D.B.; Li, Q.; Wilmer, C.E.; Botros, Y.Y.; Snurr, R.Q.; Slawin, A.M.Z.; et al. Nanoporous Carbohydrate Metal–Organic Frameworks. *J. Am. Chem. Soc.* **2012**, *134*, 406–417. [CrossRef]
4. Wang, H.; Zhu, Q.-L.; Zou, R.; Xu, Q. Metal-Organic Frameworks for Energy Applications. *Chem* **2017**, *2*, 52–80. [CrossRef]
5. Rice, A.M.; Leith, G.A.; Ejegbavwo, O.A.; Dolgopolova, E.A.; Shustova, N.B. Heterometallic Metal–Organic Frameworks (MOFs): The Advent of Improving the Energy Landscape. *ACS Energy Lett.* **2019**, *4*, 1938–1946. [CrossRef]
6. Naskar, K.; Dey, A.; Maity, S.; Ray, P.P.; Ghosh, P.; Sinha, C. Biporous Cd(II) Coordination Polymer via in Situ Disulfide Bond Formation: Self-Healing and Application to Photosensitive Optoelectronic Device. *Inorg. Chem.* **2020**, *59*, 5518–5528. [CrossRef]
7. Naskar, K.; Dey, A.; Dutta, B.; Ahmed, F.; Sen, C.; Mir, M.H.; Roy, P.P.; Sinha, C. Intercatenated Coordination Polymers (ICPs) of Carboxylato Bridged Zn(II)-Isoniazid and Their Electrical Conductivity. *Cryst. Growth Des.* **2017**, *17*, 3267–3276. [CrossRef]
8. Naskar, K.; Dey, A.; Maity, S.; Bhunia, M.K.; Ray, P.P.; Sinha, C. Novel porous polycatenated Iodo–cadmium coordination polymer for iodine sorption and electrical conductivity measurement. *Cryst. Growth Des.* **2019**, *19*, 2206–2218. [CrossRef]
9. Naskar, K.; Sil, S.; Sahu, N.; Dutta, B.; Slawin, A.M.Z.; Ray, P.P.; Sinha, C. Enhancement of Electrical Conductivity due to Structural Distortion from Linear to Nonlinear Dicarboxylato-Bridged Zn(II) 1D-Coordination Polymers. *Cryst. Growth Des.* **2019**, *19*, 2632–2641. [CrossRef]
10. Maity, S.; Naskar, K.; Bhowmik, T.; Bera, A.; Weyhermüller, T.; Sinha, C.; Ghosh, P. Coordination polymers of Ag(I) and Hg(I) ions with 2,2'-azobispyridine: Synthesis, characterization and enhancement of conductivity in the presence of Cu(II) ions. *Dalton Trans.* **2020**, *49*, 8438–8442. [CrossRef] [PubMed]
11. Maity, K.; Mukherjee, D.; Sen, M.; Biradha, K. Fluorescent Dye-Based Metal–Organic Framework Piezochromic and Multicolor-Emitting Two-Dimensional Materials for Light-Emitting Devices. *ACS Appl. Nano Mater.* **2019**, *2*, 1614–1620. [CrossRef]
12. Khatua, S.; Goswami, S.; Biswas, S.; Tomar, K.; Jena, H.S.; Konar, S. Stable Multiresponsive Luminescent MOF for Colorimetric Detection of Small Molecules in Selective and Reversible Manner. *Chem. Mater.* **2015**, *27*, 5349–5360. [CrossRef]
13. Naskar, K.; Bhanja, A.K.; Paul, S.; Pal, K.; Sinha, C. Trace Quantity Detection of $H_2PO_4^-$ by Fluorescent Metal–Organic Framework (FMOF) and Bioimaging Study. *Cryst. Growth Des.* **2020**, *20*, 6453–6460. [CrossRef]
14. Robison, L.; Zhang, L.; Drout, R.J.; Li, P.; Haney, C.R.; Brikha, A.; Noh, H.; Mehdi, B.L.; Browning, N.D.; Dravid, V.P.; et al. A Bismuth Metal–Organic Framework as a Contrast Agent for X-ray Computed Tomography. *ACS Appl. Bio Mater.* **2019**, *2*, 1197–1203. [CrossRef]
15. Zhao, H.; Hou, S.; Zhao, X.; Liu, D. Adsorption and pH-Responsive Release of Tinidazole on Metal–Organic Framework CAU-1. *J. Chem. Eng. Data* **2019**, *64*, 1851–1858. [CrossRef]
16. Lin, S.X.; Pan, W.L.; Niu, R.J.; Liu, Y.; Chen, J.X.; Zhang, W.H.; Lang, J.P.; Young, D.J. Effective loading of cisplatin into a nanoscale UiO-66 metal-organic framework with preformed defects. *Dalton Trans.* **2019**, *48*, 5308–5314. [CrossRef] [PubMed]
17. Liang, J.; Liang, Z.B.; Zou, R.Q.; Zhao, Y.L. Heterogeneous Catalysis in Zeolites, Mesoporous Silica, and Metal-Organic Frameworks. *Adv. Mater.* **2017**, *29*, 1701139. [CrossRef]
18. Dhakshinamoorthy, A.; Garcia, H. Metal–organic frameworks as solid catalysts for the synthesis of nitrogen-containing heterocycles. *Chem. Soc. Rev.* **2014**, *43*, 5750–5765. [CrossRef]
19. Abdelbaky, M.S.M.; Amghouz, Z.; Blanco, D.M.; García-Granda, S.; García, J.R. Crystal structure and characterization of a novel layered copper-lithium phosphonate with antiferromagnetic intrachain Cu(II)•••Cu(II) interactions. *J. Solid State Chem.* **2017**, *248*, 61–67. [CrossRef]

20. Jana, S.; Ray, A.; Chandra, A.; El Fallah, M.S.; Das, S.; Sinha, C. Studies on Magnetic and Dielectric Properties of Antiferromagnetically Coupled Dinuclear Cu(II) in a One-Dimensional Cu(II) Coordination Polymer. *ACS Omega* **2020**, *5*, 274–280. [CrossRef]
21. Park, K.S.; Ni, Z.; Côté, A.P.; Choi, J.Y.; Huang, R.; Uribe-Romo, F.J.; Chae, H.K.; O'Keeffe, M.; Yaghi, O.M. Exceptional chemical and thermal stability of zeolitic imidazolate frameworks. *Proc. Natl. Acad. Sci. USA* **2006**, *103*, 10186–10191. [CrossRef] [PubMed]
22. Qutaish, H.; Lee, J.; Hyeon, Y.; Han, S.A.; Lee, I.-H.; Heo, Y.-U.; Whang, D.; Moon, J.; Park, M.-S.; Kim, J.H. Design of cobalt catalysed carbon nanotubes in bimetallic zeolitic imidazolate frameworks. *Appl. Surf. Sci.* **2021**, *547*, 149134. [CrossRef]
23. Hayashi, H.; Côté, A.P.; Furukawa, H.; O'Keeffe, M.; Yaghi, O.M. Zeolite A imidazolate frameworks. *Nat. Mater.* **2007**, *6*, 501–506. [CrossRef] [PubMed]
24. Zanon, A.; Chaemchuen, S.; Verpoort, F. Zn@ZIF-67 as Catalysts for the Knoevenagel Condensation of Aldehyde Derivatives with Malononitrile. *Catal. Lett.* **2017**, *147*, 2410–2420. [CrossRef]
25. Horiuchi, Y.; Toyao, T.; Fujiwaki, M.; Dohshi, S.; Kim, T.-H.; Matsuoka, M. Zeolitic imidazolate frameworks as heterogeneous catalysts for a one-pot P–C bond formation reaction via Knoevenagel condensation and phospha-Michael addition. *RSC Adv.* **2015**, *5*, 24687–24690. [CrossRef]
26. Fan, H.; Yang, Y.; Song, J.; Ding, G.; Wu, C.; Yang, G.; Han, B. One-pot sequential oxidation and aldol-condensation reactions of veratryl alcohol catalyzed by the Ru@ZIF-8 + CuO/basic ionic liquid system. *Green Chem.* **2014**, *16*, 600–604. [CrossRef]
27. Vermoortele, F.; Ameloot, R.; Vimont, A.; Serre, C.; De Vos, D. An amino-modified Zr-terephthalate metal–organic framework as an acid–base catalyst for cross-aldol condensation. *Chem. Commun.* **2011**, *47*, 1521–1523. [CrossRef]
28. Kim, S.; Jee, S.; Choi, K.M.; Shin, D.-S. Single-atom Pd catalyst anchored on Zr-based metal-organic polyhedra for Suzuki-Miyaura cross coupling reactions in aqueous media. *Nano Res.* **2021**, *14*, 486–492. [CrossRef]
29. Gong, X.-F.; Zhang, L.-Y.; Zhang, H.-X.; Cui, Y.-M.; Jin, F.-C.; Liu, Y.; Zhai, Y.-F.; Li, J.-H.; Liu, G.-Y.; Zeng, Y.-F. Highly Active Heterogeneous $PdCl_2$/MOF Catalyst for Suzuki–Miyaura Cross-Coupling Reactions of Aryl Chloride. *Z. Anorg. Allg. Chem.* **2020**, *646*, 1336–1341. [CrossRef]
30. Nguyen, L.T.L.; Le, K.K.A.; Phan, N.T.S. A Zeolite Imidazolate Framework ZIF-8 Catalyst for Friedel-Crafts Acylation. *Chin. J. Catal.* **2012**, *33*, 688–696. [CrossRef]
31. Calleja, G.; Sanz, R.; Orcajo, G.; Briones, D.; Leo, P.; Martínez, F. Copper-based MOF-74 material as effective acid catalyst in Friedel–Crafts acylation of anisole. *Catal. Today* **2014**, *227*, 130–137. [CrossRef]
32. Nagarjun, N.; Concepcion, P.; Dhakshinamoorthy, A. MIL-101(Fe) as an active heterogeneous solid acid catalyst for the regioselective ring opening of epoxides by indoles. *Mol. Catal.* **2020**, *482*, 110628. [CrossRef]
33. Srivastava, D.; Rani, P.; Srivastava, R. ZIF-8-Nanocrystalline Zirconosilicate Integrated Porous Material for the Activation and Utilization of CO_2 in Insertion Reactions. *Chem. Asian J.* **2020**, *15*, 1132–1139. [CrossRef]
34. Lee, J.Y.; Farha, O.K.; Roberts, J.; Scheidt, K.A.; Nguyen, S.B.T.; Hupp, J.T. Metal–organic framework materials as catalysts. *Chem. Soc. Rev.* **2009**, *38*, 1450. [CrossRef]
35. Liu, J.; Chen, L.; Cui, H.; Zhang, J.; Zhang, L.; Su, C.-Y. Applications of metal–organic frameworks in heterogeneous supramolecular catalysis. *Chem. Soc. Rev.* **2014**, *43*, 6011–6061. [CrossRef]
36. Gawande, M.B.; Goswami, A.; Felpin, F.-X.; Asefa, T.; Huang, X.; Silva, R.; Zou, X.; Zboril, R.; Varma, R.S. Cu and Cu-Based Nanoparticles: Synthesis and Applications in Catalysis. *Chem. Rev.* **2016**, *116*, 3722–3811. [CrossRef] [PubMed]
37. Berg, R.; Straub, B.F. Advancements in the mechanistic understanding of the copper-catalyzed Azide—Alkyne cycloaddition. *Beilstein J. Org. Chem.* **2013**, *9*, 2715–2750. [CrossRef] [PubMed]
38. Cao, J.; Rinaldi, A.; Plodinec, M.; Huang, X.; Willinger, E.; Hammud, A.; Hieke, S.; Beeg, S.; Gregoratti, L.; Colbea, C.; et al. In situ observation of oscillatory redox dynamics of copper. *Nat. Commun.* **2020**, *11*, 3554. [CrossRef] [PubMed]
39. Fu, F.; Martinez, A.; Wang, C.; Ciganda, R.; Yate, L.; Escobar, A.; Moya, S.; Fouquet, E.; Ruiz, J.; Astruc, D. Exposure to air boosts CuAAC reactions catalyzed by PEG-stabilized Cu nanoparticles. *Chem. Commun.* **2017**, *53*, 5384–5387. [CrossRef] [PubMed]
40. Szanyi, J.; Daturi, M.; Clet, G.; Baerc, D.R.; Peden, C.H.F. Well-studied Cu-BTC still serves surprises: Evidence for facile Cu^{2+}/Cu^+ interchange. *Phys. Chem. Chem. Phys.* **2012**, *14*, 4383–4390. [CrossRef]
41. Ahmed, A.; Robertson, C.M.; Steiner, A.; Whittles, T.; Ho, A.; Dhanak, V.; Zhang, H. Cu(i)Cu(ii)BTC, a microporous mixed-valence MOF via reduction of HKUST-1. *RSC Adv.* **2016**, *6*, 8902–8905. [CrossRef]
42. Pasini, D. The Click Reaction as an Efficient Tool for the Construction of Macrocyclic Structures. *Molecules* **2013**, *18*, 9512–9530. [CrossRef]
43. Schneider, E.M.; Zeltner, M.; Kränzlin, N.; Grass, R.N.; Stark, W.J. Base-free Knoevenagel condensation catalyzed by copper metal surfaces. *Chem. Commun.* **2015**, *51*, 10695–10698. [CrossRef]
44. Pandey, R.; Singh, D.; Thakur, N.; Raj, K.K. Catalytic C–H Bond Activation and Knoevenagel Condensation Using Pyridine-2,3 Dicarboxylate-Based Metal–Organic Frameworks. *ACS Omega* **2021**, *6*, 13240–13259. [CrossRef]
45. Tehrania, A.A.; Morsalia, A.; Kubicki, M. The role of weak hydrogen and halogen bonding interactions in the assembly of a series of Hg(II) coordination polymers. *Dalton Trans.* **2015**, *44*, 5703–5712. [CrossRef] [PubMed]
46. Kouznetsov, V.V.; Robles-Castellanos, M.L.; Sojo, F.; Rojas-Ruiz, F.A.; Arvelo, F. Diverse C-6 substituted 4-methyl-2-(2-, 3- and 4-pyridinyl)quinolines: Synthesis, in vitro anticancer evaluation and in silico studies. *Med. Chem. Res.* **2017**, *26*, 551–561. [CrossRef]
47. *SMART and SAINT*; Bruker AXS Inc.: Madison, WI, USA, 1998.

48. *Bruker*; SADABS; Bruker AXS Inc.: Madison, WI, USA, 2001.
49. Sheldrick, G.M. *SHELXL 2014, SHELXL-2016/6 and SHELXL-2017/1*; Program for Crystal Structure Solution, University of Göttingen: Göttingen, Germany, 2017.
50. Sheldrick, G.M. SHELXT—Integrated space-group and crystal-structure determination. *Acta Cryst.* **2015**, *A71*, 3–8. [CrossRef]
51. Spek, A.L. Structure validation in chemical crystallography. *Acta Cryst.* **2009**, *D65*, 148–155. [CrossRef] [PubMed]
52. Wang, J.-S.; Jin, F.-Z.; Ma, H.-C.; Li, X.-B.; Liu, M.-Y.; Kan, J.-L.; Chen, G.-J.; Dong, Y.-B. Au@Cu(II)-MOF: Highly Efficient Bifunctional Heterogeneous Catalyst for Successive Oxidation–Condensation Reactions. *Inorg. Chem.* **2016**, *55*, 6685–6691. [CrossRef] [PubMed]
53. Li, Y.-A.; Yang, S.; Liu, Q.-K.; Chen, G.-J.; Ma, J.-P.; Dong, Y.-B. Pd(0)@UiO-68-AP: Chelation-directed bifunctional heterogeneous catalyst for stepwise organic transformations. *Chem. Commun.* **2016**, *52*, 6517–6520. [CrossRef] [PubMed]
54. Srivastava, S.; Aggarwal, H.; Gupta, R. Three-Dimensional Heterometallic Coordination Networks: Syntheses, Crystal Structures, Topologies, and Heterogeneous Catalysis. *Cryst. Growth Des.* **2015**, *15*, 4110–4122. [CrossRef]

Article

Synthesis, Structural and Physicochemical Characterization of a Titanium(IV) Compound with the Hydroxamate Ligand N,2-Dihydroxybenzamide

Stamatis S. Passadis [1], Sofia Hadjithoma [2], Panagiota Siafarika [3], Angelos G. Kalampounias [3,4,*], Anastasios D. Keramidas [2,*], Haralampos N. Miras [5] and Themistoklis A. Kabanos [1,*]

1. Section of Inorganic and Analytical Chemistry, Department of Chemistry, University of Ioannina, 45110 Ioannina, Greece; stamatispassadis@hotmail.com
2. Department of Chemistry, University of Cyprus, Nicosia 2109, Cyprus; hadjithoma.sofia@ucy.ac.cy
3. Physical Chemistry Laboratory, Department of Chemistry, University of Ioannina, 45110 Ioannina, Greece; p.siafarika@uoi.gr
4. Institute of Materials Science and Computing, University Research Center of Ioannina (URCI), 45110 Ioannina, Greece
5. School of Chemistry, University of Glasgow, Glasgow G12 8QQ, UK
* Correspondence: akalamp@uoi.gr (A.G.K.); akeramid@ucy.ac.cy (A.D.K.); Charalampos.moiras@glasgow.ac.uk (H.N.M.); tkampano@uoi.gr (T.A.K.)

Citation: Passadis, S.S.; Hadjithoma, S.; Siafarika, P.; Kalampounias, A.G.; Keramidas, A.D.; Miras, H.N.; Kabanos, T.A. Synthesis, Structural and Physicochemical Characterization of a Titanium(IV) Compound with the Hydroxamate Ligand N,2-Dihydroxybenzamide. *Molecules* **2021**, *26*, 5588. https://doi.org/10.3390/molecules26185588

Academic Editors: William T. A. Harrison and Alan Aitken

Received: 23 August 2021
Accepted: 9 September 2021
Published: 15 September 2021

Publisher's Note: MDPI stays neutral with regard to jurisdictional claims in published maps and institutional affiliations.

Copyright: © 2021 by the authors. Licensee MDPI, Basel, Switzerland. This article is an open access article distributed under the terms and conditions of the Creative Commons Attribution (CC BY) license (https://creativecommons.org/licenses/by/4.0/).

Abstract: The siderophore organic ligand N,2-dihydroxybenzamide (H_2dihybe) incorporates the hydroxamate group, in addition to the phenoxy group in the ortho-position and reveals a very rich coordination chemistry with potential applications in medicine, materials, and physical sciences. The reaction of H_2dihybe with $TiCl_4$ in methyl alcohol and KOH yielded the tetranuclear titanium oxo-cluster (TOC) [$Ti^{IV}_4(\mu\text{-}O)_2(HOCH_3)_4(\mu\text{-Hdihybe})_4(Hdihybe)_4$]$Cl_4 \cdot 10H_2O \cdot 12CH_3OH$ (1). The titanium compound was characterized by single-crystal X-ray structure analysis, ESI-MS, ^{13}C, and ^{1}H NMR spectroscopy, solid-state and solution UV–Vis, IR vibrational, and luminescence spectroscopies and molecular orbital calculations. The inorganic core $Ti_4(\mu\text{-}O)_2$ of 1 constitutes a rare structural motif for discrete Ti^{IV}_4 oxo-clusters. High-resolution ESI-MS studies of 1 in methyl alcohol revealed the presence of isotopic distribution patterns which can be attributed to the tetranuclear clusters containing the inorganic core {$Ti_4(\mu\text{-}O)_2$}. Solid-state IR spectroscopy of 1 showed the presence of an intense band at ~800 cm^{-1} which is absent in the spectrum of the H_2dihybe and was attributed to the high-energy $\nu(Ti_2\text{-}\mu\text{-}O)$ stretching mode. The $\nu(C=O)$ in 1 is red-shifted by ~10 cm^{-1}, while the $\nu(N\text{-}O)$ is blue-shifted by ~20 cm^{-1} in comparison to H_2dihybe. Density Functional Theory (DFT) calculations reveal that in the experimental and theoretically predicted IR absorbance spectra of the ligand and Ti-complex, the main bands observed in the experimental spectra are also present in the calculated spectra supporting the proposed structural model. ^{1}H and ^{13}C NMR solution (CD$_3$OD) studies of 1 reveal that it retains its integrity in CD$_3$OD. The observed NMR changes upon addition of base to a CD$_3$OD solution of 1, are due to an acid–base equilibrium and not a change in the Ti^{IV} coordination environment while the decrease in the complex's lability is due to the improved electron-donating properties which arise from the ligand deprotonation. Luminescence spectroscopic studies of 1 in solution reveal a dual narrow luminescence at different excitation wavelengths. The TOC 1 exhibits a band-gap of 1.98 eV which renders it a promising candidate for photocatalytic investigations.

Keywords: titanium(IV) oxo-clusters; band gap modification; multinuclear NMR; ESI-MS studies; photoluminescence

1. Introduction

The design, synthesis, and physicochemical characterization of polyoxo-titanium clusters (PTCs) have been an active research area over the last decade, due to their interesting electronic properties for various applications in nanotechnology [1–6], photocatalytic hydrogen production [7–9], degradation of environmental pollutants [10,11], catalysis [12–17], solar energy conversion [18,19], and copolymerization of carbon dioxide [20]. At this point, it is worth noting that according to an excellent review, published very recently, the chemistry of group IV elements is underexplored [21].

The knowledge of the structural features of PTCs from single-crystal X-ray structure analysis is of fundamental importance to predict the binding modes of various ligands to TiO_2 since the PTCs are considered solution-processable molecular analogs of TiO_2. Moreover, the 3.20 eV band-gap of TiO_2 limits its applications in photocatalysis [22,23]. However, the use of strong organic chelators allows the modulation of the band-gap with subsequent Vis-NIR absorption by PTCs to appropriate values, which is a fundamentally important parameter for practical applications [24].

Siderophores are low molecular weight organic compounds that are produced by microorganisms and plants suffering from iron deficiency [25]. Siderophores have received much attention in recent years due to their potential applications in environmental research [26]. Siderophores are divided into three main families depending on the characteristic functional group, i.e., hydroxamates, catecholates, and carboxylates [27].

Metal-hydroxamate has greater resistance to hydrolysis [28] in comparison to carboxylic acids and greater electronic coupling over carboxylic and phosphonic acids that ensures efficient electron transport [29,30]. The main binding modes of hydroxamate with metal ions are shown in Figure S1 [28].

The organic molecule N,2-dihydroxybenzamide (H_2dihybe) (Scheme 1A) incorporates a hydroxamate group in addition to the phenoxy group in the ortho-position and exhibits a rich coordination chemistry with many transition metals [31–40] with potential applications in various fields ranging from medicine [41–43] to materials [44], and physical sciences [40,45].

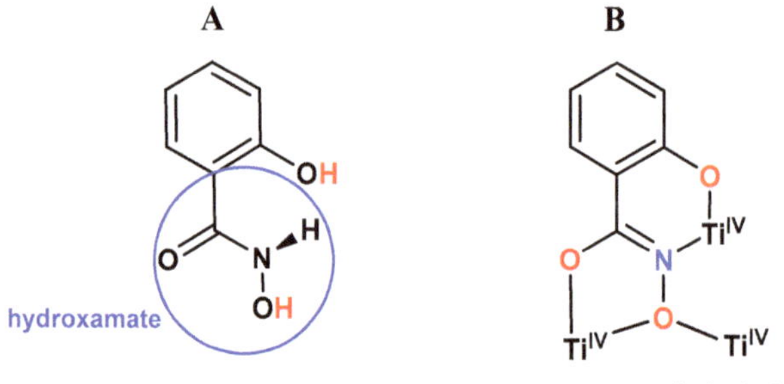

Scheme 1. The ligand used in this study (**A**) and the μ_3-η^1:η^2:η^1:η^1 binding mode of the ligand H_2dihybe, in its iminol form, in the reported [46] three Ti^{IV}/H_2dihybe TOCs (**B**).

The ligand H_2dihybe has been used previously in the synthesis of titanium(IV) oxoclusters [47] under hydrothermal conditions to give three TOCs, namely: [$Ti_6(\mu$-O)(μ_3-O)$_2$ (O^{i}Pr)$_{10}$(OOCCH$_3$)$_2$(dihybe)$_2$]; [$Ti_7(\mu_3$-O)$_2$(OEt)$_{18}$(dihybe)$_2$]; and [$Ti_{12}(\mu$-O)$_4(\mu_3$-O)$_4$ (OEt)$_{20}$ (dihybe)$_4$]. The ligand in these three TOCs interacts with the titanium(IV) in its iminol tri-deprotonated μ_3-η^1:η^2:η^1:η^1 form (Scheme 1B).

The formation of various metal–siderophore complexes (nuclearity-binding modes of siderophore) among other factors greatly depends on the pH of the reaction mixture since there is a competition between free protons and metal ions for the free siderophore ligands.

To further explore the coordination chemistry of N,2-dihydroxybenzamide (H_2dihybe) (Scheme 1) with titanium(IV), we explored the interaction of $Ti^{IV}Cl_4$ with H_2dihybe at low pH under mild conditions (room temperature). Herein, we report the synthesis, structural, and physicochemical characterization of the tetranuclear TOC [$Ti^{IV}_4(\mu$-O)$_2$(HOCH$_3$)$_4$ (μ-Hdihybe)$_4$(Hdihybe)$_4$]Cl$_4$·10H$_2$O·12CH$_3$OH (**1**). The cluster **1** constitutes a rare example of a discrete TOC containing the inorganic core {$Ti_4(\mu$-O)$_2$}. Spectroscopic studies in the solid state revealed a reduced bandgap value of 1.98 eV. In addition, this compound produced dual emission because of changes in the emitting light with variations in the excitation wavelength. The materials with dual emission properties such as **1** can be applied in OLED devices, allowing tuning of the color of the emitting light depending on the voltage applied to the device [47].

2. Results and Discussion

2.1. Synthesis of **1** and Comparison with the Reported Higher Nuclearity TOCs/H_2dihybe

The synthesis of the TOC **1** takes place according to equation 1 and the produced HCl is responsible for the low pH (1.5) of the system and presumably for the mono-deprotonation of the eight ligands of **1**.

$$4TiCl_4 + 8H_2dihybe + 8KOH \rightarrow [Ti_4(\mu\text{-}O)_2(Hdihybe)_8]Cl_4 + 8KCl + 4HCl + 6H_2O(l) \quad (1)$$

In Equation (1), the molar ratio (mr) of $TiCl_4/H_2$dihybe is 1:2, while in the reported [46] Ti^{IV}/H_2dihybe TOCs {[$Ti_6(\mu\text{-}O)_2(\mu_3\text{-}O)_2(O^iPr)_{10}(OOCCH_3)_2(dihybe)_2$]; [$Ti_7(\mu_3\text{-}O)_2(OEt)_{18}$(dihybe)$_2$]; and [$Ti_{12}(\mu\text{-}O)_4(\mu_3\text{-}O)_4(OEt)_{20}(dihybe)_4$]} the mr of $Ti(O^iPr)_4/H_2$dihybe is four and this means that more positions of the coordination sphere of titanium(IV), in our case, are occupied by the donor atoms of the ligand which precludes the formation of bigger clusters. On the other hand, the use of $Ti(O^iPr)_4$ with the very strong base $^-O^iPr$, [deprotonated HO^iPr (pKa = 16.5) is a stronger base than H_2dihybe (pKa = 9.57), thus, in these solutions, the ligand was deprotonated, $^-$Hdihybe], the high temperature and high pressure under the hydrothermal conditions lead to the formation of the tri-deprotonated ligand which is capable of bridging more metals and thus, leading to the formation of higher nuclearity TOCs.

From all the above, it is clear that the low molar ratio of titanium(IV)/H_2dihybe, low pH, and room temperature lead to the formation of low nuclearity TOCs. The ligation of the ligand to Ti^{IV} makes the hydroxamic proton [~C(O)NHO-**H**] more acidic, and thus, despite the low pH, results in the deprotonation of the ligand.

2.2. Description of the Structure

Interatomic distances and bond angles relevant to the Ti(1) coordination sphere are listed in Table 1. The molecular structure of the cation [$Ti^{IV}_4(\mu\text{-}O)_2(HOCH_3)_4(\mu\text{-}\eta^1,\eta^2\text{-}Hdihybe\text{-}O,O')_4(\eta^1,\eta^1\text{-}Hdihybe\text{-}O,O')_2$]$^{4+}$ of **1** is presented in Figure 1 which is composed of two dinuclear [$Ti^{IV}_2(HOCH_3)_2(\mu\text{-}\eta^1,\eta^2\text{-}Hdihybe\text{-}O,O')_2(\eta^1,\eta^1\text{-}Hdihybe\text{-}O,O')_2$]$^{4+}$ (Figure 2A) units interlinked through two μ-bridging oxygen atoms (Figure 2B). Each titanium(IV) atom in **1** is bonded to two mono-deprotonated Hdihybe$^-$ligands, one of which acts as a bidentate-O,O' chelate through the carbonyl and the deprotonated hydroxamate oxygen atoms (see Scheme 2a) and the other one as a chelate-bridging-O,O' through the same oxygen atoms (see Scheme 2b). All the titanium centers in **1** are seven-coordinate, with an O_7 donor set, in a pentagonal bipyramidal environment and are sharing one of their edge (Figure 2B) in the {Ti_2} structural unit and through a corner in the Ti(1)-O(15)-Ti(1)$'''$ unit (Figure 2B). The Ti(1)···Ti(1)$'$ and Ti(1)···Ti(1)$''$ distances within the two dimeric {Ti_2} units are 3.515(1) and 3.561(1) Å, respectively, while the Ti(1)-O(15)-Ti(1)$'''$ angle is 165.8(1)°. The inorganic core {$Ti_4(\mu\text{-}O)_2$} constitutes a rare example of such a structural motif for discrete {Ti^{IV}_4} oxo-clusters. The other two examples which have been reported are the following: $(NH_4)_6[Ti^{IV}_4(\mu\text{-}O)_2(C_2H_2O_3)_4(C_2H_3O_3)_2(O_2)_4]$ [48,49] and [$Ti^{IV}_4(\mu\text{-}O)_2(\mu\text{-}OEt)_4(\kappa^3\text{-}tbop)_4$] [50].

Table 1. Interatomic distances (Å) and angles (°) relevant to the Ti(1) coordination sphere for the tetranuclear titanium cluster **1**.

Bond Distances			
Ti(1)-O(1)	2.112(3)	Ti(1)-O(4)	1.978(3)
Ti(1)-O(1)'	2.087(3)	Ti(1)-O(9)	2.114(3)
Ti(1)-O(2)	2.043(3)	Ti(1)-O(15)	1.7945(10)
Ti(1)-O(3)	2.027(3)		
Bond Angles			
Ti(1)'-O(1)-Ti(1)	113.65(13)	O(15)-Ti(1)-O(1)	91.07(13)
Ti(1)"-O(15)-Ti(1)	165.8(2)	O(4)-Ti(1)-O(1)	138.26(12)
O(15)-Ti(1)-O(4)	97.77(10)	O(3)-Ti(1)-O(1)	144.82(12)
O(15)-Ti(1)-O(3)	94.60(15)	O(2)-Ti(1)-O(1)	72.89(11)
O(4)-Ti(1)-O(3)	75.20(12)	O(1)'-Ti(1)-O(1)	63.64(13)
O(15)-Ti(1)-O(2)	90.87(10)	O(15)-Ti(1)-O(9)	176.42(13)
O(4)-Ti(1)-O(2)	146.94(12)	O(4)-Ti(1)-O(9)	85.45(13)
O(3)-Ti(1)-O(2)	72.33(12)	O(3)-Ti(1)-O(9)	87.73(2)
O(15)-Ti(1)-O(1)'	93.56(14)	O(2)-Ti(1)-O(9)	87.23(12)
O(4)-Ti(1)-O(1)'	75.11(12)	O(1)'-Ti(1)-O(9)	85.73(12)
O(3)-Ti(1)-O(1)'	150.00(12)	O(1)-Ti(1)-O(9)	85.47(12)
O(2)-Ti(1)-O(1)'	136.35(12)		

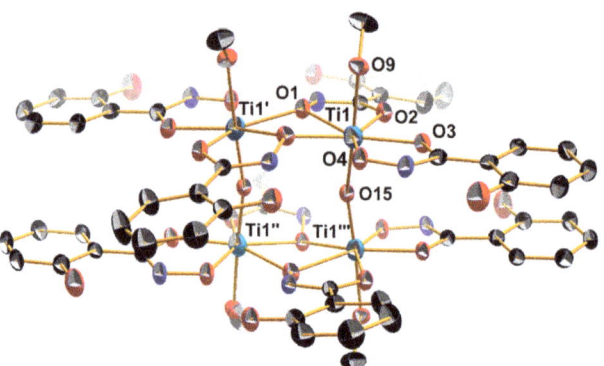

Figure 1. ORTEP plot (50% probability level) of the cation [Ti$^{IV}_4$(μ-O)$_2$(HOCH$_3$)$_4$(μ-η^1,η^2-Hdihybe-O,O')$_4$(η^1,η^1-Hdihybe-O,O')$_4$]$^{4+}$ of **1** with a partial labeling scheme. Hydrogen atoms are omitted for clarity.

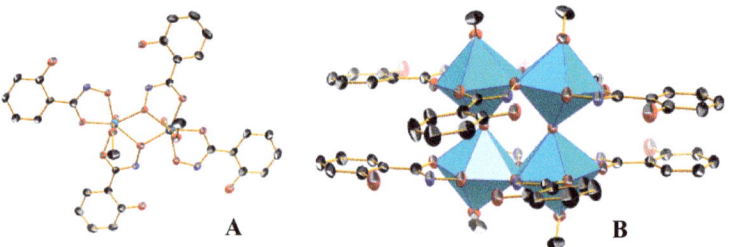

Figure 2. Ball-and-stick plot of the structural unit [Ti$^{IV}_2$(HOCH$_3$)$_2$(μ-η^1,η^2-Hdihybe-O,O')$_2$(η^1,η^1-Hdihybe-O,O')$_2$] (**A**). Polyhedral/ball-and-stick representation of the cation of **1** (**B**).

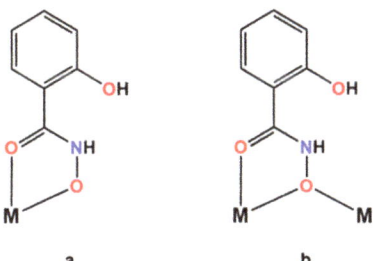

Scheme 2. The two bonding modes of the mono-deprotonated ligand Hdihybe⁻ in the tetranuclear titanium cluster **1**. Chelating (**a**) and bridging-chelating (**b**).

2.3. ESI-MS Spectrometry

In an effort to further characterize the tetranuclear titanium(IV) cluster in solution, we employed high-resolution ESI-MS to unambiguously determine the structural integrity and composition of the titanium-based species in solution [51,52]. The low m/z region of the negative ion mass spectrum of **1** exhibits two characteristic sets of isotopic distribution patterns (Figure 3) which can be attributed to the dinuclear and trinuclear fragments of **1** and are centered at: ca. (a) 698.98, 736.93, and 768.95 m/z with the formulae of $\{Ti^{III}_2O_2(OCH_3)_3(C_7H_5NO_3)_3(OH_2)H_6\}^-$, $\{Ti^{III}_2O_2(C_7H_5NO_3)_4H_5\}^-$ and $\{Ti^{III}_2O_2(OCH_3)(C_7H_5NO_3)_4H_6\}^-$ for the dinuclear fragment and at (b) 895.911 and 931.90 m/z with the formulae of $\{Ti^{III}_3O_2(OCH_3)(C_7H_5NO_3)_4(HOCH_3)_2(OH_2)H_3\}^-$ and $\{Ti^{III}_3O_2(OCH_3)(C_7H_5NO_3)_4(HOCH_3)_2(OH_2)_3H_3\}^-$ for the trinuclear fragments, respectively. The observation of different oxidation states for the metal centers and the presence of other fragments during the studies is due to the ionization and transfer process and has been observed in numerous occasions [53–57].

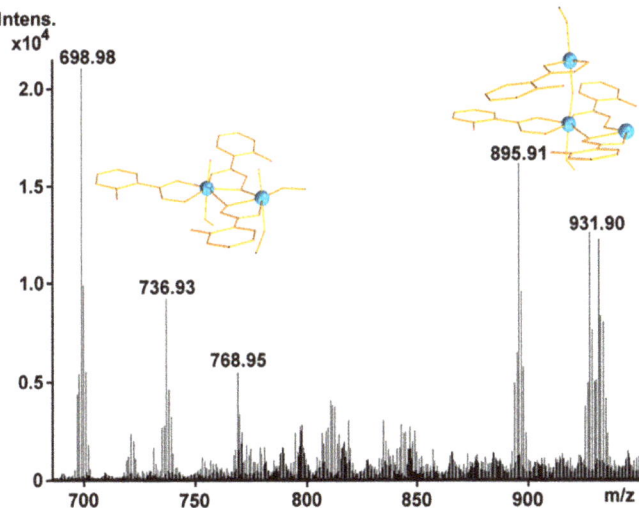

Figure 3. Negative mode of the low m/z region electrospray ionization mass spectrum (ESI-MS) of the $\{Ti_4\}$ (**1**) cluster in CH$_3$OH.

The high m/z region of the negative ion mass spectrum of **1** (Figure 4) exhibits characteristic isotopic distribution patterns which can be attributed to the tetranuclear cluster containing the inorganic core $\{Ti_4O_2\}$ and are centered in the region ca. 1078–1315 m/z. See Table 2 for the assigned species. The partial fragmentation during the ESI-MS studies provides additional information on the relevant stability of the fragments that can exist in solution. Thus, this can provide an indication of the potential assembly pathway followed during the formation of the tetranuclear species which can be formed by the combination of smaller dimeric fragments, e.g., $2 \times \{Ti_1\} \rightarrow 2 \times \{Ti_2\} \rightarrow \{Ti_4\}$.

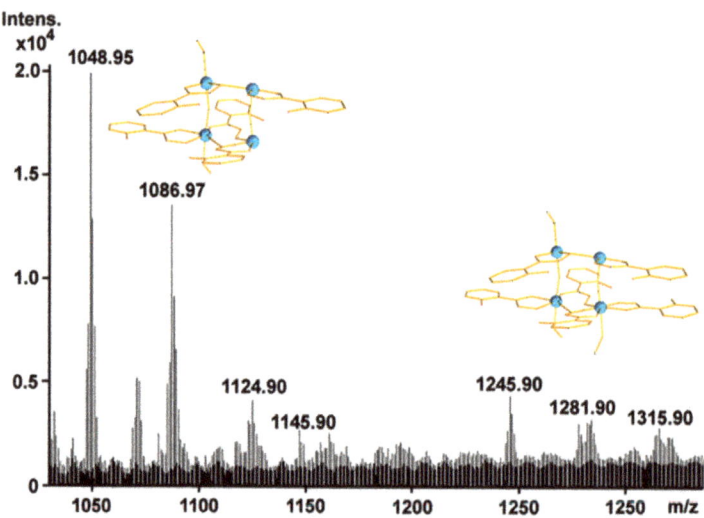

Figure 4. Negative mode of the high m/z region electrospray ionization mass spectrum (ESI-MS) of the {Ti$_4$} (**1**) cluster in CH$_3$OH.

Table 2. Representation of the experimentally identified and simulated m/z values of distribution patterns of {Ti$_4$} cluster.

Experimental	Theoretical	Charge	Formula
698.98	698.09	−1	{Ti$^{III}_2$O$_2$(OCH$_3$)$_3$(C$_7$H$_5$NO$_3$)$_3$(OH$_2$)H$_6$}$^-$
736.93	737.03	−1	{Ti$^{III}_2$O$_2$(C$_7$H$_5$NO$_3$)$_4$H$_5$}$^-$
768.95	769.06	−1	{Ti$^{III}_2$O$_2$(OCH$_3$)(C$_7$H$_5$NO$_3$)$_4$H$_6$}$^-$
895.91	896.04	−1	{Ti$^{III}_3$O$_2$(OCH$_3$)(C$_7$H$_5$NO$_3$)$_4$(HOCH$_3$)$_2$(OH$_2$)H$_3$}$^-$
931.90	932.10	−1	{Ti$^{III}_3$O$_2$(OCH$_3$)(C$_7$H$_5$NO$_3$)$_4$(HOCH$_3$)$_2$(OH$_2$)$_3$H$_3$}$^-$
1048.95	1049.05	−1	{Ti^{II}Ti$^{III}_3$O$_2$(OCH$_3$)(C$_7$H$_5$NO$_3$)$_5$(H$_2$O)$_2$H$_3$}$^-$
1086.90	1086.97	−1	{Ti$^{II}_3$Ti^{III}O$_2$(OCH$_3$)(C$_7$H$_5$NO$_3$)$_5$(H$_2$O)$_4$H$_5$}$^-$
1124.90	1125.03	−1	{Ti^{IV}Ti$^{III}_3$O$_2$(OCH$_3$)$_4$(C$_7$H$_5$NO$_3$)$_5$(OH$_2$)H$_4$}$^-$
1245.90	1246.05	−1	{Ti$^{III}_2$Ti$^{IV}_2$O$_2$(OCH$_3$)$_3$(C$_7$H$_5$NO$_3$)$_6$(OH$_2$)H$_5$}$^-$
1281.90	1281.06	−1	{Ti^{III}Ti$^{IV}_3$O$_2$(OCH$_3$)$_3$(C$_7$H$_5$NO$_3$)$_6$(OH$_2$)$_3$H$_3$}$^-$
1315.90	1316.05	−1	{Ti^{III}Ti$^{IV}_3$O$_2$(OCH$_3$)$_3$(C$_7$H$_5$NO$_3$)$_6$(OH$_2$)$_5$H$_3$}$^-$

2.4. IR Spectroscopy

Quantitative agreement between experimental and theoretical spectra predicted by ab initio DFT/B3LYP calculations with the Los Alamos National Laboratory 2 double zeta (LanL2DZ) split-valence basis set. The specific basis set is an ideal choice for quantum mechanical calculations for complexes whose centers are first-row transition metals, such as titanium. The calculations of such molecules with the LANL2DZ basis set and the fact that the complex has D_2 point group symmetry are characterized by a relatively short computational time. Considering the LANL2DZ basis set, a reason for the reduction in the computational time is that ECP (Effective Core Potential) plus double zeta on Na-Bi is used in this basis set. ECP describes the inner electron orbitals and so no basis functions are required for them. Another advantage of this basis set is that it includes relativistic effects, but does not include polarization functions. These results add detailed confidence to our present understanding of the chemistry of the systems. The experimental spectra of the ligand and complex, denoted as H$_2$dihybe and **1**, respectively, in the solid state at ambient conditions are given in Figure 5, allowing for direct comparison. At a first glance, we observe that the spectrum of the H$_2$dihybe is less complicated compared to the spectrum of the complex **1**. Furthermore, its energy minimum is lower as expected for a simpler structure. In the high-frequency spectral region, only the N–H, OH, and C–H stretching modes are detected for both spectra and do not provide any significant information concerning complexation. Thus, we have chosen to focus our attention on the low-frequency region, the so-called fingerprint region. This part of the spectrum is more informative in relation to the comprehensive understanding of the structural features of the titanium complex. In this region, a larger number of bands are present. Besides the fact that these bands are sharper and of higher intensity, also evidenced is a strong band overlapping that implies the inherent structural complexity of the studied complex. An intense band at ~800 cm^{-1} is observed in the spectrum of the complex which is absent in the spectrum of the H$_2$dihybe. This band is attributed to the titanium complex formation and more specifically is assigned to the high-energy ν(Ti$_2$–μ-O) stretching modes. The low-frequency band of H$_2$dihybe observed at ~780 is red-shifted to 770 cm^{-1} in the spectrum of complex, while the band of H$_2$dihybe at ~744 cm^{-1} remains in the same frequency with much lower absorbance. Additional relative absorbance changes are observed between the two spectra. The formation of the complex is expected to affect the vibrational frequency and/or absorbance of the C=O, C-O, C-N, and N-O bonds. Indeed, the frequency of C=O is red-shifted ~10 cm^{-1}. This band is observed at ~1577 cm^{-1} and at ~1567 cm^{-1} in the spectrum of H$_2$dihybe and complex, respectively. The bands at ~1250 cm^{-1} and ~1360 cm^{-1}, attributed to C-O and to C-N, respectively, exhibit only a significant absorbance decrease, while there is no frequency shift upon formation of the complex formation. On the contrary, the band assigned to the N-O vibration is blue-shifted from ~1050 cm^{-1} to ~1070 cm^{-1} without any additional absorbance variation after complex formation. The experimental and theoretically predicted by ab initio DFT/B3LYP/LANL2DZ IR absorbance spectra of the H2dihybe and Ti-complex **1** are presented in Figure S2a,b, respectively. All the main bands detected in the experimental spectra are also present in the calculated spectra supporting our proposed structural model. Any differences observed in intensities and frequencies are reasonable, considering that the calculation was performed in the vapor state without the presence of any additional interactions. The IR absorbance data are indicative of a structural rearrangement and complex formation which further support the structural information revealed by the rest of the experimental techniques utilized in the present study.

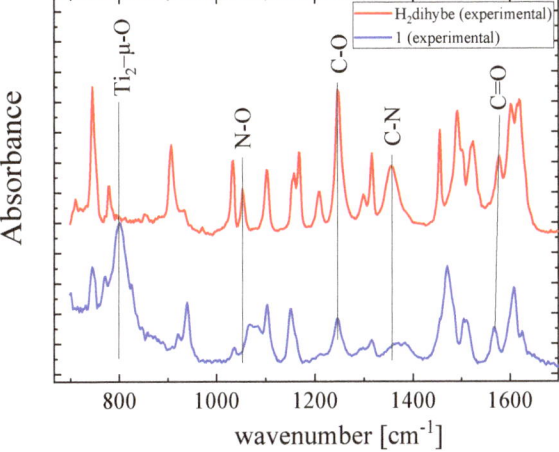

Figure 5. Experimental IR absorbance spectra of the ligand *N*,2-Dihydroxybenzamide denoted as H$_2$dihybe and titanium(IV) complex with *N*,2-dihydroxybenzamide denoted as **1**.

2.5. NMR Spectroscopy

The ^{1}H and ^{13}C NMR chemical shifts of the CD$_3$OD solutions of the H$_2$dihybe and **1** are collected in Table 3. The ^{1}H NMR spectrum of a CD$_3$OD solution of the H$_2$dihybe (Figure 6A) gave two doublets of doublets at 7.669, 6.928 ppm and two triplets of doublets at 7.338, 6.897 ppm assigned to the protons attached to carbon atoms C(d), C(a), and C(b), C(c), respectively.

Table 3. ^{1}H and ^{13}C NMR chemical shifts (ppm) for the ligand H$_3$dihybe, the tetranuclear titanium compound **1**, and the shielding/deshielding effect ($\Delta\delta$) upon complexation $^\alpha$.

	1		H$_2$dihybe			
	^{13}C	^{1}H	^{13}C	^{1}H	^{13}C ($\Delta\delta$, ppm) $^\beta$	^{1}H ($\Delta\delta$, ppm) $^\beta$
C(d)$^\gamma$	129.9	7.837	128.2	7.669	1.7	0.168
C(c)$^\gamma$	119.9	7.004	119.4	6.897	0.5	0.107
C(b)$^\gamma$	134.3	7.436	133.6	7.338	0.7	0.098
C(a)$^\gamma$	115.5	7.017	117.2	6.928	−1.7	0.089
C(g)$^\gamma$	163.3		167.2		−3.9	
C(f)$^\gamma$	156.6		159.2		−2.6	
C(e)$^\gamma$	111.1		114.1		−3.0	

$^\alpha$ The chemical shifts of the protons are the center of multiplets. $^\beta$ $\Delta\delta$ is the chemical shift difference between the chemical shift of the NMR peaks (^{13}C or ^{1}H) of the complex **1** and the respective peaks of the ligand. $^\gamma$ The carbon atoms of the ligand H$_2$dihybe shown in Figure 6.

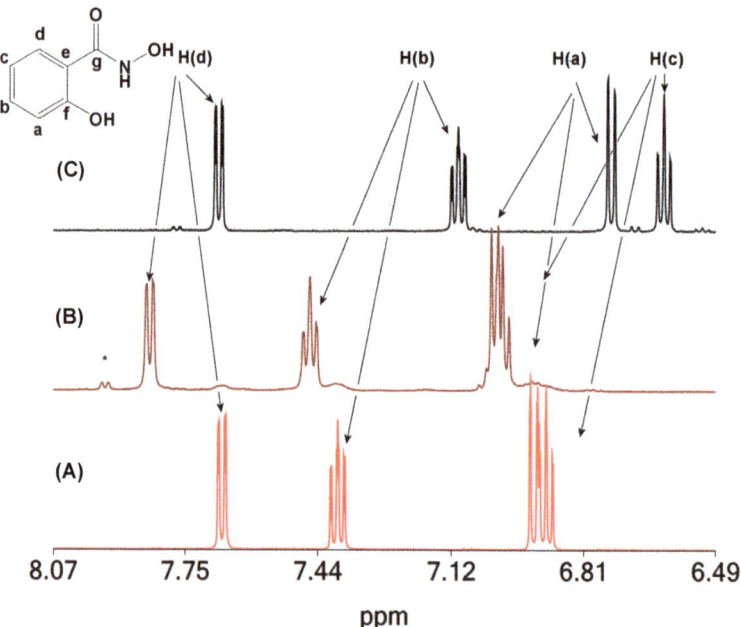

Figure 6. ^{1}H NMR spectra of the solutions (CD$_3$OD) of the ligand H$_3$dihybe (**A**); of **1** (**B**); and of **1** after the addition of two equivalents of But$_4$NOH (**C**). H(a)–H(d) are the protons of the ligand shown at the top of the Figure. * The H(d)-^{1}H NMR peaks of the minor titanium complex **2** (See Supplementary Material: Table S1).

The ^{1}H NMR spectrum of the CD$_3$OD solution of **1** shows peaks of the same multiplicity with H$_2$dihybe, however, the peaks were shifted to lower field suggesting ligation of the ligand to Ti^{IV}; with $\Delta\delta$ for a, b, c, d protons 0.10, 0.07, 0.11, 0.16 ppm, respectively (Figure 6B). The ^{1}H NMR spectrum of **1** shows only one set of peaks for the complex.

However, the crystal structure of **1** (Figure 6) shows two mono-deprotonated Hdihybe⁻ ligands bound to each titanium atom with two different modes of ligation (Scheme 2), and this means that either the complex **1** in solution does not have the same structure as the solid state or there is a fast-chemical exchange between the bridged and the non-bridged Hdihybe⁻ ligands. ESI MS experiments indicate that the complex retains its solid-state structure in CD$_3$OD, therefore, the bridged and the non-bridged Hdihybe⁻ ligands cannot be distinguished by ^{1}H NMR because they exchange fast. In addition, the chemical shifts of the bridged and the non-bridged Hdihybe⁻ ligands are not expected to be very different, thus, any exchange will result in their coalescence. Chemical exchange between the two coordinated Hdihybe⁻ ligands is supported by the much broader ^{1}H peaks of **1** than the respective peaks of H$_2$dihybe. The carbonylic ^{13}C NMR peak of the CD$_3$OD solution of **1**, determined by 2D grHSQC and gr HMBC spectroscopies (Figures S3–S6), shows the largest shift compared to the respective peak of the ligand, $\Delta\delta = -3.9$ ppm, suggesting coordination of Ti^{IV} to the carbonylic oxygen atom [24,58–61].

In addition to the peaks originated from **1**, the ^{1}H NMR spectrum gave peaks originated from a minor species (10%) and the free ligand (12%) (Figure 6). The chemical shifts of ^{1}H and ^{13}C, as they have been found from 2D {^{1}H} grCOSY (Figure S4) and 2D {^{1}H, ^{13}C} grHSQC (Figure S5), are the same as **1** except C(d)-H proton which is shifted to the lower field (0.110 ppm) than the respective proton of **1** (Table 3). Apparently, the minor species **2** and **1** have a similar structure. Looking at the structure of **1**, one can suggest various isomers. The most possible one is that with the non-bridged Hdihybe⁻ ligated to Ti^{IV} at different orientations (Scheme 3).

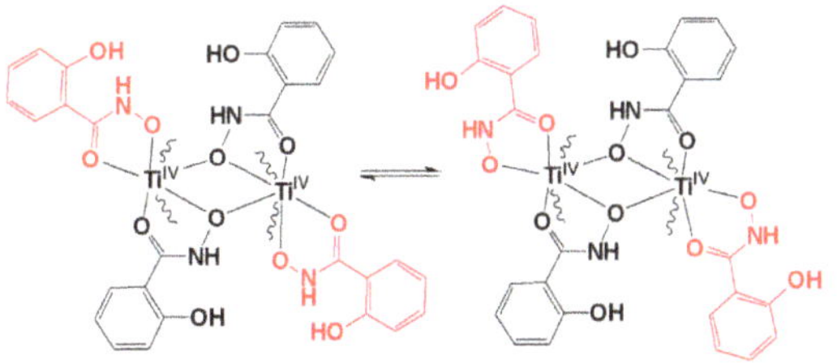

Scheme 3. Two possible isomers that might be formed in CD$_3$OD solutions of **1**.

The 2D {^{1}H} NOESY-EXSY spectrum (Figure 7) of the CD$_3$OD solution of **1** gave NOESY cross peaks between the neighboring aromatic protons and EXSY cross-peaks between **1** and the free ligand, the minor species and ligand as well as between **1** and the minor species. The intensity ratios of the cross-peaks vs the diagonal were similar for all exchange processes suggesting that the conversion of **1** to the other isomer (Scheme 3) is intermolecular.

In addition, the 2D {^{1}H} NOESY-EXSY spectrum showed two NOESY cross-peaks between protons H(**b**) and H(**d**) (See Figure 6) having the same phase with the diagonal peaks (Figure 7, blue circles). Despite the opposite phase than the expected one, these peaks should be originated from NOESY interactions; the possibility of these peaks originating from a chemical exchange is improbable. The phase of this peak is attributed to the fast exchange between the bridged and non-bridged Hdihybe⁻. These NOESY signals are assigned to the interactions between the H(**b**) and H(**d**) protons of different non-bridged Hdihybe⁻ ligands, each of which belongs to one of the two parallel planes, defined from two Ti^{IV} and four Hdihybe⁻ ligands (Scheme 4). Apparently, this supports the complex in retaining its tetranuclear structure in agreement with the ESI measurements.

The 2D {^{1}H} NOESY-EXSY spectrum (Figure 8) of the CD$_3$OD solution of **1** gave NOESY cross-peaks between the neighboring aromatic protons and EXSY cross-peaks between **1** and the free ligand. In addition, it showed and two NOESY cross-peaks between protons **d** and **b** (Figure 8, blue circles).

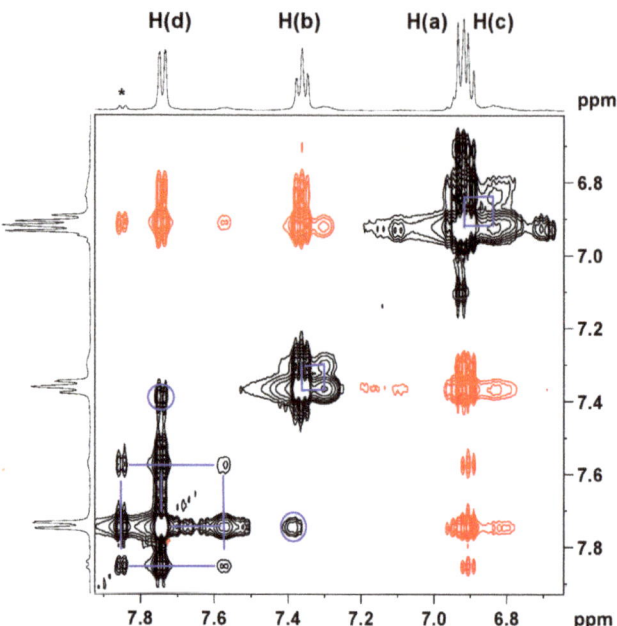

Figure 7. Two-dimensional {^{1}H} grNOESY of **1**. The blue lines show the chemical exchange between the titanium complexes and the free ligand. The red lines show the chemical exchange between the titanium complexes. In the blue circles, NOESY peaks between H(**d**) and H(**b**) protons of two chelating Hdihybe$^-$ ligands (Scheme 2). H(**a**)–H(**d**) the protons of the ligand shown in Figure 6. * The H(**d**)-^{1}H NMR peaks of the minor titanium complex **2** (See Supplementary Material; Table S1).

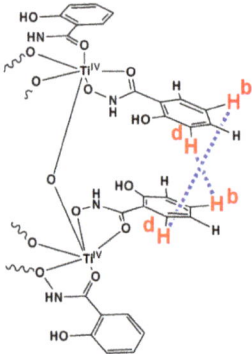

Scheme 4. NOESY interactions between H(**d**) and H(**b**) protons of the two chelating Hdihybe$^-$ ligands which belong to two parallel planes. Each plane is defined by two Ti^{IV} atoms and four Hdihybe$^-$ ligands (Figure 1).

In contrast to the spectrum of **1** without a base, the peaks now have the expected negative phase, assigned to the significant decrease in the complex's lability after the addition of the base to the solution. The NOESY interactions between protons **b** and **d** suggest that the complex retains its tetra-nuclear structure after the addition of the base, permitting interactions between the protons of the parallel aromatic rings (Scheme 4). Another observation that supports a similar tetranuclear structure before and after the addition of the base, is the fact that the NMR spectra of **1** after the addition either of one or four equivalents per equivalent of **1** are the same (Figures S7 and S8). Thus, supporting that the NMR changes observed after the addition of the base in the CD$_3$OD solution of **1**, is a result of acid–base equilibrium and not a change in the Ti^{IV} coordination environment, and the decrease in the complex's lability is due to the better electron-donating properties of the ligand after its deprotonation as shown in Scheme 5.

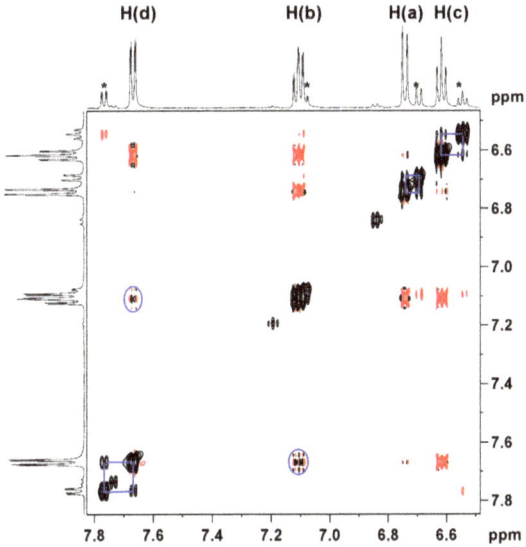

Figure 8. 2D {^{1}H} grNOESY **1** + 4 equivalents of But$_4$NOH + 0.1 equivalents of H$_2$dihybe. Blue lines show the chemical exchange between the tetranuclear titanium complex and the free ligand. In the blue circles correspond the NOESY peaks between H(**d**) and H(**b**) protons of two chelating dihybe^{2-} ligands (Scheme 5). H(**a**)–H(**d**) are the protons of the ligand shown on the top of Figure 6. * The protons of the free ligand.

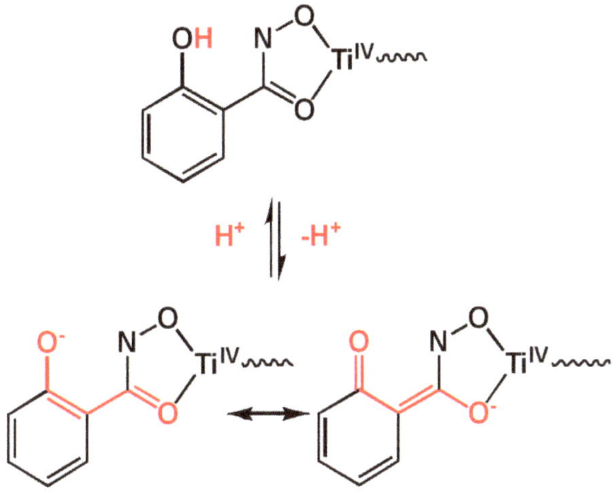

Scheme 5. Deprotonation of the phenoxy group of Hdihybe$^-$ and resonance structures showing the better electron-donating properties of dihybe^{2-} than Hdihybe$^-$.

2.6. Solution UV–Vis and Luminescence Spectroscopies

Figure 9 shows the solution UV–Vis spectra of the MeOH solutions of compound **1** without and with the presence of a base (But$_4$NOH), The MeOH solution of **1** gave two peaks at 310 nm (38 cm^{-1}M^{-1}) and 242 nm (71 cm^{-1}M^{-1}). The peaks are attributed to n-π* and π-π* electronic transitions. The peak at 310 nm is characteristic for phenolate groups. After the addition of But$_4$NOH, the peaks remain the same, but their intensity increased significantly. The results support that the structure of **1** remains the same in solution after the addition of the base, whereas the increase in the intensity of the peak at 310 nm is assigned to deprotonation of the phenol, in line with the NMR experiment.

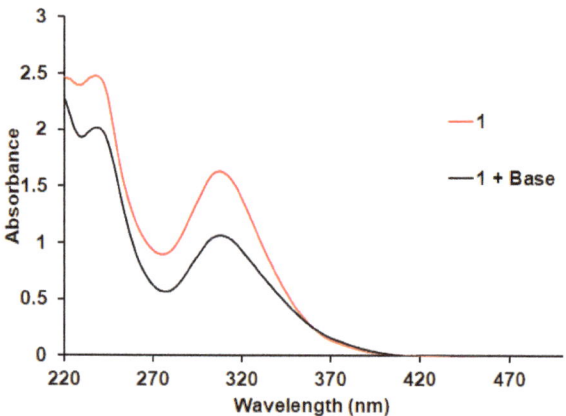

Figure 9. UV–Vis spectra of a MeOH solution of **1** (0.028 mM) (black line) and of **1** (0.028 mM) + But$_4$NOH (0.028 mM).

Figure 10 shows the luminescence spectra of the MeOH solutions of compound **1** in the presence of a base (But$_4$NOH), The MeOH solution of **1** gave dual-luminescence, emitting light at 620 nm for excitation wavelength 524 nm and at 573 nm for excitation wavelength 490 nm. The emitting peaks are relatively sharp with linewidths ~30 nm. The intensity of the emitting light was doubled after the addition of an equimolar quantity of base in the MeOH solution of **1**, however, the excitation and emission wavelengths remain the same suggesting that the structure in solution remains the same after the addition of the base. The increase in the intensity is assigned to the phenol deprotonation.

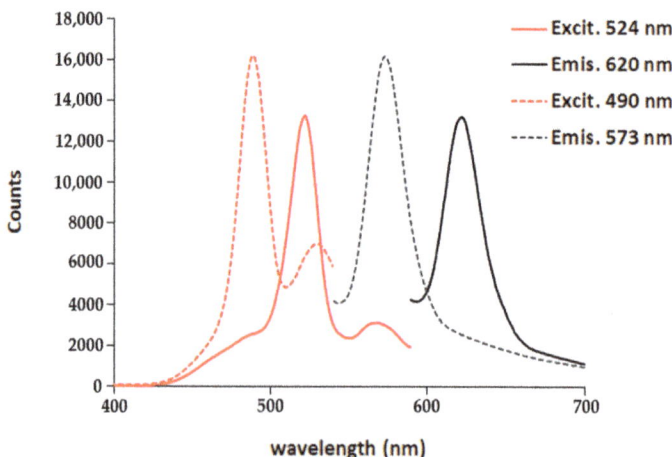

Figure 10. Luminescence spectra of a solution of **1** (2.00 mM) + But$_4$NOH (2.00 mM) at excitation wavelengths 524 nm (solid line) and 490 nm (dashed line).

2.7. Solid-State UV–Vis Spectroscopy

Figure 11 shows the solid-state UV–Vis spectra of the compound **1** and the ligand H$_2$dihybe. The band gap for the compound **1** was found to be 1.98 eV and was calculated from the solid-state spectrum by the Kubelka–Munk method [62] (Figure S9). This low band gap value for compound **1** reveals its potential use as a semiconducting photocatalyst.

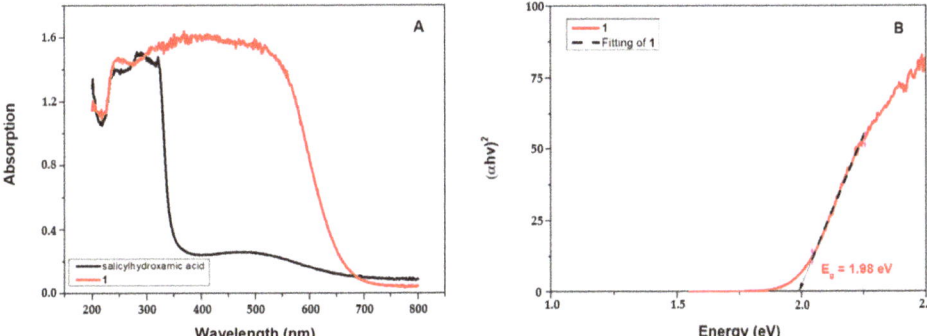

Figure 11. Solid-state UV–Vis spectra of the tetranuclear titanium(IV) compound **1** and the ligand H$_2$dihybe (**A**). Tauc plot of compound **1** (**B**).

3. Materials and Methods

3.1. Experimental Details

All chemicals and solvents were purchased from Sigma-Aldrich and Merck (Saint Louis, MO, USA), were of reagent grade, and were used without further purification, except TiCl$_4$, which was distilled under high vacuum just prior to use. C, H, and N analyses were conducted by the microanalytical service of the School of Chemistry, the University of Glasgow. FT-IR transmission spectra of the compounds, in KBr pellets, were acquired using a Bruker Alpha spectrophotometer (Bruker, Billerica, MA, USA) in the 4000–400 cm^{-1} range. The UV–Vis diffuse reflectance spectra were recorded at room temperature on an Agilent Cary 60 UV–Vis spectrophotometer (Agilent Technologies, Santa Clara, CA, USA). The UV–Vis and the luminescence solution spectra were acquired on a Shimadzu UV-2600i UV–Vis Spectrophotometer (Shimadzu, Nagoya, Japan) and on a Jasco Spectrofluorometer FP-8300 (JASCO, Mary's Court Easton, MD 21601, USA), respectively, at room temperature.

3.2. Synthesis of $[Ti^{IV}_4(\mu\text{-}O)_2(HOCH_3)_4(\mu\text{-}\eta^1,\eta^2\text{-}Hdihybe\text{-}O,O')_4(\eta^1,\eta^1\text{-}hdihybe\text{-}O,O')_4]$ $Cl_4 \cdot 10H_2O \cdot 12CH_3OH$ (**1**)

To a stirred methyl alcohol solution (4 mL) were successively added N,2-dihydroxybenzamide (H$_2$dihybe) (139.7 mg, 0.912 mmol) and TiCl$_4$ (0.05 mL, 86.5 mg, 0.456 mmol). The colorless solution of the ligand turned orange upon the addition of TiCl$_4$. Then, solid KOH (51.1 mg, 0.912 mmol) was added in one portion. The solution was filtered, and the orange filtrate (pH = 1.5) was kept at ≈4 °C for 9–10 days during which period 90.0 mg of orange crystals of compound **1** were formed. The crystals were filtered off and dried at an ambient atmosphere (≈20 °C). (Yield: 35%, based on TiCl$_4$). Elemental anal. calc. for (C$_{72}$H$_{132}$N$_8$O$_{52}$Cl$_4$Ti$_4$, M_r = 2275.112 g mol^{-1}): C, 38.01; H, 5.85; N, 4.92; found: C, 37.98; H, 5.81; N, 4.95.

3.3. X-ray Crystallographic Details

A suitable single crystal was selected and mounted onto a rubber loop using Fomblin oil. Single-crystal X-ray diffraction data of **1** was recorded on a Bruker Apex II Quazar CCD diffractometer (Bruker, Bremen, Germany) (λ (MoK$_\alpha$) = 0.71073 Å) at 150 K equipped with a graphite monochromator. Structure solution and refinement were carried out with SHELXS-97 [63] and SHELXL-97 [64] using the WinGX software package [65]. Data collection and reduction were performed using the Apex2 software package. Corrections for the incident and diffracted beam absorption effects were applied using empirical absorption corrections [66]. All the atoms and most of the carbon atoms were refined anisotropically. Solvent molecule sites were found and included in the refinement of the structures. Final unit cell data and refinement statistics for compounds **1** are collated in Table 4. The crystallographic data for compound **1** (CCDC 1: 2096669) can be obtained free of charge from the Cambridge Crystallographic Data Centre, 12, Union Road, Cambridge, CB2 1EZ; fax:(+44)-1223-336-033, deposit@ccdc.cam.ac.uk.

Table 4. Crystal data and details of the structure determination and refinement for the tetranuclear titanium cluster **1**.

Formula	$C_{72}H_{132}N_8O_{52}Cl_4Ti_4$	
Formula weight	2275.112 g mol^{-1}	
Temperature	150(2) K	
Wavelength	0.71073 Å	
Crystal system	Orthorhombic	
Space group	F d d d	
Unit cell dimensions	a = 20.040(4) Å b = 26.118(5) Å c = 36.745(11) Å	a = 90° b = 90° g = 90°
Volume	19232(8) Å^3	
Z	16	
Density (calculated)	1.453 Mg/m^3	
Absorption coefficient	0.587 mm^{-1}	
F(000)	8624	
Crystal size	0.200 × 0.170 × 0.100 mm^3	
Theta range for data collection	1.396 to 26.515°.	
Index ranges	$-24 \leq h \leq 25, -32 \leq k \leq 32, -45 \leq l \leq 45$	
Reflections collected	45,786	
Independent reflections	4976 [R(int) = 0.1227]	
Completeness to theta = 25.242°	100.0%	
Absorption correction	Empirical	
Max. and min. transmission	0.7454 and 0.6576	
Refinement method	Full-matrix least-squares on F^2	
Data/restraints/parameters	4976/5/304	
Goodness-of-fit on F^2	0.967	
Final R indices [I>2sigma(I)]	R1 = 0.0657, wR2 = 0.1703	
R indices (all data)	R1 = 0.1101, wR2 = 0.2058	
Extinction coefficient	n/a	
Largest diff. peak and hole	1.117 and -0.473 e.E^{-3}	

3.4. ESI MS Experimental Details

All MS data were collected using a Bruker Q-trap, time-of-flight MS (Maxis Impact MS, Bremen, Germany) instrument supplied by Bruker Daltonics Ltd. The detector was a time-of-flight, micro-channel plate detector and all data was processed using the Bruker Daltonics Data Analysis 4.1 software, whilst simulated isotope patterns were investigated using Bruker Isotope Pattern software and Molecular Weight Calculator 6.45. The calibration solution used was the Agilent ES tuning mix solution, Recorder No. G2421A, enabling calibration between approximately 100 *m/z* and 3000 *m/z*. This solution was diluted 60:1 with MeCN. Samples were dissolved in MeOH and introduced into the MS via direct injection at 180 µL h^{-1}. The ion polarity for all MS scans recorded was negative, at 180 °C, with the voltage of the capillary tip set at 4000 V, endplate offset at −500 V, funnel 1 RF at 300 Vpp, and funnel 2 RF at 400 Vpp.

3.5. FT-IR Spectroscopy

The FT-IR spectra of the ligand and the complex were recorded in the 4000–400 cm^{-1} mid-infrared range on a Bruker Apha FT-IR spectrophotometer (Bruker, Billerica, MA, USA) with 256 scans at a resolution of 2 cm^{-1}. All samples in the solid form were ground with spectroscopic grade potassium bromide (KBr) powder (2 mg of sample per 200 mg dry KBr) and then pressed into pellets with a thickness of 1 mm.

3.6. Ab Initio Modeling of Ligand and Ti-Complex

Based on the crystal structures of the ligand N,2-Dihydroxybenzamide (H$_2$dihybe) and titanium(IV) complex with N,2-Dihydroxybenzamide (**1**), we calculated the corresponding vibrational properties. All calculations were performed with the Gaussian 09 W Revision D.01 package [67]. The initial structure of the complex used in the calculations emerged after its study by X-ray crystallography, while the structure of the H$_2$dihybe was obtained from the electronic library of chemical compounds from PubChem [68]. The Density Functional Theory (DFT) using hybrid functional B3LYP, Becke's three-parameter exchange functional with the Lee–Yang–Parr correlation functional [69,70], was chosen for all calculations. In addition, the basis set used was the 3–21 G split valence basis set. All calculations were performed without the effect of solvent, in the gaseous phase. The vibrational frequencies were calculated and scaled by a vibrational scaling factor of 0.965 to attain an acceptable agreement between the theoretical and experimental values. This is reasonable since the 3–21 G basis set used in the calculation is relatively simple and provides larger inter-atomic distances and shifted vibrational frequencies. Nevertheless, the predicted geometry for the complex resulted in reasonable parameters. No imaginary frequencies were observed in the results of all calculations indicating that the structures correspond to minimal points on the potential energy surface.

4. Conclusions

In conclusion, we synthesized a tetranuclear TOC **1** through the reaction of the hydroxamate ligand H$_2$dihybe with TiCl$_4$ and KOH in methyl alcohol at a pH of 1.5. The X-ray structure analysis of **1** revealed that it constitutes a rare example containing an {Ti$_4$(μ-O)$_2$} inorganic core with an almost square planar arrangement of the {Ti$_4$} unit. The low molar ratio of Ti^{IV}/H$_2$dihybe, low pH, and room temperature lead to the formation of low nuclearity TOCs.

^{1}H and ^{13}C NMR solution (CD$_3$OD) studies of **1** show its structural integrity in solution which is in good agreement with the high-resolution ESI-MS studies which revealed characteristic isotopic distribution envelopes attributed to the intact tetranuclear clusters containing the inorganic core {Ti$_4$(μ-O)$_2$}. The observed NMR, UV–Vis, and luminescence changes after the addition of the base to the CD$_3$OD solution of **1**, are a result of acid–base equilibrium and not a change in the Ti^{IV} coordination sphere. Moreover, the decrease in the complex's lability is due to the improved electron-donating properties of the ligand dihybe^{2-} associated with the deprotonation of its phenoxy group.

The structural features of **1** have also been investigated by means of vibrational spectroscopy revealing a ν(C=O) red-shift by ~10 cm^{-1}, and a ν(N-O) blue-shift by ~20 cm^{-1} upon complexation in comparison to the free ligand H$_2$dihybe.

The solid-state spectroscopic studies of **1** revealed a band gap of 1.98 eV (band gap of TiO$_2$ 3.20 eV) demonstrating not only the ability of the siderophore H$_2$dihybe to stabilize rare metallic cores but to also modulate their electronic structure with potential uses in semiconducting photocatalytic applications. The origin of the dual-luminescence properties of the cluster **1** is currently under investigation.

Supplementary Materials: The following are available online, Figure S1: Possible modes for hydroxamate binding; Figure S2: Experimental and calculated in vacuum by ab initio DFT/B3LYP/LANL2DZ modeling spectra for the H2dihybe (a) and **1** (b); Figure S3: 2D {^{1}H,^{13}C} HMBC of H$_2$dihybe; Figure S4: 2D {^{1}H} grCOSY of **1**; Figure S5: 2D {^{1}H,^{13}C} grHSQC of **1**; Figure S6: 2D {^{1}H,^{13}C} grHMBC of **1**; Figure S7: 2D {^{1}H,^{13}C} grHSQC of **1** + 4 eq But$_4$NOH + 0.1 eq H$_2$dihybe; Figure S8: 2D {^{1}H,^{13}C} grHMBC of **1** + 4 eq But$_4$NOH + 0.1 eq H$_2$dihybe; Figure S9: Tauc plot of compound **1**; Table S1: ^{1}H and ^{13}C NMR chemical shifts (ppm) for the ligand H3dihybe, the minor titanium species **2** and the shielding/deshielding effect ($\Delta\delta$) upon complexation; Table S2: ^{1}H and ^{13}C NMR chemical shifts (ppm) for **1** and H2dihybe after addition of three equivalents of But$_4$NOH per **1** and the shielding/deshielding effect ($\Delta\delta$, ppm) upon complexation. The following are available online, Supporting Information (pdf) containing crystal data. CIF files containing crystal data for complex **1**.

Author Contributions: Conceptualization, T.A.K. and S.S.P.; synthesis of the titanium cluster **1** and solid-state UV–Vis spectroscopy, S.S.P.; ESI-MS spectrometry and crystallography, H.N.M.; NMR spectroscopy, S.H. and A.D.K.; IR spectroscopy and DFT calculations, P.S. and A.G.K.; writing-original draft preparation, T.A.K., H.N.M., A.D.K. and A.G.K.; writing—review and editing, T.A.K., H.N.M., A.D.K. and A.G.K.; supervision of all contributions, T.A.K., H.N.M. and A.D.K. All authors have read and agreed to the published version of the manuscript.

Funding: The research work was supported by the Hellenic Foundation for Research and Innovation (HFRI) under the HFRI PhD Fellowship grant (Fellowship Number: 1553). This work was co-funded by the European Regional Development Fund and the Republic of Cyprus through the Research and Innovation Foundation (Project: EXCELLENCE/1216/0515). The APC was funded by EPSRC. We would like to thank EPSRC (EP/S017046/1) and the University of Glasgow for supporting this work.

Data Availability Statement: Not applicable.

Conflicts of Interest: The authors declare no conflict of interest.

Sample Availability: Samples of compound **1** is available from the authors upon request.

References

1. Liu, P.Y.; Cui, L.N.; Yang, L.; Shu, X.P.; Zhu, Q.Y.; Dai, J. Bio-compatible fluorescent nano TiO materials prepared from titanium-oxo-cluster precursors. *Chem. Commun.* **2019**, *55*, 12360–12363. [CrossRef]
2. Chen, X.; Selloni, I.A. Introduction: Titanium dioxide (TiO$_2$) nanomaterials. *Chem. Rev.* **2014**, *114*, 9281–9282. [CrossRef]
3. Rajh, T.; Dimitrijevic, N.M.; Bissonnette, M.; Koritarov, T.; Konda, V. Titanium Dioxide in the Service of the Biomedical Revolution. *Chem. Rev.* **2014**, *114*, 10177–10216. [CrossRef] [PubMed]
4. Bai, J.; Zhou, B. Titanium Dioxide Nanomaterials for Sensor Applications. *Chem. Rev.* **2014**, *114*, 10131–10176. [CrossRef]
5. Bai, Y.; Mora-Sero', I.; Angelis, F.D.; Bisquert, J.; Wang, P. Titanium dioxide nanomaterials for photovoltaic applications. *Chem. Rev.* **2014**, *114*, 10095–10130. [CrossRef] [PubMed]
6. Ma, Y.; Wang, X.; Jia, Y.; Chen, X.; Han, H.; Li, C. Titanium Dioxide-Based Nanomaterials for Photocatalytic Fuel Generations. *Chem. Rev.* **2014**, *114*, 9987–10043. [CrossRef]
7. Fan, X.; Wang, J.H.; Wu, K.F.; Zhang, L.; Zhang, J. Isomerism in Titanium-oxo Clusters: Molecular Anatase Model with Atomic Structure and Improved Photocatalytic Activity. *Angew. Chem. Int. Ed.* **2019**, *58*, 1320–1323. [CrossRef] [PubMed]
8. Wang, C.; Liu, C.; Li, L.J.; Sun, Z.M. Synthesis, Crystal Structures, and Photochemical Properties of a Family of Heterometallic Titanium Oxo Clusters. *Inorg. Chem.* **2019**, *58*, 6312–6319. [CrossRef] [PubMed]
9. Wang, X.; Yu, Y.; Wang, Z.; Zheng, J.; Bi, Y.; Zheng, Z. Thiacalix[4]arene-Protected Titanium–Oxo Clusters: Influence of Ligand Conformation and Ti−S Coordination on the Visible-Light Photocatalytic Hydrogen Production. *Inorg. Chem.* **2020**, *59*, 7150–7157. [CrossRef] [PubMed]
10. Wang, S.B.; Wang, X.C. Multifunctional Metal-Organic Frameworks for Photocatalysis. *Small* **2015**, *11*, 3097–3112. [CrossRef]
11. Janek, M.; Radtke, A.; Muzioł, T.M.; Jerzykiewicz, M.; Piszczek, P. Tetranuclear Oxo-Titanium Clusters with Different Carboxylate Aromatic Ligands: Optical Properties, DFT Calculations, and Photoactivity. *Materials* **2018**, *11*, 1661. [CrossRef]
12. Pearce, A.J.; Harkins, R.P.; Reiner, B.R.; Wotal, A.C.; Dunscomb, R.J.; Tonks, I.A. Multicomponent Pyrazole Synthesis from Alkynes, Nitriles, and Titanium Imido Complexes via Oxidatively Induced N−N Bond Coupling. *J. Am. Chem. Soc.* **2020**, *142*, 4390–4399. [CrossRef]

13. Tuskaev, V.A.; Gagieva, S.C.; Kurmaev, D.A.; Melnikova, E.K.; Zubkevich, S.V.; Buzin, M.I.; Nikiforova, G.G.; Vasil'ev, V.G.; Saracheno, D.; Bogdanov, V.S.; et al. Olefin polymerization behavior of titanium(IV) alkoxo complexes with fluorinated diolate ligands: The impact of the chelate ring size and the nature of organoaluminum compounds. *Appl. Organomet. Chem.* **2020**, *34*, e5953. [CrossRef]
14. Robinson, S.G.; Wu, X.; Jiang, B.; Sigman, M.S.; Lin, S. Mechanistic Studies Inform Design of Improved Ti(salen) Catalysts for Enantioselective [3 + 2] Cycloaddition. *J. Am. Chem. Soc.* **2020**, *142*, 18471–18482. [CrossRef] [PubMed]
15. Ayla, E.Z.; Potts, D.S.; Bregante, D.T.; Flaherty, D.W. Alkene Epoxidations with H2O2 over Groups 4−6 Metal-Substituted BEA Zeolites: Reactive Intermediates, Reaction Pathways, and Linear Free-Energy Relationships. *ACS Catal.* **2021**, *11*, 139–154. [CrossRef]
16. Engler, H.; Lansing, M.; Gordon, C.P.; Neudörfl, J.M.; Schafer, M.; Schlörer, N.E.; Copéret, C.; Berkessel, A. Olefin Epoxidation Catalyzed by Titanium–Salalen Complexes: Synergistic H_2O_2 Activation by Dinuclear Ti Sites, Ligand H-Bonding, and π-Acidity. *ACS Catal.* **2021**, *11*, 3206–3217. [CrossRef]
17. Zhang, T.; Solé-Daura, A.; Fouilloux, H.; Poblet, J.M.; Proust, A.; Carbó, J.J.; Guillemot, G. Reaction Pathway Discrimination in Alkene Oxidation Reactions by Designed Ti-Siloxy-Polyoxometalates. *ChemCatChem* **2021**, *13*, 1220–1229. [CrossRef]
18. Hou, J.L.; Huo, P.; Tang, Z.Z.; Cui, L.N.; Zhu, Q.Y.; Dai, J. A Titanium Oxo Cluster Model Study of Synergistic Effect of Cocoordinated Dye Ligands on Photocurrent Responses. *Inorg. Chem.* **2018**, *57*, 7420–7427. [CrossRef]
19. Zou, D.H.; Cui, L.N.; Liu, P.Y.; Yang, S.; Zhu, Q.Y.; Dai, J. Molecular Model of Dye Sensitized Titanium Oxides Based on ArylAmine Dye Anchored Titanium Oxo Clusters. *Inorg. Chem.* **2019**, *58*, 9246–9252. [CrossRef]
20. Lakshmi Suresh, L.; Ralte Lalrempuia, R.; Ekeli, J.B.; Gillis-D'Hamers, F.; Törnroos, K.W.; Jensen, V.R.; Le Roux, E. Unsaturated and Benzannulated N-Heterocyclic Carbene Complexes of Titanium and Hafnium: Impact on Catalysts Structure and Performance in Copolymerization of Cyclohexene Oxide with CO_2. *Molecules* **2020**, *25*, 4364. [CrossRef] [PubMed]
21. Zhang, Y.; Azambuja, F.; Parac-Vogt, T.N. The forgotten chemistry of group(IV) metals: A survey on the synthesis, structure, and properties of discrete Zr(IV), Hf(IV), and Ti(IV) oxo clusters. *Coord. Chem. Rev.* **2021**, *438*, 213886. [CrossRef]
22. Wang, J.F.; Fang, W.H.; Li, D.S.; Zhang, L.; Zhang, J. Cocrystal of {Ti4} and {Ti6} Clusters with Enhanced Photochemical Properties. *Inorg. Chem.* **2017**, *56*, 2367–2370. [CrossRef]
23. Wang, C.; Liu, C.; Tian, H.R.; Li, L.J.; Sun, Z.M. Designed Cluster Assembly of Multidimensional Titanium Coordination Polymers: Syntheses, Crystal Structure and Properties. *Chem.-Eur. J.* **2018**, *24*, 2952–2961. [CrossRef] [PubMed]
24. Passadis, S.S.; Papanikolaou, M.G.; Elliott, A.; Tsiafoulis, C.G.; Tsipis, A.C.; Keramidas, A.D.; Miras, H.N.; Kabanos, T.A. Synthesis, Structural, and Physicochemical Characterization of a Ti6 and a Unique Type of Zr6 Oxo Clusters Bearing an Electron-Rich Unsymmetrical {OON} Catecholate/Oxime Ligand and Exhibiting Metalloaromaticity. *Inorg. Chem.* **2020**, *59*, 18345–18357. [CrossRef] [PubMed]
25. Schwyn, B.; Neilands, J.B. Universal chemical assay for the detection and determination of siderophores. *Anal. Biochem.* **1987**, *160*, 47–56. [CrossRef]
26. Ahmed, E.; Holmström, S.J.M. Siderophores in environmental research: Roles and applications. *Microb. Biotechnol.* **2014**, *7*, 196–208. [CrossRef] [PubMed]
27. Boukhalfa, H.; Lack, J.; Reilly, S.D.; Hersman, L.; Neu, M.P. Siderophore production and facilitated uptake of iron and plutonium in *P. putida*. *Proc. AIP Conf. Proc.* **2003**, *673*, 343.
28. Brennan, B.J.; Chen, J.; Rudshteyn, B.; Chaudhuri, S.; Mercado, B.Q.; Batista, V.S.; Crabtree, R.H.; Brudvig, G.W. Molecular titanium–hydroxamate complexes as models for TiO_2 surface binding. *Chem. Commun.* **2016**, *52*, 2972–2975. [CrossRef] [PubMed]
29. Brewster, T.P.; Konezny, S.J.; Sheehan, S.W.; Martini, L.A.; Schmuttenmaer, C.A.; Batista, V.S.; Crabtree, R.H. Hydroxamate Anchors for Improved Photoconversion in Dye-Sensitized Solar Cells. *Inorg. Chem.* **2013**, *52*, 6752–6764. [CrossRef]
30. McNamara, W.R.; Snoeberger, R.C.; Li, G.; Richter, C.; Allen, L.J.; Milot, R.L.; Schmuttenmaer, C.A.; Crabtree, R.H.; Brudvig, G.W.; Batista, V.S. Hydroxamate anchors for water-stable attachment to TiO_2 nanoparticles. *Energy Environ. Sci.* **2009**, *2*, 1173–1175. [CrossRef]
31. Zaleski, C.M.; Kampf, J.W.; Mallah, T.; Kirk, M.L.; Pecoraro, V.L. Assessing the Slow Magnetic Relaxation Behavior of $Ln^{III}_4Mn^{III}_6$ Metallacrowns. *Inorg. Chem.* **2007**, *46*, 1954–1956. [CrossRef] [PubMed]
32. Boron, T.T., III; Kampf, J.W.; Pecoraro, V.L. A Mixed 3d-4f 14-Metallacrown-5 Complex That Displays Slow Magnetic Relaxation through Geometric Control of Magnetoanisotropy. *Inorg. Chem.* **2010**, *49*, 9104–9106. [CrossRef]
33. Zaleski, C.M.; Depperman, E.C.; Kampf, J.W.; Kirk, M.L.; Pecoraro, V.L. Synthesis, Structure, and Magnetic Properties of a Large Lanthanide−Transition-Metal Single-Molecule Magnet. *Angew. Chem. Int. Ed.* **2004**, *43*, 3912–3914. [CrossRef] [PubMed]
34. Deb, A.; Boron, T.T., III; Itou, M.; Sakurai, Y.; Mallah, T.; Pecoraro, V.L.; Penner-Hahn, J.E. Understanding Spin Structure in Metallacrown Single-Molecule Magnets using Magnetic Compton Scattering. *J. Am. Chem. Soc.* **2014**, *136*, 4889–4892. [CrossRef] [PubMed]
35. Chow, C.Y.; Trivedi, E.R.; Pecoraro, V.; Zaleski, C.M. Heterometallic Mixed 3d-4f Metallacrowns: Structural Versatility, Luminescence, and Molecular Magnetism. *Comments Inorg. Chem.* **2015**, *35*, 214–253. [CrossRef]
36. Ostrowska, M.; Toporivska, Y.; Golenya, I.A.; Shova, S.; Fritsky, I.O.; Pecoraro, V.L.; Gumienna-Kontecka, E. Explaining How α-Hydroxamate Ligands Control the Formation of Cu(II)-, Ni(II)-, and Zn(II)-Containing Metallacrowns. *Inorg. Chem.* **2019**, *58*, 16642–16659. [CrossRef]

37. Eliseeva, S.V.; Salerno, E.V.; Bermudez, B.A.L.; Petoud, S.; Pecoraro, V.L. Dy^{3+} White Light Emission Can Be Finely Controlled by Tuning the First Coordination Sphere of Ga^{3+}/Dy^{3+} Metallacrown Complexes. *J. Am. Chem. Soc.* **2020**, *142*, 16173–16176. [CrossRef]
38. Salerno, E.V.; Eliseeva, S.V.; Schneider, B.L.; Kampf, J.W.; Petoud, S.; Pecoraro, V.L. Visible, Near-Infrared, and Dual-Range Luminescence Spanning the 4f Series Sensitized by a Gallium(III)/Lanthanide(III) Metallacrown Structure. *J. Phys. Chem. A* **2020**, *124*, 10550–10564. [CrossRef]
39. Alaimo, A.A.; Koumousi, E.S.; Cunha-Silva, L.; McCormick, L.J.; Teat, S.J.; Psycharis, V.; Raptopoulou, C.P.; Mukherjee, S.; Li, C.; Gupta, S.D.; et al. Structural Diversities in Heterometallic Mn−Ca Cluster Chemistry from the Use of Salicylhydroxamic Acid: $\{Mn^{III}_4Ca_2\}$, $\{Mn^{II/III}_6Ca_2\}$, $\{Mn^{III/IV}_8Ca\}$, and $\{Mn^{III}_8Ca_2\}$ Complexes with Relevance to Both High and Low-Valent States of the Oxygen-Evolving Complex. *Inorg. Chem.* **2017**, *56*, 10760–10774. [CrossRef]
40. Sun, O.; Chen, P.; Li, H.F.; Gao, T.; Yan, P.F. Wheel-like $\{Ln_6\}$ luminescent lanthanide complexes covering the visible and near-infrared domains. *CrystEngComm* **2020**, *22*, 5200–5206. [CrossRef]
41. Pathak, A.; Blair, V.L.; Ferrero, R.L.; Junk, P.C.; Tabora, R.F.; Andrews, P.C. Synthesis and structural characterisation of bismuth(III) hydroxamates and their activity against Helicobacter pylori. *Dalton Trans.* **2015**, *44*, 16903–16913. [CrossRef] [PubMed]
42. Doua, M.; Yanga, H.; Zhaoa, X.; Zhangb, Z.; Lia, D.; Dou, J. A novel "sawtooth-like" heterometallic Sr-Mo 18-metallacrown-6 complex: Synthesis, structure and anticancer activity. *Inorg. Chem. Commun.* **2020**, *119*, 108127–108130. [CrossRef]
43. Wang, B.; Luo, X. A first-principles study on potential chelation agents and indicators of Alzheimer's disease. *RSC Adv.* **2020**, *10*, 35574–35581. [CrossRef]
44. Gao, D.D.; Gao, Q.; Chen, Y.M.; Li, Y.H.; Li, W. Syntheses, Structures, and Luminescent Properties of the Zn-II and Cd-II 1-D Chain Polymers Assembled by Salicylhydroxamic Acid. *Chin. J. Struct. Chem.* **2015**, *34*, 1371–1378.
45. Sun, O.; Chen, P.; Li, H.F.; Gao, T.; Yan, P.F. Structural, photophysical and magnetic studies of $\{Ln_2\}$ assembled about oxime. *Inorg. Chem. Commun.* **2020**, *114*, 107841–197844. [CrossRef]
46. Chen, S.; Fang, W.H.; Zhang, L.; Zhang, J. Synthesis, Structures, and Photocurrent Responses of PolyoxoTitanium Clusters with Oxime Ligands: From Ti_4 to Ti_{18}. *Inorg. Chem.* **2018**, *57*, 8850–8856. [CrossRef] [PubMed]
47. Ito, W.; Hattori, S.; Kondo, M.; Sakagami, H.; Kobayashi, O.; Ishimoto, T.; Shinozaki, K. Dual emission from an iridium(iii) complex/counter anion ion pair. *Dalton Trans.* **2021**, *50*, 1887–1894. [CrossRef]
48. Hong, Q.M.; Wang, S.Y.; An, D.L.; Li, H.Y.; Zhou, J.M.; Deng, Y.F.; Zhou, Z.H. Transformations of dimeric and tetrameric glycolato peroxotitanates and their thermal decompositions for the preparations of anatase and rutile oxides. *J. Solid State Chem.* **2019**, *277*, 169–174. [CrossRef]
49. Tomita, K.; Petrykin, V.; Kobayashi, M.; Shiro, M.; Yoshimura, M.; Kakihana, M. A Water-Soluble Titanium Complex for the Selective Synthesis of Nanocrystalline Brookite, Rutile, and Anatase by a Hydrothermal Method. *Angew. Chem. Int. Ed.* **2006**, *45*, 2378–2381. [CrossRef]
50. Janas, Z.; Jerzykiewicz, L.; Przybylak, K.; Sobota, P.; Szczegot, K. Titanium Complexes Stabilized by a Sulfur-Bridged Chelating Bis(aryloxo) Ligand as Active Catalysts for Olefin Polymerization. *Eur. J. Inorg. Chem.* **2004**, 1639–1645. [CrossRef]
51. Miras, H.N.; Stone, D.; Long, L.; McInnes, E.J.L.; Kögerler, P.; Cronin, L. Exploring the Structure and Properties of Transition Metal Templated $\{VM_{17}(VO_4)_2\}$ Dawson-Like Capsules. *Inorg. Chem.* **2011**, *50*, 8384–8391. [CrossRef]
52. Xu, F.; Scullion, R.A.; Yan, J.; Miras, H.N.; Busche, C.; Scandurra, A.; Pignataro, B.; Long, D.L.; Cronin, L. A Supramolecular Heteropolyoxopalladate $\{Pd_{15}\}$ Cluster Host Encapsulating a $\{Pd_2\}$ Dinuclear Guest: $[Pd^{II}_2 \subset \{H_7Pd^{II}_{15}O_{10}(PO_4)_{10}\}]^{9-}$. *J. Am. Chem. Soc.* **2011**, *133*, 4684–4686. [CrossRef]
53. Miras, H.N.; Sorus, M.; Hawkett, J.; Sells, D.O.; McInnes, E.J.L.; Cronin, L. Oscillatory Template Exchange in Polyoxometalate Capsules: A Ligand-Triggered, Redox-Powered, Chemically Damped Oscillation. *J. Am. Chem. Soc.* **2012**, *134*, 6980–6983. [CrossRef]
54. Miras, H.N.; Wilson, E.F.; Cronin, L. Unravelling the complexities of inorganic and supramolecular self-assembly in solution with electrospray and cryospray mass spectrometry. *Chem. Commun.* **2009**, *11*, 1297–1311. [CrossRef] [PubMed]
55. Zang, H.; Surman, A.; Long, D.; Cronin, L.; Miras, H.N. Exploiting the equilibrium dynamics in the self-assembly of inorganic macrocycles based upon polyoxothiometalate building blocks. *Chem. Commun.* **2016**, *52*, 9109–9112. [CrossRef]
56. Zang, H.Y.; Chen, J.J.; Long, D.L.; Cronin, L.; Miras, H.N. Assembly of Thiometalate-Based $\{Mo_{16}\}$ and $\{Mo_{36}\}$ Composite Clusters Combining $[Mo_2 O_2S_2]^{2+}$ Cations and Selenite Anions. *Adv. Mater.* **2013**, *25*, 6245–6249. [CrossRef]
57. Miras, H.N.; Zang, H.Y.; Long, D.L.; Cronin, L. Direct Synthesis and Mass Spectroscopic Observation of the $\{M_{40}\}$ Polyoxothiometalate Wheel. *Eur. J. Inorg. Chem.* **2011**, *33*, 5105–5111. [CrossRef]
58. Drouza, C.; Hadjithoma, S.; Nicolaou, M.; Keramidas, A.D. Structural characterization, hydrolytic stability, and dynamics of cis-$Mo^{VI}O_2^{2+}$ hydroquinonate/phenolate complexes. *Polyhedron* **2018**, *152*, 22–30. [CrossRef]
59. Drouza, C.; Stylianou, M.; Keramidas, A.D. NMR characterization and dynamics of vanadium(V) complexes with tripod (hydroquinonate/phenolate) iminodiacetate ligands in aqueous solution. *Pure Appl. Chem.* **2009**, *81*, 1313–1321. [CrossRef]
60. Drouza, C.; Stylianou, M.; Papaphilippou, P.; Keramidas, A.D. Structural and electron paramagnetic resonance (EPR) characterization of novel vanadium(V/IV) complexes with hydroquinonate-iminodiacetate ligands exhibiting "noninnocent" activity. *Pure Appl. Chem.* **2013**, *85*, 329–342. [CrossRef]

61. Passadis, S.S.; Hadjithoma, S.; Kalampounias, A.G.; Tsipis, A.C.; Sproules, S.; Miras, H.N.; Keramidas, A.D.; Kabanos, T.A. Synthesis, structural and physicochemical characterization of a new type Ti6-oxo cluster protected by a cyclic imide dioxime ligand. *Dalton Trans.* **2019**, *48*, 5551–5559. [CrossRef]
62. Shi, C.R.; Zhang, M.; Hang, X.; Bi, Y.; Huang, L.; Zhou, K.; Xu, Z.; Zheng, Z. Assembly of thiacalix[4]arene-supported high-nuclearity Cd_{24} cluster with enhanced photocatalytic activity. *Nanoscale* **2018**, *10*, 14448–14454. [CrossRef]
63. Sheldrick, G.M. Phase Annealing in SHELX-90: Direct Methods for Larger Structures. *Acta Crystallogr. Sect. A* **1990**, *46*, 467. [CrossRef]
64. Sheldrick, G.M. A short history of SHELX. *Acta Crystallogr. Sect. A* **2008**, *64*, 112. [CrossRef]
65. Farrugia, L.J. WinGX Program Features. *J. Appl. Cryst.* **1999**, *32*, 837. [CrossRef]
66. Clark, R.C.; Reid, J.S. The Analytical Calculation of Absorption in Multifaceted Crystals. *Acta Crystallogr. Sect. A* **1995**, *51*, 887. [CrossRef]
67. Frisch, M.J.; Trucks, G.W.; Schlegel, H.B.; Scuseria, G.E.; Robb, M.A.; Cheeseman, J.R.; Scalmani, G.; Barone, V.; Mennucci, B.; Petersson, G.A.; et al. *Gaussian 09, Revision D.01*; Gaussian, Inc.: Wallingford, CT, USA. Available online: https://gaussian.com/g09citation/ (accessed on 15 July 2021).
68. Available online: https://pubchem.ncbi.nlm.nih.gov (accessed on 15 July 2021).
69. Becke, A.D. Density-functional thermochemistry. III. The role of exact exchange. *J. Chem. Phys.* **1993**, *98*, 5648. [CrossRef]
70. Lee, C.; Yang, W.; Parr, R.G. Development of the Colic-Salvetti correlation-energy formula into a functional of the electron density. *Phys. Rev. B* **1988**, *37*, 785. [CrossRef]

Article

Synthesis of Indoloquinolines: An Intramolecular Cyclization Leading to Advanced Perophoramidine-Relevant Intermediates

Craig A. Johnston, David B. Cordes, Tomas Lebl, Alexandra M. Z. Slawin and Nicholas J. Westwood *

School of Chemistry and Biomedical Sciences Research Complex, University of St. Andrews and EaStCHEM, North Haugh, St. Andrews KY16 9ST, UK; craig.johnston85@gmail.com (C.A.J.); dbc21@st-andrews.ac.uk (D.B.C.); tl12@st-andrews.ac.uk (T.L.); amzs@st-andrews.ac.uk (A.M.Z.S.)
* Correspondence: njw3@st-andrews.ac.uk; Tel.: +44-1334-463816

Abstract: The bioactive natural product perophoramidine has proved a challenging synthetic target. An alternative route to its indolo[2,3-b]quinolone core structure involving a N-chlorosuccinimde-mediated intramolecular cyclization reaction is reported. Attempts to progress towards the natural product are also discussed with an unexpected deep-seated rearrangement of the core structure occurring during an attempted iodoetherification reaction. X-ray crystallographic analysis provides important analytical confirmation of assigned structures.

Keywords: perophoramidine; natural product; Claisen rearrangement; indoloquinoline; intramolecular cyclization; X-ray structure

1. Introduction

Attempts to prepare complex bioactive natural products often test synthetic methodology under very challenging circumstances. In addition, unexpected outcomes, in what initially look like simple transformations, frequently occur and provide interesting analytical conundrums. One approach to solving these structural questions is the use of small-molecule X-ray crystallography which continues to play an essential role in developing routes to complex molecules. Through collaborations, such as ours with Professor Alexandra Slawin, difficult analytical challenges are frequently solved with apparent ease. For anyone who has had the pleasure to work with someone with Professor Slawin's level of expertise, the phrase "Oh of course Alex, now you've pointed it out that has to be the structure" will probably be familiar!

The structure of the natural product perophoramidine **1** (Scheme 1) was first reported in 2002 by Ireland [1] although earlier biosynthetic proposals, synthetic and small molecule X-ray crystallographic work had sparked interest in closely related alkaloids [2–10]. To date, a number of elegant total syntheses of perophoramidine **1** have been reported [11–16] along with a range of other attempts [17,18]. In addition, in their initial report, Ireland et al. reported the dehalogenation of perophoramidine **1** under hydrogenation conditions using HCO$_2$NH$_4$ and Pd/C in MeOH. This led to the formation of a compound that they named dehaloperophoramidine **2** (Scheme 1) [1]. Additional syntheses of compound **2** have been reported [19–22], with one of these routes to **2** being developed in our laboratory [21]. As our studies to **2** progressed, investigations into the synthesis of the halogen-containing **1** were also carried out.

Here, we report the successful synthesis of the halogen-containing analogue **3** of compound **4**, which was a key intermediate in our synthesis of **2** (Scheme 1) [21]. Attempts to progress forward from **3** towards **1**, using an alternative route to that reported by us for the synthesis of **2** [21], are also described. In addition, we include the structural assignment of two of the prepared compounds by small-molecule X-ray crystallographic analysis, bringing to 18 the total number of times this technique has guided this overall program

of work (for the previous small molecule X-ray crystallographic structures see CCDC 737646-737648, 1486344, 1478152-4, 1582892-7 and 1811882-4).

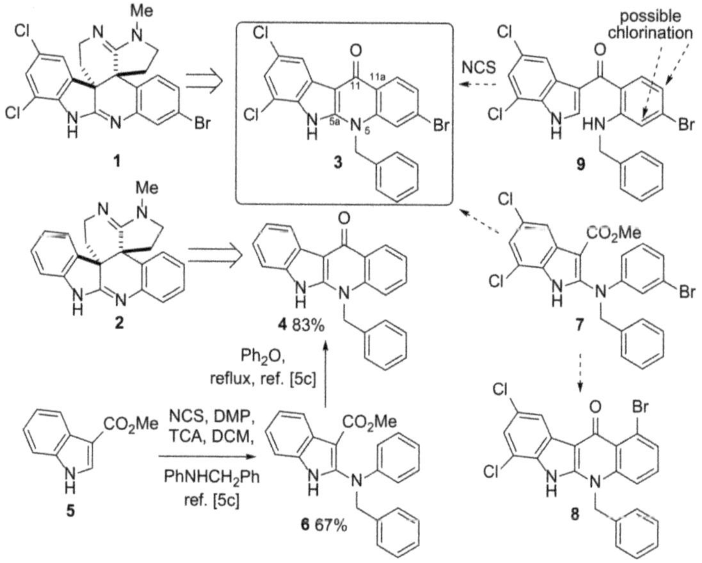

Scheme 1. Structures of the natural product perophoramidine **1** and the product prepared by Ireland et al. [1] following dehalogenation, dehaloperophoramidine **2**. Previous reports [5c] described the synthesis of **4** via a NCS-mediated coupling reaction of **5** and subsequent cyclization of **6**. **4** was then converted to **2** [21]. In this work, access to the analogous intermediate **3** was explored with two possible routes either via **7** or an intramolecular NCS-mediated cyclization of **9** considered. An alternative product, compound **8**, from a potential cyclization reaction of **7** was possible. Chlorination of the aniline ring in **9** was also a potential side reaction. NCS = N-chlorosuccinimide; DMP = N,N′-dimethylpiperazine; TCA = trichloroacetic acid.

2. Results

2.1. Proposed Route to Halogenated Intermediate 3

We have previously reported [21] a robust and scalable route to the indolo[2,3-b]quinoline core structure of **4** using a N-chlorosuccinimide (NCS)-mediated coupling of methyl indole-3-carboxylate (**5**) with N-benzylaniline followed by cyclization of the product **6** in refluxing diphenyl ether (Scheme 1) [23]. This reaction sequence proved a suitable starting point for the synthesis of dehaloperophoramidine **2** [21]. Whilst this approach was relatively straightforward and high yielding in the case of **4**, it seemed likely that the analogous reaction using substrate **7** with the required halogenation pattern to prepare perophoramidine **1** would lead to the formation of regioisomers **3** and **8** (Scheme 1). It was envisaged that separation of **3** and **8** would prove challenging as these type of compounds exhibit low solubility in organic solvents.

To avoid this issue, an alternative approach was proposed involving a NCS-mediated cyclization reaction of substituted indole **9** to form the N5–C5a bond in **3** after the C11–C11a bond. This is an intramolecular version of the reaction used to form **6** (Scheme 1). NCS-mediated intramolecular cyclization reactions have recently been used to form a C–N bond at the indole 2-position in the total synthesis of (−)-chaetominine (**10**) [8] with tetrahydro-1H-pyrido[2,3-b]indole (**11**) being prepared from substituted indole **12** using NCS (1.3 equiv.) and Et$_3$N (4 equiv.) in DCM in 52% yield (Scheme 2). One possible challenge with this approach in our system was competing chlorination in the aniline ring in **9** (Scheme 1). Given difficulties in predicting the result of this competition a priori, it was decided to investigate this proposed route to **3**.

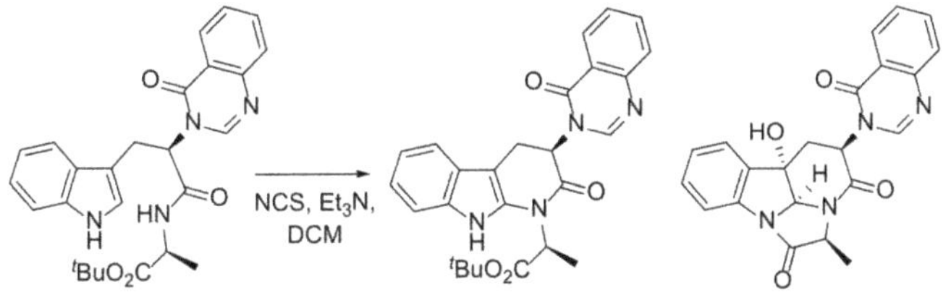

Scheme 2. Reported synthesis of **11** via a NCS-mediated formation of the C–N bond at the indole 2-position during the preparation of the natural product (−)-chaetominine (**10**) [24].

2.2. Synthesis of Halogenated Intermediate 3

Whilst several methods are available for the synthesis of 5,7-dichloroindole (**13**) [24–28] including electrochemical-mediated cyclization [25] and gold-catalyzed annulation of the corresponding 2-alkynylaniline [26], it was decided to use the Bartoli indole synthesis [27,28] as multi-gram quantities of **13** were required (Scheme 3). Reaction of 2,4-dichloronitrobenzene (**14**) with vinylmagnesium chloride [28] gave indole **13** with key steps in the process involving a [3,3]-sigmatropic rearrangement followed by cyclization onto the resulting aldehyde to form the 5-membered ring. The yield of this reaction was relatively low (48%, in line with the literature precedent [28]); however, the starting materials were readily available and the reaction could be carried out to give almost 9 g of **13**. Indole **13** was subsequently converted to aldehyde **15** using a Vilsmeier-Haack reaction [29] and Boc protection of the indole nitrogen using di-*tert*-butyl dicarbonate (Boc$_2$O) in the presence of 4-dimethylaminopyridine (DMAP) formed **16** in high yield (Scheme 3).

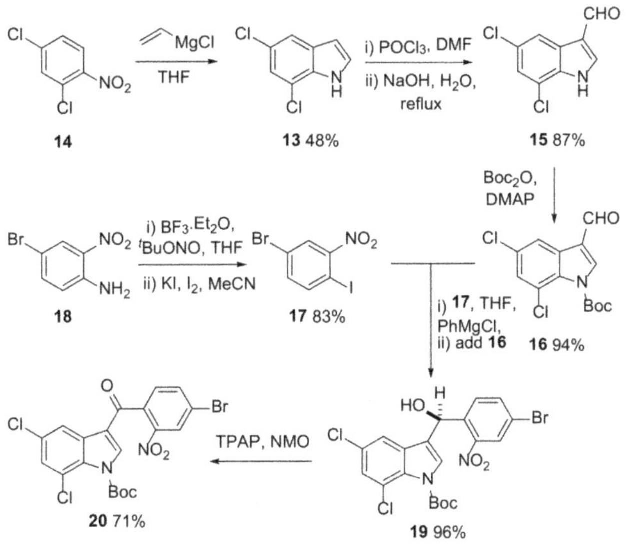

Scheme 3. Synthesis of ketone **20** involving generation of indole **13** using a Bartoli reaction [27,28] and a Knochel-type coupling reaction. TPAP = tetrapropylammonium perruthenate; NMO = N-methylmorpholine N-oxide.

Aryl iodide **17** was then synthesized from **18** in 83% yield using a literature procedure (Scheme 3) [30]. Having prepared intermediates **16** and **17**, a Grignard-mediated coupling reaction was attempted. This reaction was based on a method reported by Knochel et al. in which aryl rings containing an iodine atom *ortho* to a nitro group can undergo I-Mg exchange when treated with phenylmagnesium chloride [31,32]. The resulting Grignard reagent can then react with an electrophile to form a new carbon–carbon bond. Interestingly, more reactive Grignard reagents such as methyl magnesium chloride react with the nitro group, leading to complex mixtures of products. The reaction is also reported to be unsuccessful when *meta*- or *para*-iodonitrobenzenes are used [31,32], leading to the proposal that chelation of the nitro group to the magnesium atom stabilizes the *ortho*-substituted Grignard reagent. The treatment of iodide **17** with phenylmagnesium chloride at −40 °C for 1 h followed by reaction with aldehyde **16** gave alcohol **19** (racemate, Scheme 3) in excellent yield, even on a multiple-gram scale. Alcohol **19** was then oxidized to ketone **20** using tetrapropylammonium perruthenate (TPAP, 5 mol%) and co-oxidant N-methylmorpholine N-oxide (NMO) (Scheme 3) [33].

Attempts were made to reduce the nitro group to the corresponding amine and remove the Boc protecting group in one step by refluxing **20** in acetic acid and ethanol in the presence of iron powder. However, this reaction produced a mixture of the expected product **21** and Boc protected **22**. A two step protocol using iron in acetic acid and ethanol followed by treatment of purified **22** with TFA in DCM gave **21** in an excellent yield over the two steps (Scheme 4). The required reductive N-benzylation of **21** proved more challenging. When **21** was heated with benzaldehyde in refluxing toluene for six hours followed by the addition of sodium triacetoxyborohydride (STAB) the required product **9** was obtained in only moderate yield presumably due to the relatively poor nucleophilicity of the aniline nitrogen in **21** disfavouring initial imine formation. An alternative improved procedure based on the report of Boros et al. [34] was eventually found. This involved the additional use of TFA and so enabled telescoping of the conversion of **22** to **9** which after optimization of this reaction (a second aliquot of the STAB/TFA solution after 30 min) enabled the formation of **9** from **22** to 68% with a small quantity of **21** also being obtained (Scheme 4).

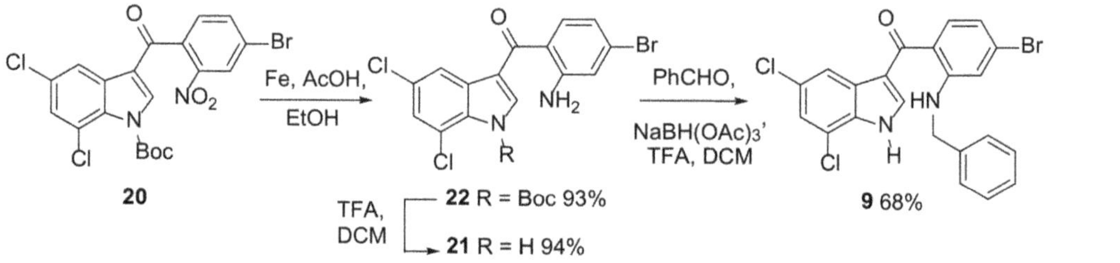

Scheme 4. Nitro group reduction and reductive amination.

After carrying out a series of model studies (see Supplementary Material), a NCS-mediated cyclization of **9** was attempted. The initial conditions used were NCS (2.0 equiv.) and DMP (0.56 equiv.) in DCM at room temperature overnight. After stirring for 28 h, it was observed that a small amount of precipitate was formed. The precipitate was isolated by filtration, washed with DCM and analyzed by ^{1}H NMR in d6-DMSO (Figure S2). The NMR and mass spectrometric analysis was consistent with the presence of two compounds with m/z values equal to the required product **3** and a chlorinated analogue of **3** (assigned structure **23** (Scheme 5 and Figure S3), vide infra). Analysis of the filtrate and washings from the reaction indicated that again two main compounds were present, one of which was unreacted **9**. The second compound was assigned as **24** (Scheme 5), which was presumably formed by NCS-mediated chlorination at the aniline 4-position rather than at the indole 3-position of **9**. The isolation of **24** also led to the proposal that the unidentified chlorinated analogue of **3** present in the precipitate was in fact **23**, the cyclized version of **24**. To confirm this, a small-scale reaction of **24** with NCS and base was carried out, giving **23** (Scheme 5, Figure S3 and Supplementary Material for protocols and analytical data for **23** and **24**).

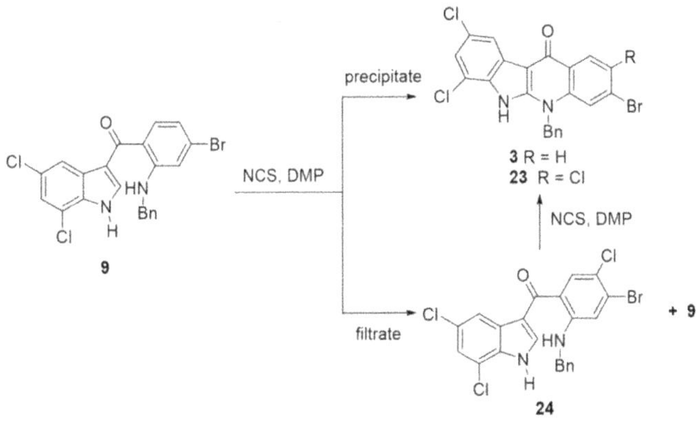

Scheme 5. Preliminary attempts at NCS-mediated cyclization of **9** using NCS (2.0 equiv.), DMP (0.56 equiv.) in DCM at room temperature for 28 h led to formation of a precipitate that contained the desired cyclized product **3** and a second product assigned as **23** (Figure S3, see Supplementary Material for analytical data for **23** and **24**). **23** was independently prepared from an isolated sample of **24** (NCS (1.0 equiv.), DMP (0.56 equiv.) in DCM at room temperature for 16 h). Significant optimization of this reaction was required.

Optimization of this reaction started with a solvent screen based on literature precedent [23,24,35,36]. The reaction of **9** with NCS (1.0 equiv.) and DMP (0.56 equiv.) was carried out in seven common solvents (DCM, MeOH, acetone, THF, CH_3CN, hexane and toluene) and the amount of precipitate isolated and the ratio of **9**:**24** in the filtrates was determined (Table S1). Acetonitrile was judged as the preferred solvent as the largest amount of precipitate was formed (cyclized products, Table S1, Entry 5). As related reactions have been reported using a range of bases [23,35–40], it was decided to react **9** with NCS (1.2 equiv.) in acetonitrile using six different bases (DMP, NaH, Et_3N, DMAP, DIPEA and pyridine, Table S2). In brief, the use of Et_3N gave the highest product yield (48%) with only the required **3** being present in the precipitate and starting material **9** being the dominant product in the filtrate (Table S2, Entry 3). Further optimization of this reaction found that when 2.4 equivalents of both NCS and Et_3N were used at room temperature for 18 h, **3** could be isolated in 55% yield with **9** (21%) recovered on purification by column chromatography of the filtrate (Scheme 6 and Supplementary Material demonstrating the reproducibility of this reaction up to a 2 g scale).

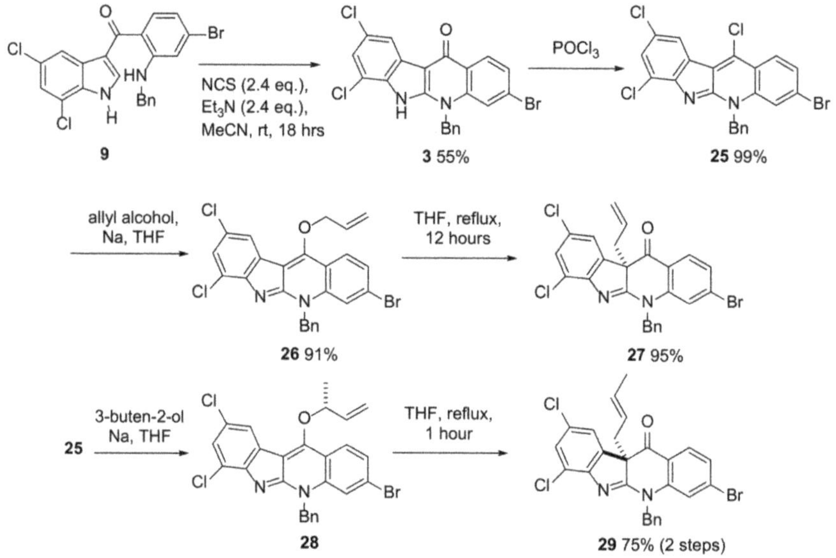

Scheme 6. Optimized conditions for the conversion of **9** to **3** and instalment of an all-carbon quaternary center in compounds **27** and **29**. Insights into the impact of the halogen substituents on both the addition-elimination and Claisen rearrangement steps were gained.

2.3. Installation of the First All-Carbon Quaternary Center and X-ray Structure Determination of Advanced Intermediate 27

After developing the route to ketone **3**, the instalment of the first of the all-carbon quaternary centers (Schemes 1 and 6) present in perophoramidine **1** was achieved over 3 steps. Reaction of **3** with POCl$_3$ gave **25** in almost quantitative yield. Substitution of one of the chlorines in **25** using sodium allyloxide gave allyl ether **26** (Scheme 6) which underwent Claisen rearrangement to form ketone **27** on heating at reflux in THF for 12 h. Recrystallization of **27** provided crystals suitable for X-ray crystallographic analysis, which enabled confirmation of its structure (Figure 1). This three-step sequence was also successful when the alkoxide formed from the reaction of racemic 3-buten-2-ol with sodium was used. As observed previously in the dehalo system (Scheme S4) [21], the Claisen rearrangement of the crotyl-containing **28** generated from **25** was feasible at room temperature (unlike allyl-containing **26** which does not rearrange at all at room temperature) and as a result it was not possible to isolate **28** in pure form. Heating the obtained sample of predominantly **28** at reflux in THF only required 1 h for full conversion to **29** in 75% yield over the two steps (c.f. the 12 h at reflux in THF required for complete conversion of **26** to **27** and the 5 h required in the analogous reaction in the crotyl-dehalo series, Scheme S4). It is clear that the presence of the halogens in **28** accelerated the Claisen rearrangement. In this sequence it was also interesting to note that the alkoxide addition-elimination reactions of **25** to give **26** or **28** proceeded much faster than the corresponding reaction of dehalo analogue which required 18 h to convert fully [23]. This is likely due to the electron-withdrawing effect of the halogen substituents in **25** promoting nucleophilic addition of the alkoxide. Although the synthesis of intermediate **29** was carried out in such a way that racemic product was obtained, the use of enantiomerically enriched (R)-3-buten-2-ol would enable the formation of the C-10b quaternary center with the correct absolute stereochemistry required for an asymmetric synthesis of perophoramidine **1** [23].

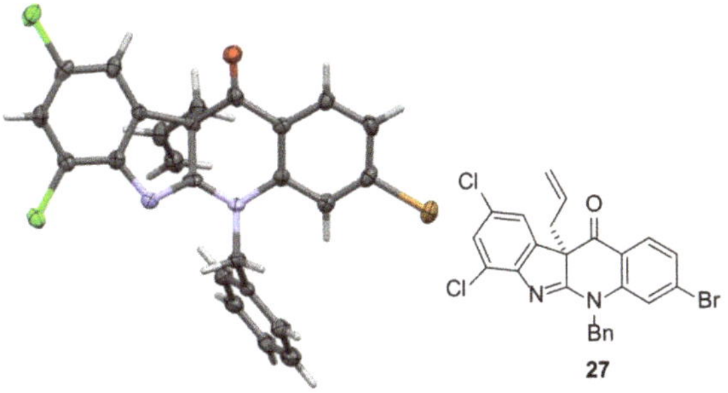

Figure 1. A view of the X-ray crystal structure of **27** (ellipsoids drawn at the 50 % probability level) used to confirm its structure.

2.4. Attempted Progress towards Perophoramidine 1

The second half of this report briefly describes two approaches that were attempted to progress from advanced synthetic intermediates **27** and **29** towards perophoramidine **1**. Whilst ultimately unsuccessful, these approaches did provide additional insights into the inherent reactivity of this complex system. One of the initial strategies employed by others [17,41] and us [42] for the synthesis of **1** and **2** involved the preparation of a C11-ester-containing intermediate, in our case, of general structure **30** (Scheme 7 and Scheme S5). The approach attempted here for installing a C11-ester involved reaction of the previously prepared **29** with in situ-generated chloromethyl lithium to form epoxide **31** in good yield (Scheme 7). The relative stereochemistry of **31** was tentatively assigned based on the absence of a nOe correlation between the CH_2 of the epoxide and any of the protons of the crotyl chain. In addition, the assigned stereochemistry in **31** was expected due to the attack of the chloromethyl lithium from the least hindered face of the molecule as was observed in previous studies in which proof of structure had been obtained through small-molecule X-crystal structure analysis of the analogous dehalo-epoxide [21]. Reductive opening of **31** to give **32** was then achieved using excess boron trifluoride and sodium cyanoborohydride in moderate yield. It was believed that this reaction proceeded with inversion of configuration at the C-11 stereocenter due to hydride attack occurring from the least hindered face, again consistent with previously reported studies in which small-molecule X-ray analysis of a related dehalo-analogue had been achieved [42]. Unfortunately, the results obtained from the Jones oxidation [43] of **32** were different to those of the previously reported non-halogenated alcohols [42]. Following the reaction by LC–MS showed that after just 5 min reaction time, **32** was almost completely consumed with a number of different product peaks being observed in the LC–MS spectrum. From the complex mixture of products obtained at this time point, evidence for the presence of **33** was obtained. After an increased reaction time (5 h), the initially complex mixture simplified with the major product being assigned as the decomposition product **34** (Scheme 7 and Figure S4, observed m/z = 501.17 $[M+H]^+$; theoretical m/z for formation of **34**, $C_{23}H_{13}{}^{81}Br^{35}Cl_2N_2O_2$ $[M+H]^+$ 500.96). No evidence for the presence of **33** in the reaction mixture at the 5 h time point was found.

As conversion of **32** to **34** clearly involved loss of the crotyl group (for one possible mechanism see Scheme S6), it was next decided to develop the crotyl chain in **29** (or in fact the allyl chain in **27**, Scheme 8) by reacting the double bond. Building on an approach previously reported by us in the dehalo series [44], **35** became the new target molecule. Subsequent conversion of **35** to **36** could provide an approach to perophoramidine **1** by differentiating between the two allyl groups as seen in a related system [21].

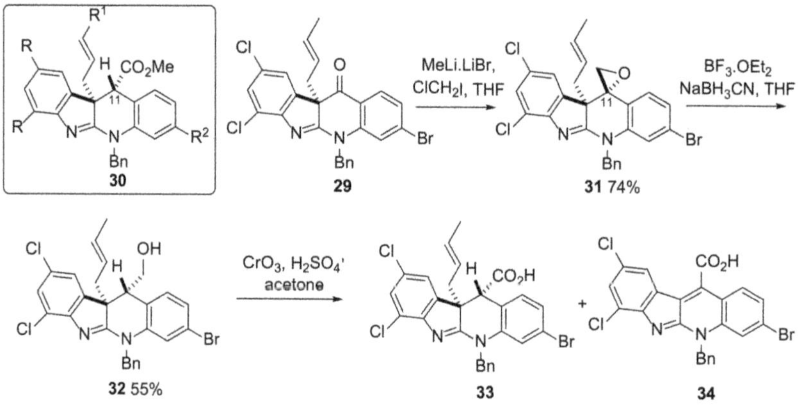

Scheme 7. Attempted formation of ester **30** (R = Cl, R¹ = Me, R² = Br) to enable ester alkylation as a method of incorporating the second all-carbon quaternary center. Reaction of **32** under Jones oxidation conditions gave **34** rather than the desired **33**.

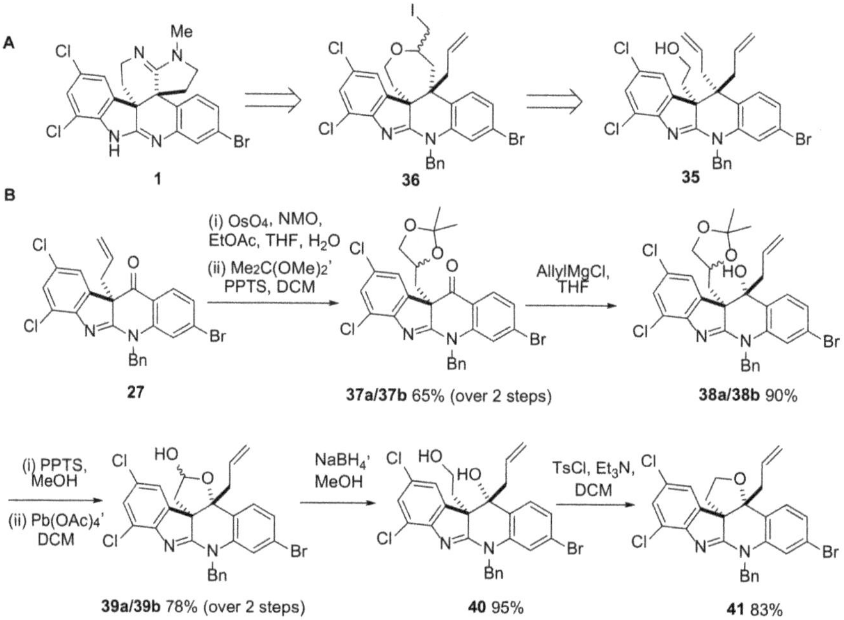

Scheme 8. A. One possible approach to perophoramidine **1** via diallyl-containing compound **35**. Differentiation between the two allyl groups in **35** could potentially be achieved through an iodoetherification protocol. **B.** Synthesis of cyclic ether **41**, a precursor to **35**. PPTS = pyridinium p-toluenesulfonate, TsCl = p-toluenesulfonyl chloride.

After dihydroxylation of the allyl group in **27**, the corresponding diastereomeric acetonides **37a/37b** (1:1 mixture) were formed in 65% yield over the two steps. Reaction of **37a/37b** with allylmagnesium chloride proceeded in high yield to give only two of the possible diastereomers (assigned structures as shown in **38a/38b** based on expected stereochemical outcome of addition to the ketone). Deprotection of the acetonide group in **38a/38b** was achieved under relatively mild conditions using PPTS in methanol [45] and the resulting crude reaction mixture was treated with lead tetraacetate, leading to lactols **39a/39b** still as a 1:1 mixture of diastereomers. Reduction of **39a/39b** with sodium borohydride in methanol gave **40** which could be converted to cyclic ether **41** in excellent yield over the two steps. Importantly, after dealing with a series of diastereomeric mixtures for several steps in this sequence, recrystallization of **41** by slow evaporation of a solution of **41** in ethyl acetate provided crystals suitable for small-molecule X-ray crystallography. From the data obtained, it was clear that the stereochemistry at the two stereogenic centers in **41** was as planned. This places the ether ring almost perpendicular to the tetracyclic core of **41**, with the allyl substituent on the opposite face to the ether ring (Figure 2).

Figure 2. Two views of the X-ray crystal structure of **41** (ellipsoids drawn at the 50 % probability level) used to confirm its structure.

Opening of the cyclic ether in **41** proceeded smoothly with **35** being obtained in 79% yield after reaction with TiCl$_4$ and allyltrimethylsilane in DCM at −78 °C for 4.5 h. The presence of the halogen substituents in **35** had little effect on this robust reaction which has previously been reported to proceed in a similar manner with the non-halogenated analogue **42** to give **43** (87% yield [44]). Previous attempts to differentiate between the two allyl groups present in **43** using an iodoetherification reaction resulted in the formation of **44**, the structural assignment of which (especially the relative stereochemistry) relied heavily on X-ray crystallographic analysis [44]. It was proposed that the presence of the chlorine substitutents in **35** may steer the outcome of this reaction away from that observed with **43** as a reduction in the electron density in the aryl ring due to the presence of the net deactivating chlorines would be expected to slow the proposed reaction of this ring with an intermediate iodonium ion central to the formation of **44** from **43** (Scheme 9 and Scheme S7).

Reaction of **35** with excess NIS in freshly base-washed CDCl$_3$ was followed by purification of the crude reaction mixture by column chromatography using triethylamine treated silica. Mass spectrometric analysis of the product indicated that one iodine atom had been added to the molecule, possibly consistent with the desired formation of **36**. However, detailed NMR analysis indicated that the structure of the major product from this reaction was not the planned iodoether **36** or an analogue of **44**. The characterization data obtained suggested that a major rearrangement of the molecule had taken place, leading to the formation of **45** (Scheme 9, Schemes S8–S10 and Figure S5 for a more detailed discussion of the data used in this assignment). In this case, it was not possible for us to prepare suitable quality crystals of **45**, emphasizing again how, in the absence of X-ray crystallographic analysis, certainty in structural assignment and especially stereochemistry can prove difficult in such complex systems (compound **45** is a single diastereomer but at present the relative stereochemistry at the centers marked * is unknown).

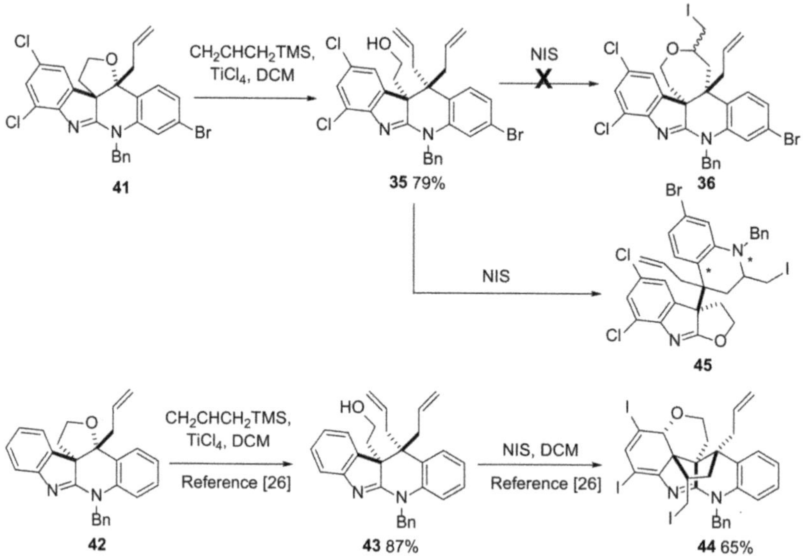

Scheme 9. Attempted conversion of **35** to **36** led instead to formation of **45**.

3. Experimental

3.1. General Experimental Details

Nuclear magnetic resonance (NMR) spectra were recorded on a Bruker Advance 300 (^{1}H, 300; ^{13}C, 75 MHz), Bruker Advance II 400 (^{1}H, 400; ^{13}C, 101 MHz) or Bruker Ascend 500 (^{1}H 500; ^{13}C 126 MHz). ^{13}C NMR spectra were recorded using the PENDANT pulse sequence. Peaks were assigned where possible with the aid of the two-dimensional NMR spectroscopic techniques COSY, HSQC, and HMBC. All NMR spectra were acquired using the deuterated solvent as the lock and the residual solvent as the internal reference. Melting points were recorded in open capillaries using an Electrothermal 9100 melting point apparatus. Values are quoted to the nearest 1 °C and are uncorrected. Infrared spectra were recorded on a Perkin Elmer Paragon 1000 FT spectrometer. Absorption maxima are reported in wavenumbers (cm^{-1}). Mass spectra were recorded using either atmospheric pressure chemical ionization (APCI) or electrospray (ES) ionization methods in the positive or negative ionization mode by Mrs Caroline Horsburgh in the University of St Andrews School of Chemistry mass spectrometry service or via the EPSRC Mass Spectrometry Service Centre (Swansea, UK).

3.2. X-ray Structure Determination for **27** and **41**

X-ray diffraction data for compounds **27** and **41** were collected at 93 K using a Rigaku MM007 High Brilliance RA generator/confocal optics and Mercury70 CCD system [Mo Kα radiation (λ = 0.71073 Å)]. Intensity data were collected using both ω and φ steps accumulating area detector images spanning at least a hemisphere of reciprocal space. Data for all compounds analyzed were collected and processed (including correction for Lorentz, polarization and absorption) using CrystalClear [46]. Structures were solved by Patterson methods (PATTY [47]) and refined by full-matrix least-squares against F^2 (SHELXL-2018/3 [48]). Non-hydrogen atoms were refined anisotropically, and hydrogen atoms were refined using a riding model. All calculations were performed using the CrystalStructure [49] interface. Selected crystallographic data are presented in Table S3. Deposition numbers 2109434 and 2109435 contain the supplementary crystallographic data for this paper. These data are provided free of charge by the joint Cambridge Crystallographic Data Center and Fachinformationszentrum Karlsruhe Access Structures service www.ccdc.cam.ac.uk/structures.

3.3. Preparation of Selected Compounds

3.3.1. N-Boc-5,7-dichloroindole-3-carbaldehyde 16

To a suspension of **15** (5.00 g, 23.36 mmol) and Boc$_2$O (5.61 g, 25.69 mmol) in DCM (100 mL) was added DMAP (0.342 g, 2.80 mmol). After stirring for 15 min, a saturated solution of NaHCO$_3$ (aq.) (100 mL) was added and the reaction mixture extracted with DCM (2 × 50 mL). The organic phase was washed with 0.5 M HCl (aq.) (100 mL) before being dried (MgSO$_4$), filtered and the solvent removed under reduced pressure to give **16** as a pale yellow solid that required no further purifiazion (6.93 g, 94%). m.p. 102–103 °C; I.R. (KBr) ν_{max} 2924, 1723, 1601, 1465, 1453, 1253, 735 cm^{-1}; ^{1}H NMR (400 MHz, CDCl$_3$) δ 10.07 (s, 1H, CHO), 8.30 (d, J = 2.0 Hz, 1H, C4-H), 8.20 (s, 1H, C2-H), 7.46 (d, J = 2.0 Hz, 1H, C6-H), 1.71 (s, 9H, (CH$_3$)$_3$); ^{13}C NMR (100 MHz, CDCl$_3$) 184.8 (CHO), 147.5 (CO$_2^t$Bu), 140.0 (C2), 131.3 (C7a), 130.9 (C3a), 129.9 (C5), 128.0 (C6), 120.7 (C7), 120.6 (C4), 120.2 (C3), 86.9 (C(CH$_3$)$_3$), 27.8 (C(CH$_3$)$_3$); HR MS [ES$^+$]: m/z calcd. for C$_{14}$H$_{13}$^{35}Cl$_2$NO$_3$Na 336.0170, found 336.0164 [M+Na]$^+$.

3.3.2. tert-Butyl 3-((4-bromo-2-nitrophenyl)(hydroxy)methyl)-5,7-dichloro-1H-indole-1-carboxylate 19

To a solution of **17** (5.16 g, 15.70 mmol) in THF (23 mL) at −40 °C was added 2 M solution of phenylmagnesium chloride in THF (8.40 mL, 16.80 mmol), followed 15 min later by a solution of **16** (3.30 g, 10.50 mmol) in THF (23 mL). The mixture was stirred for 1 h at −40 °C before being allowed to warm to room temperature for a further 1 h. A saturated solution of NH$_4$Cl (aq.) (50 mL) was then added followed by water (75 mL) and ethyl acetate (75 mL). The mixture was separated and the organic phase washed with a saturated solution of NaCl (aq.) (50 mL), dried (MgSO$_4$) and the solvent removed under reduced pressure. The crude product was purified by column chromatography (10–20% EtOAc/hexanes), giving **19** as a yellow solid (5.21 g, 96%). m.p. 56–58 °C; I.R. (KBr) ν_{max} 3386, 1737, 1531, 1346, 1150 cm^{-1}; ^{1}H NMR (400 MHz, CDCl$_3$) δ 8.08 (d, J = 2.0 Hz, 1H, C3'-H), 7.70 (dd, J = 8.4, 2.0 Hz, 1H, C5'-H), 7.57 (d, J = 8.4 Hz, 1H, C6'-H), 7.30 (s, 1H, C2-H), 7.29 (d, J = 1.9 Hz, 1H, ArC-H), 7.26 (d, J = 1.9 Hz, 1H, ArC-H), 6.50 (d, J = 4.4 Hz, 1H, CHOH), 2.84 (d, J = 4.4 Hz, 1H, OH), 1.56 (s, 9H, (CH$_3$)$_3$); ^{13}C NMR (100 MHz, CDCl$_3$) δ 148.5 (CO$_2^t$Bu), 148.3 (C2'), 136.8 (C5'), 135.8 (C1'), 132.2 (ArC), 131.3 (ArC), 130.6 (C6'), 129.0 (ArC), 128.8 (C2), 128.0 (C3'), 126.8 (ArC), 122.4 (C4'), 121.2 (ArC), 119.9 (ArC), 118.1 (ArC), 85.5 (C(CH$_3$)$_3$), 65.0 (CHOH), 27.8 (C(CH$_3$)$_3$); HR MS [APCI$^+$]: m/z calcd. for C$_{20}$H$_{17}$^{79}Br^{35}Cl$_2$N$_2$O$_5$NH$_4$ 532.0036, found 532.0034 [M+NH$_4$]$^+$.

3.3.3. tert-Butyl 3-(4-bromo-2-nitrobenzoyl)-5,7-dichloro-1H-indole-1-carboxylate 20

To a solution of **19** (3.25 g, 6.30 mmol) and N-methylmorpholine-N-oxide (1.48 g, 12.60 mmol) in DCM (35 mL) in the presence of 3Å molecular sieves at 0 °C was added tetrapropylammonium perruthenate (0.110 g, 0.310 mmol). The mixture was stirred for 30 min at 0 °C followed by a further 6 h at room temperature before the mixture was filtered and the solvent removed under reduced pressure. The crude product was purified by column chromatography (5–25% EtOAc/hexanes), giving **20** as a white solid (2.31 g, 71%). m.p. 126-127 °C; I.R. (KBr) ν_{max} 2924, 1745, 1654, 1544, 1348 cm^{-1}; ^{1}H NMR (400 MHz, CDCl$_3$) δ 8.42 (d, J = 1.9 Hz, 1H, C4-H), 8.39 (d, J = 1.8 Hz, 1H, C3'-H), 7.95 (dd, J = 8.1, 1.9 Hz, 1H, C5'-H), 7.64 (s, 1H, C2-H), 7.48 (d, J = 2 Hz, 1H, C6-H), 7.47 (d, J = 8.1 Hz, 1H, C6'-H), 1.66 (s, 9H, C(CH$_3$)$_3$); ^{13}C NMR (100 MHz, CDCl$_3$) δ 186.2 (C=O), 147.9 (CO$_2^t$Bu), 147.3 (C2'), 137.5 (C2), 137.0 (C5'), 134.5 (C1'), 131.1 (C7a), 131.1 (ArC), 130.9 (ArC), 130.0 (C6'), 128.0 (C6), 128.0 (C3'), 124.5 (C4'), 121.1 (C4), 120.7 (ArC), 118.6 (C3a), 87.5 (C(CH$_3$)$_3$), 27.7 (C(CH$_3$)$_3$); HR MS [ES$^+$]: m/z calcd. for C$_{20}$H$_{15}$^{79}Br^{35}Cl$_2$N$_2$O$_5$ 534.9439, found 534.9432 [M+H]$^+$.This was prepared as in 3.3.1. using benzotriazole (14.69 g, 123.3 mmol), thionyl chloride (3.67 g, 2.24 mL, 30.8 mmol) and (E)-3-(2-methylphenyl)prop-2-enoic acid (5.00 g,

30.8 mmol). Drying and evaporation followed by recrystallization of the residue (CH_2Cl_2) gave **33** (1.77 g, 22%) as a white powder, m.p. 124–125 °C (Lit. [21] 127–129 °C); δ_H 8.47 (1 H, d, *J* 15, COC*H*=CH), 8.42 (1 H, dt, *J* 8, 1, H-4 or 7 of Bt), 8.16 (1 H, dt, *J* 8, 1, H-4 or 7 of Bt), 8.07 (1 H, d, *J* 15, COCH=C*H*), 7.86 (1 H, d, *J* 8), 7.69 (1 H, ddd, *J* 8, 7, 1, H-5 or 6 of Bt), 7.54 (1 H, ddd, *J* 8, 7, 1, H-5 or 6 of Bt), 7.40–7.27 (3 H, m) and 2.56 (3 H, s, Me).

3.3.4. (2-Amino-4-bromophenyl)(5,7-dichloro-1*H*-indol-3-yl)methanone **21**

To a solution of **22** (800 mg, 1.85 mmol) in DCM (9 mL) was added trifluoroacetic acid (3 mL) and the solution was stirred at room temperature for 18 h. The solvent was removed under reduced pressure before a saturated solution of $NaHCO_3$ (aq.) (20 mL) was added and the mixture extracted with DCM (3 × 20 mL). The organic extracts were dried ($MgSO_4$), filtered and the solvent removed under reduced pressure to give **21** as an orange solid (673 mg, 94%) that required no further purification. m.p. 191–193 °C; I.R. (KBr) ν_{max} 3446, 3346, 1691, 1597, 1428, 1189 cm^{-1}; ^{1}H NMR (400 MHz, d^6-DMSO) δ 12.65 (s, 1H, NH), 8.09 (d, *J* = 1.9 Hz, 1H, C4-H), 7.94 (d, *J* = 3.1 Hz, 1H, C2-H), 7.53 (d, *J* = 8.4 Hz, 1H, C6'-H), 7.46 (d, *J* = 1.8 Hz, 1H, C6-H), 7.03 (d, *J* = 1.9 Hz, 1H, C3'-H), 6.73 (dd, *J* = 8.4, 1.9 Hz, 1H, C5'-H), 6.64 (br. s, 2H, NH_2); ^{13}C NMR (100 MHz, d^6-DMSO) δ 190.3 (C=O), 151.0 (C2'), 136.0 (C2), 133.5 (C6'), 132.6 (C7a), 129.2 (C3a), 126.7 (C5), 126.6 (C3), 126.3 (C4'), 122.2 (C6), 119.6 (C4), 119.1 (C1'), 118.5 (C3'), 117.6 (C5') 116.8 (C7); HR MS (ES$^-$) *m/z* calcd. for $C_{15}H_8{}^{79}Br{}^{35}Cl_2N_2O$ 380.9197, found 380.9192 [M-H]$^-$.

3.3.5. tert-Butyl 3-(2-amino-4-bromobenzoyl)-5,7-dichloro-1*H*-indole-1-carboxylate **22**

A suspension of **20** (1.00 g, 1.945 mmol) and iron powder (0.543 g, 9.725 mmol) in acetic acid (10 mL) and ethanol (10 mL) was stirred for 18 h at room temperature. The mixture was filtered through a plug of Celite. The celite was washed with DCM (50 mL) and combined with the existing filtrate. Water (50 mL) was added. The mixture was partitioned and the aqueous layer further extracted with DCM (2 × 25 mL). The combined organic layers were concentrated under reduced pressure before being re-dissolved in DCM (50 mL), washed with a saturated solution of $NaHCO_3$ (aq.) (100 mL), dried ($MgSO_4$), filtered and the solvent removed under reduced pressure. The crude product was purified by column chromatography (10–25% EtOAc/hexanes) to give **22** as a yellow solid (0.876 g, 93%). m.p. 143–144 °C; I.R. (KBr) ν_{max} 3442, 3336, 1735, 1645, 1605 cm^{-1}; ^{1}H NMR (400 MHz, d^6-DMSO) δ 8.01 (d, *J* = 2.0 Hz, 1H, C4-H), 7.80 (s, 1H, C2-H), 7.44 (d, *J* = 8.5 Hz, 1H, C6'-H), 7.35 (d, *J* = 1.9 Hz, 1H, C6-H), 6.87 (d, *J* = 1.8 Hz, 1H, C3'-H), 6.75 (dd, *J* = 8.4, 1.9 Hz, 1H, C5'-H), 5.87 (br. s, 2H, NH_2), 1.58 (s, 9H, $C(CH_3)_3$); ^{13}C NMR (100 MHz, d^6-DMSO) δ 190.8 (C=O), 150.9 (C2'), 147.0 ($\underline{C}O_2{}^tBu$), 135.7 (C2), 133.7 (C6'), 132.4 (C7a), 130.3 (C), 130.0 (C), 128.9 (C4'), 127.6 (C6), 120.8 (C), 120.6 (C4), 119.5 (C3'), 119.3 (C5'), 119.3 (C), 118.6 (C1'), 86.6 ($\underline{C}(CH_3)_3$), 27.8 ($C(\underline{C}H_3)_3$); HR MS [ES$^+$]: *m/z* calcd. for $C_{20}H_{17}{}^{79}Br{}^{35}Cl_2N_2O_3Na$ 504.9697, found 504.9693 [M+Na]$^+$.

3.3.6. (2-Benzylamino)-4-bromophenyl)(5,7-dichloro-1*H*-indol-3-yl)methanone **9**

To a solution of **22** (1.00 g, 2.07 mmol) in TFA (8 mL) was added sodium triacetoxyborohydride (0.875 g, 4.13 mmol). After 5 min, a solution of benzaldehyde (0.219 g, 2.07 mmol) in DCM (8 mL) was added dropwise and the solution was stirred for 1 h at room temperature. A second aliquot of sodium triacetoxyborohydride (0.875 g, 4.13 mmol) in TFA (4mL) was added and the mixture was stirred for a further 1 h at room temperature. The mixture was then poured into an ice cold saturated solution of $NaHCO_3$ (aq.) (20 mL) and extracted with DCM (3 × 10 mL). The combined organic extracts were dried ($MgSO_4$), filtered and the solvent removed under reduced pressure before the crude product was purified by column chromatography (5-30% EtOAc/hexanes), giving **22** as an orange-yellow solid (0.558 g, 68%). m.p. 81–82 °C; I.R. (KBr) ν_{max} 1691 (C=O), 1602, 1419, 1185 cm^{-1}; ^{1}H NMR (400 MHz, $CDCl_3$) δ 8.81 (br. s, 1H, NH), 8.28 (br. t, *J* = 5.3 Hz, 1H, NHBn), 8.05 (d, *J* = 1.5 Hz, 1H, C4-H), 7.54 (d, *J* = 2.9 Hz, 1H, C2-H), 7.48 (d, *J* = 8.4 Hz, 1H, C6'-H), 7.30–7.19 (m, 6H, C6-H and 5 × ArC-H), 6.83 (d, *J* = 1.8 Hz, 1H, C3'-H), 6.68

(dd, *J* = 8.4, 1.9 Hz, 1H, C5'-H), 4.34 (d, *J* = 5.5 Hz, 2H, CH$_2$); ^{13}C NMR (100 MHz, CDCl$_3$) δ 191.6 (C=O), 151.0 (C2'), 137.9 (ArC), 134.1 (C6'), 132.6 (C2), 132.1 (C7a), 129.2 (C4'), 128.8 (ArCH), 128.4 (C3a), 128.3 (C3), 127.5 (ArCH), 127.3 (ArCH), 123.5 (C6), 120.5 (C4), 118.9 (ArC), 118.5 (C1'), 117.8 (C5'), 117.3 (C), 114.7 (C3'), 47.2 (CH$_2$); HR MS [ES$^-$]: *m/z* calcd. for C$_{22}$H$_{14}$^{79}Br^{35}Cl$_2$N$_2$O 470.9667, found 470.9673 [M-H]$^-$. **21** was also obtained as an orange solid (0.151 g, 19%).

3.3.7. 5-Benzyl-3-bromo-7,9-dichloro-5*H*-indolo[2,3-*b*]quinolin-11(6*H*)-one **3**

To a solution of **9** (250 mg, 0.527 mmol) in MeCN (6 mL) was added *N*-chlorosuccinimide (169 mg, 1.265 mmol) and Et$_3$N (176 µL, 1.265 mmol). The mixture was stirred for 24 h at room temperature before the precipitate was collected by filtration, giving **3** as a cream-colored solid (137 mg, 55%). m.p. >320 °C (dec.); I.R. (KBr) ν$_{max}$ 1706, 1603, 1536, 1181 cm^{-1}; ^{1}H NMR (400 MHz, d6-DMSO) δ 12.45 (br. s, 1H, NH), 8.27 (d, *J* = 8.5 Hz, 1H, C1-H), 8.14 (d, *J* = 1.9 Hz, 1H, C10-H), 7.78 (d, *J* = 1.6 Hz, 1H, C4-H), 7.54 (dd, *J* = 8.5, 1.6 Hz, 1H, C2-H), 7.46 (d, *J* = 1.9 Hz, 1H, C8-H), 7.36–7.26 (m, 3H, 3 × ArC-H), 7.18–7.16 (m, 2H, 2 × ArC-H), 5.98 (s, 2H, CH$_2$); ^{13}C NMR (100 MHz, d6-DMSO) δ 171.4 (C=O), 148.0 (C5a), 139.6 (C4a), 135.6 (ArC), 130.6 (C6a), 129.0 (ArCH), 127.9 (C1), 127.5 (ArCH), 126.5 (ArC), 126.3 (ArC), 126.0 (ArCH), 125.6 (C2), 125.5 (C3), 123.8 (C11a), 122.4 (C8), 118.8 (C4), 118.0 (C10), 116.2 (ArC), 49.0 (CH$_2$); HR MS [ES$^+$]: *m/z* calcd. for C$_{22}$H$_{14}$^{35}Cl$_2$^{81}BrN$_2$O 470.9661, found 470.9659 [M+H]$^+$.

3.3.8. 5-Benzyl-3-bromo-7,9,11-trichloro-5*H*-indolo[2,3-*b*]quinoline **25**

A solution of **3** (1.50 g, 3.18 mmol) in POCl$_3$ (15 mL) was heated to reflux for 1 h before removing the solvent under reduced pressure. DCM (100 mL) and a saturated solution of NaHCO$_3$ (aq.) (250 mL) were added and the mixture partitioned before the organic layer was dried (MgSO$_4$), filtered and concentrated to give **25** in sufficient purity without further purification (1.54 g, 99%). m.p. 260–263 °C; I.R. (KBr) ν$_{max}$ 1705, 1602, 1484, 1175, 847 cm^{-1}; ^{1}H NMR (400 MHz, CDCl$_3$) δ 8.31 (d, *J* = 2.0 Hz, 1H, C10-H), 8.28 (d, *J* = 8.8 Hz, 1H, C1-H), 7.87 (d, *J* = 1.7 Hz, 1H, C4-H), 7.59 (d, *J* = 1.7 Hz, 1H, C2-H), 7.58 (d, *J* = 2.0 Hz, 1H, C8-H), 7.36–7.27 (m, 5H, 5 × Ar-H), 6.20 (s, 2H, CH$_2$); ^{13}C NMR (101 MHz, CDCl$_3$) δ 155.4 (C5a), 150.3 (C6a), 137.2 (ArC), 134.8 (ArC), 129.3 (C8), 129.1 (ArCH), 128.1 (ArCH), 127.7 (C1), 127.0 (ArCH), 126.9 (ArC), 126.3 (C2), 125.9 (ArC), 125.6 (ArC), 124.4 (ArC), 123.2 (ArC), 122.0 (C10), 118.5 (C4), 118.3 (ArC), 50.0 (CH$_2$); HR MS (ES$^+$) *m/z* calcd. for C$_{22}$H$_{12}$N$_2$^{35}Cl$_3$^{81}BrNa 512.9127, found 512.9130 [M+Na]$^+$.

3.3.9. 11-(Allyloxy)-5-benzyl-3-bromo-7,9-dichloro-5*H*-indolo[2,3-*b*]quinoline **26**

To a suspension of sodium (77 mg, 3.36 mmol) in THF (2 mL) at room temperature was added allyl alcohol (0.76 mL, 11.21 mmol) dropwise to maintain a steady reaction. Once all of the sodium had dissolved, the alkoxide solution was added via cannula to a stirred solution of **25** (550 mg, 1.12 mmol) in THF (11 mL) and the mixture was stirred at room temperature for 2 h. A saturated solution of NH$_4$Cl (aq.) (10 mL) was added. The solvent was removed under reduced pressure and the crude residue was extracted with DCM (3 × 30 mL). The combined organic extracts were dried (MgSO$_4$) filtered and concentrated under reduced pressure, giving **26** in sufficient purity without any further purification as an orange solid (525 mg, 91%). m.p. 225–227 °C; I.R. (KBr) ν$_{max}$ 2925, 1641, 1561, 1487, 1177, 858 cm^{-1}; ^{1}H NMR (300 MHz, CDCl$_3$) δ 8.18 (d, *J* = 8.7 Hz, 1H, Ar-H), 7.99 (d, *J* = 2.0 Hz, 1H, Ar-H), 7.86 (d, *J* = 1.6 Hz, 1H, Ar-H), 7.56 (d, *J* = 1.9 Hz, 1H, Ar-H), 7.53 (dd, *J* = 8.7, 1.7 Hz, 1H, Ar-H), 7.37–7.27 (m, 5H, 5 × Ar-H), 6.32–6.20 (m, 1H, CH$_2$=C<u>H</u>), 6.19 (s, 2H, CH$_2$N), 5.56 (dq, *J* = 17.1, 1.4 Hz, 1H, C<u>H$_2$</u>=CH), 5.43 (dq, *J* = 10.4, 1.1 Hz, 1H, CH$_2$=CH), 4.99 (dt, *J* = 5.7, 1.3 Hz, 2H, CH$_2$); ^{13}C NMR (75 MHz, CDCl$_3$) δ 158.7 (C5a), 158.4 (C11), 149.4 (ArC), 148.9 (ArC), 138.6 (C4a), 135.2 (ArC), 132.0 (CH=CH$_2$), 129.1 (ArCH), 128.0 (ArCH), 127.9 (ArCH), 127.0 (ArCH), 126.5 (C), 126.1 (ArCH), 125.6 (ArC), 125.2 (ArC), 124.7 (ArC), 121.3 (ArCH), 119.8 (CH=CH$_2$), 118.5 (ArCH), 117.0 (ArC), 116.0 (ArC), 76.1

(CH_2-C=C), 49.7 (CH_2N); HR MS (ES$^+$) m/z calcd. for $C_{25}H_{18}{}^{79}Br^{35}Cl_2N_2O$ 510.9980, found 510.9979 [M+H]$^+$.

3.3.10. 10b-Allyl-5-benzyl-3-bromo-7,9-dichloro-5H-indolo[2,3-b]quinolin-11(10bH)-one 27

A solution of **26** (500 mg, 0.98 mmol) in toluene (10 mL) was heated to reflux for 1.5 h before removing the solvent under reduced pressure. The crude product was purified by column chromatography (10–25 % EtOAc/hexanes) to give **27** as an orange-yellow solid (477 mg, 95%). Recrystallization by slow evaporation of a solution of **27** in EtOAc/Hexane provided crystals suitable for X-ray analysis. m.p. 240 °C (dec.); I.R. ν_{max} 1695, 1543, 1196, 854 cm^{-1}; ^{1}H NMR (400 MHz, CDCl$_3$) δ 7.80 (d, J = 8.8 Hz, 1H, C1-H), 7.59 (d, J = 2.0 Hz, 1H, C10-H), 7.43–7.37 (m, 5H, 5 × ArC-H), 7.33 (m, 1H, ArC-H), 7.29–7.26 (m, 2H, 2 × ArC H), 5.98 (d, J = 16.4 Hz, 1H, 1 × C$\underline{H}_2$N), 5.38 (ddt, J = 16.9, 10.1, 7.3 Hz, 1H, C$\underline{H}$=CH$_2$), 5.14–5.03 (m, 2H, 1 × C$\underline{H}_2$N, 1 × CH=C$\underline{H}_2$), 4.90 (dd, J = 16.8, 1.3 Hz, 1H, CH=C$\underline{H}_2$), 2.87 (dd, J = 13.4, 6.9 Hz, 1H, CH$_2$), 2.53 (dd, J = 13.4, 7.7 Hz, 1H, CH$_2$); ^{13}C NMR (101 MHz, CDCl$_3$) δ 190.4 (C11), 172.2 (C5a), 149.0 (C6a), 145.1 (C4a), 135.7 (C10a), 135.4 (C), 131.1 (C3), 129.7 (C1), 129.2 (ArCH), 129.1 (ArCH), 129.1 (ArC), 128.9 (C10b^{II}), 128.0 (ArCH), 126.8 (ArCH), 126.4 (ArCH), 123.9 (ArC), 123.5 (C10), 121.6 (C10b^{III}), 118.9 (ArCH), 117.6 (ArC), 67.0 (C10b), 49.9 (C10b^I), 44.8 (CH$_2$N); **HR MS** (ES$^+$) m/z calcd. for $C_{25}H_{17}{}^{79}Br^{35}Cl^{37}ClN_2ONa$ 534.9769, found 534.9767 [M+Na]$^+$.

3.3.11. (E)-5-Benzyl-3-bromo-10b-(but-2-en-1-yl)-7,9-dichloro-5H-indolo[2,3-b]quinolin-11(10bH)-one 29

To a suspension of sodium (205 mg, 8.93 mmol) in THF (5 mL) at room temperature was added (±)-3-buten-2-ol (2.58 mL, 29.76 mmol) dropwise to maintain a steady reaction. Once all of the sodium had dissolved, the alkoxide solution was added via a cannula to a stirred solution of **25** (1.46 g, 2.98 mmol) in THF (30 mL) and the mixture was stirred at room temperature for 2 h. A saturated solution of NH$_4$Cl (aq.) (30 mL) was added before the organic solvent was removed under reduced pressure and the crude residue extracted with DCM (3 × 100 mL). The combined organic extracts were dried (MgSO$_4$), filtered and concentrated under reduced pressure to give a crude product that contained **28** and **29**. This residue was redissolved in THF (30 mL) and the solution was heated at reflux for 1 h before removing the solvent under reduced pressure to give the crude product. Purification by column chromatography (10–25% EtOAc/hexanes) gave **29** as a yellow solid (1.17 g, 75%). m.p. 140–142 °C; ν_{max} 1698, 1590, 1545, 1196, 855 cm^{-1}; ^{1}H NMR (500 MHz, CDCl$_3$) δ 7.71 (d, J = 8.6 Hz, 1H, C1-H), 7.48 (d, J = 2.0 Hz, 1H, C10-H), 7.36–7.14 (m, 8H, 8 × ArC-H), 5.91 (d, J = 16.3 Hz, 1H, CH$_2$N), 5.21 (dq, J = 12.6, 5.8 Hz, 1H, C10b^{II}-H), 5.01–4.88 (m, 2H, CH$_2$N + C10b^{III}-H), 2.74 (dd, J = 13.2, 6.4 Hz, 1H, C10b^I-H$_2$), 2.35 (dd, J = 13.2, 8.1 Hz, 1H, C10b^I-H$_2$), 1.49 (d, J = 6.2 Hz, 3H, C10b^{IV}-H$_3$); ^{13}C NMR (126 MHz, CDCl$_3$) δ 190.5 (C11), 172.4 (C5a), 148.9 (C6a), 145.1 (C4a), 136.0 (C10a), 135.5 (C), 132.9 (C10b^{II}), 131.0 (C3), 129.6 (C1), 129.1 (ArCH), 129.0 (ArCH), 128.9 (ArC), 128.0 (ArCH), 126.8 (ArCH), 126.3 (ArCH), 123.8 (ArC), 123.6 (C10), 121.4 (C10b^{III}), 118.8 (C8), 117.8 (C11a), 67.4 (C10b), 49.9 (CH$_2$N), 44.3 (C10b^I), 17.9 (C10b^{IV}); HR MS (ES$^-$) m/z calcd. for $C_{26}H_{18}N_2O^{35}Cl_2{}^{81}Br$ 524.9959, found 524.9971 [M − H]$^-$.

3.3.12. 5-Benzy-3-bromo-10b-((E)-but-2-en-1-yl)-7,9-dichloro-5,10b-dihydrospiro[indolo[2,3-b]quinoline-11,2′-oxirane] 31

To a solution of **29** (200 mg, 0.380 mmol) and chloroiodomethane (42 μL, 0.570 mmol) in THF (4 mL) at −78 °C was added methyl lithium—lithium bromide complex (1.5 M in THF, 0.38 mL, 0.570 mmol) dropwise over 5 min. The mixture was stirred at −78 °C for a further 30 min before removing the cold bath and stirring for a further 18 h at room temperature. A solution of saturated NH$_4$Cl (aq.) (10 mL) was added and the mixture extracted with DCM (3 × 10 mL) before the combined organic extracts were dried (MgSO$_4$), filtered and concentrated. The crude product was purified by column chromatography

(10–20 % EtOAc/hexanes) to give **31** as an orange solid (142 mg, 74%). m.p. 90–93 °C; I.R. (KBr) ν_{max} 2921, 1709, 1595, 1550, 1486, 729 cm^{-1}; ^{1}H NMR (400 MHz, CDCl$_3$) δ 7.41–7.35 (m, 4H, 4 × ArC-H), 7.33–7.28 (m, 2H, 2 × ArC-H), 7.19 (dd, J = 8.1, 1.7 Hz, 1H, C2-H), 7.12 (d, J = 1.7 Hz, 1H, C4-H), 7.11–7.05 (m, 2H, 2 × ArC-H), 5.89 (d, J = 16.2 Hz, 1H, CH$_2$N), 5.37–5.22 (m, 1H, C10b^{III}-H), 4.97–4.86 (m, 2H, CH$_2$N + C10b^{II}-H), 3.03 (d, J = 5.3 Hz, 1H, CH$_2$O), 2.76 (ddt, J = 13.9, 6.1, 1.3 Hz, 1H, C10b^{I}-H$_2$), 2.59 (ddt, J = 13.9, 6.1, 1.3 Hz, 1H, C10b^{I}-H$_2$), 2.52 (d, J = 5.3 Hz, 1H, CH$_2$O), 1.50 (d, J = 6.0 Hz, 3H, CH$_3$); ^{13}C NMR (101 MHz, CDCl$_3$) δ 172.6 (C5a), 151.0 (C6a), 142.0 (C4a), 137.2 (C10a), 136.1 (ArC), 130.8 (C10b^{III}), 129.1 (ArCH), 129.0 (ArCH), 128.4 (ArC), 127.7 (ArCH), 126.9 (ArCH), 126.5 (ArCH), 125.2 (ArCH), 123.3 (ArC), 123.1 (ArC), 122.9 (C10b^{II}), 122.1 (C11a), 121.5 (ArCH), 118.4 (ArCH), 59.5 (C11), 56.4 (C10b), 53.2 (CH$_2$O), 50.0 (CH$_2$N), 37.1 (C10b^{I}), 17.8 (C10b^{IV}); HR MS [APCI$^+$]: m/z calcd. for C$_{27}$H$_{22}$^{79}Br^{35}Cl$_2$N$_2$O 539.0287, found 539.0275 [M+H]$^+$.

3.3.13. (10bR,11R)-5-Benzyl-3-bromo-10b-((E)-but-2-en-1-yl)-7,9-dichloro-10b,11-dihydro-5H-indolo[2,3-b]quinolin-11-yl)methanol **32**

To a solution of **31** (90 mg, 0.167 mmol) in THF (3 mL) at -78 °C was added sodium cyanoborohydride (26 mg, 0.416 mmol) and boron trifluoride diethyl etherate (82µL, 0.666 mmol). The mixture was slowly allowed to warm to room temperature over a period of 6 h before a saturated solution of NaHCO$_3$ (aq.) (6 mL) was added. The mixture was extracted with DCM (3 × 10 mL) before the combined organic extracts were dried (MgSO$_4$), filtered and concentrated to give the crude product. The crude reaction mixture was purified by column chromatography (20–30 % EtOAc/hexanes) to give **32** as a white solid (45 mg, 50%). m.p. 85–86 °C, I.R. (KBr) ν_{max} 3419 (OH), 2931, 1545, 1486, 1422, 1206 cm^{-1}; ^{1}H NMR (300 MHz, CDCl$_3$) δ 7.54 (dd, J = 8.3, 1.0 Hz, 1H, C2-H), 7.38–7.24 (m, 6H, 6 × ArC-H), 7.23–7.16 (m, 2H, 2 × ArC-H), 7.10 (d, J = 1.9 Hz, 1H, ArC-H), 5.75 (d, J = 16.2 Hz, 1H, CH$_2$N), 5.19 (dq, J = 13.0, 6.5 Hz, 1H, CH=CHCH$_3$), 4.80 (d, J = 16.2 Hz, 1H, CH$_2$N), 4.72–4.58 (m, 1H, CH=CHCH$_2$), 4.53 (dd, J = 11.2, 2.4 Hz, 1H, CH$_2$OH), 4.39–4.26 (m, 1H, CH$_2$OH), 3.00 (d, J = 6.2 Hz, 1H, C11-H), 2.42–2.26 (m, 2H, CH$_2$), 2.14 (br., s, 1H, OH), 1.40 (d, J = 6.5 Hz, 3H, CH$_3$); ^{13}C NMR (75 MHz, CDCl$_3$) δ 174.4 (C5a), 151.9 (C6a), 142.3 (C4a), 140.1 (C10a), 136.7 (ArC), 130.6 (C10b^{III}), 129.2 (2 × overlapping ArCH), 129.1 (C1), 128.0 (ArCH), 127.5 (ArCH), 126.9 (C2), 125.0 (ArC), 124.5 (C11a), 123.8 (C10b^{II}), 123.1 (ArC), 122.9 (ArCH), 122.1 (C3), 119.1 (C4), 61.7 (CH$_2$OH), 56.2 (C10b), 50.8 (CH$_2$N), 44.0 (C11), 33.9 (C10b^{I}), 18.2 (C10b^{IV}); HR MS [ES$^+$]: m/z calcd. for C$_{27}$H$_{23}$^{79}Br^{35}Cl^{37}ClN$_2$ONa 565.0239, found 565.0220 [M+Na]$^+$.

3.3.14. (10bR,11R)-11-Allyl-5-benzyl-3-bromo-7,9-dichloro-10b-(2-hydroxyethyl)-10b,11-dihydro-5H-indolo[2,3-b]quinolin-11-ol **40**

For details of the synthesis and analysis of the diastereomeric mixtures **37a/37b**, **38a/38b** and **39a/39b** see SI. To a solution of **39a/39b** (230 mg, 0.413 mmol) in MeOH (10 mL) was added NaBH$_4$ (31 mg, 0.826 mmol) and the mixture was stirred at room temperature for 1 h. A saturated solution of NH$_4$Cl (aq.) (10 mL) was added and the organic solvent was removed under reduced pressure before the mixture was extracted with DCM (3 × 10 mL). The combined organic extracts were dried (MgSO$_4$), filtered and concentrated in vacuo. The crude product was purified by column chromatography (15-25% EtOAc/hexanes) to give **40** as a white solid (220 mg, 95%). m.p. = 151–153 °C; I.R. (KBr) ν_{max} 3163, 2931, 1545, 1481, 1418, 1206, 847, 729 cm^{-1}; ^{1}H NMR (500 MHz, CDCl$_3$) δ 7.40–7.31 (m, 6H, C8-H, C10-H, 4 × ArC-H), 7.30–7.25 (m, 2H, C1-H, ArC-H), 7.19 (dd, J = 8.1, 1.7 Hz, 1H, C2-H), 7.13 (d, J = 1.7 Hz, 1H, C4-H), 5.56 (d, J = 16.1 Hz, 1H, CH$_2$N), 5.38 (ddt, J = 17.3, 10.0, 7.3 Hz, 1H, C11^{II}-H), 5.22 (d, J = 16.0 Hz, 1H, CH$_2$N), 4.98–4.92 (m, 1H, C11^{III}-H$_2$), 4.73 (dt, J = 17.0, 1.6 Hz, 1H, C11^{III}-H$_2$), 3.60 (ddd, J = 12.0, 8.7, 3.6 Hz, 1H, C10b^{II}-H$_2$), 3.52–3.45 (m, 1H, C10b^{II}-H$_2$), 2.41–2.32 (m, 1H, C10b^{I}-H$_2$), 2.12 (d, J = 7.0 Hz, 2H, C11^{I}-H$_2$), 1.78–1.69 (m, 1H, C10b^{I}-H$_2$); ^{13}C NMR (126 MHz, CDCl$_3$) δ 175.4 (C5a), 150.9 (C6a), 138.9 (C4a), 137.9 (C10a), 135.8 (ArC), 131.6 (C11^{II}), 129.2 (C8), 128.9 (2 ×

overlapping carbons ArCH, C11a), 128.4 (ArCCl), 128.2 (C1 or ArCH), 127.9 (C1 or ArCH), 127.3 (ArCH), 126.4 (C2), 122.9 (ArCCl), 122.7 (C10), 122.3 (C3), 120.0 (C11^{III}), 118.6 (C4), 75.3 (C11), 61.3 (C10b), 58.6 (C10b^{II}), 49.4 (CH$_2$N), 40.6 (C11^{I}), 35.0 (C10b^{I}). HR MS [ES$^+$]: m/z calcd. for C$_{27}$H$_{24}$^{79}Br^{35}Cl$_2$N$_2$O$_2$ 557.0393, found 557.0392 [M+H]$^+$.

3.3.15. (3a*R*,13b*R*)-13b-Allyl-9-benzyl-11-bromo-5,7-dichloro-2,3,9,13b-tetrahydrofuro[3,2-*c*]indolo[2,3-*b*]quinoline 41

To a solution of **40** (50 mg, 0.090 mmol) in DCM (2 mL) was added *p*-toluenesulfonyl chloride (26 mg, 0.134 mmol) and triethylamine (124 µL, 0.896 mmol) and the mixture was stirred at room temperature for 1 h. The mixture was then heated to reflux for a further 5 h before cooling to room temperature and addition of a saturated solution of NH$_4$Cl (aq.) (5 mL). The mixture was extracted with DCM (3 × 5 mL) and the combined organic extracts were dried (MgSO$_4$), filtered and concentrated in vacuo. The crude product was purified by column chromatography (5–15% EtOAc/hexanes) to give **41** as a white solid (40 mg, 83%). Recrystallization of **41** by slow evaporation from ethyl acetate gave crystals of suitable quality for X-ray crystallographic analysis. m.p. = 177–179 °C; I.R. (KBr) ν_{max} 2926, 1548, 1484, 1425, 1204, 1071, 1044, 850 cm^{-1}; ^{1}H NMR (500 MHz, CDCl$_3$) δ 7.29–7.15 (m, 7H, C6-H, C13-H, 5 × ArC-H), 7.14 (d, *J* = 2.0 Hz, 1H, C4-H), 7.10 (dd, *J* = 8.2, 1.8 Hz, 1H, C12-H), 7.00 (d, *J* = 1.8 Hz, 1H, C10-H), 5.62 (d, *J* = 16.1 Hz, 1H, CH$_2$N), 5.06–4.95 (m, 2H, CH$_2$N, C15-H), 4.67–4.60 (m, 1H, C16-H$_2$), 4.39 (dq, *J* = 16.9, 1.3 Hz, 1H, C16-H$_2$), 4.12 (q, *J* = 8.3 Hz, 1H, C2-H$_2$), 3.74 (ddd, *J* = 10.1, 8.6, 3.4 Hz, 1H, C2-H$_2$), 2.49 (ddd, *J* = 12.6, 10.1, 8.2 Hz, 1H, C3-H$_2$), 2.38–2.35 (m, 2H, 14-H$_2$), 2.03 (ddd, *J* = 12.2, 8.2, 3.4 Hz, 1H, C3-H$_2$); ^{13}C NMR (126 MHz, CDCl$_3$) δ 171.6 (C8a), 149.2 (C7a), 139.2 (C3b), 139.0 (C9a), 134.9 (ArC), 128.9 (C15), 127.9 (C6), 127.9 (ArCH), 127.7 (C13 or ArCH), 127.2 (ArCCl), 126.7 (C13 or ArCH), 126.1 (ArCH), 125.5 (C12), 124.8 (C13a), 122.3 (ArCCl), 121.8 (C11), 120.1 (C4), 119.2 (C16), 117.2 (C10), 85.0 (C13b), 63.3 (C2), 59.5 (C3a), 48.3 (CH$_2$N), 40.5 (C14), 36.8 (C3); HR MS [ES$^+$]: m/z calcd. for C$_{27}$H$_{22}$^{79}Br^{35}Cl$_2$N$_2$O 539.0287, found 539.0284 [M+H]$^+$.

3.3.16. (*R*)-2-(11,11-Diallyl-5-benzyl-3-bromo-7,9-dichloro-10b,11-dihydro-5*H*-indolo[2,3-*b*]quinolin-10b-yl)ethanol 35

To a solution of **41** (40 mg, 0.074 mmol) in DCM (1.5 mL) at −78 °C was added allyltrimethylsilane (42 mg, 0.370 mmol) and TiCl$_4$ (70 mg, 0.370 mmol) and the mixture was stirred at −78 °C for 4.5 h. Methanol (0.5 mL) was added and the mixture was stirred for an additional 10 min before the cold bath was removed and a saturated solution of NH$_4$Cl (aq.) (5 mL) was added. The mixture was extracted with DCM (3 × 5 mL) and the combined organic extracts were dried (MgSO$_4$), filtered and concentrated in vacuo. The crude product was purified by column chromatography (15–30% EtOAc/hexanes) to give **35** as a pale yellow solid (34 mg, 79%). m.p. = 73–75 °C; I.R. (KBr) ν_{max} 3389, 3074, 2921, 1607, 1548, 1486, 1422, 1206, 909, 847, 798, 729, 702 cm^{-1}; ^{1}H NMR (400 MHz, CDCl$_3$) δ 7.47–7.39 (m, 2H, 2 × ArC-H), 7.36–7.27 (m, 4H, C8-H, 3 × ArC-H), 7.23 (d, *J* = 1.9 Hz, 1H, C4-H), 7.19 (d, *J* = 1.9 Hz, 1H, C10-H), 7.15 (dd, *J* = 8.3, 1.9 Hz, 1H, C2-H), 7.03 (d, *J* = 8.3 Hz, 1H, C1-H), 6.09 (ddt, *J* = 17.1, 9.6, 4.9 Hz, 1H, C11^{IIa}-H), 5.44 (d, *J* = 15.9 Hz, 1H, CH$_2$N), 5.38 (d, *J* = 17.1 Hz, 1H, C11^{IIIa}-H$_2$), 5.28 (d, *J* = 15.9 Hz, 1H, CH$_2$N), 5.17 (d, *J* = 10.2 Hz, 1H, C11^{IIIa}-H$_2$), 5.14–5.03 (m, 1H, C11^{IIIb}-H), 4.82 (d, *J* = 10.0 Hz, 1H, C11^{IIIb}-H$_2$), 4.62 (d, *J* = 16.2 Hz, 1H, C11^{IIIb}-H$_2$), 3.17–3.05 (m, 2H, C10b^{II}-H$_2$, C11^{Ia}-H$_2$), 2.95 (ddd, *J* = 10.6, 8.5, 5.3 Hz, 1H, C10b^{II}-H2), 2.75 (dd, *J* = 16.9, 9.2 Hz, 1H, C11^{Ia}-H$_2$), 2.34 (ddd, *J* = 13.4, 8.2, 5.3 Hz, 1H, C10b^{I}-H$_2$), 2.14–2.03 (m, 2H, C10b^{I}-H$_2$, C11^{Ib}-H$_2$), 1.80–1.72 (m, 1H, C11^{Ib}-H$_2$); ^{13}C NMR (101 MHz, CDCl$_3$) δ 173.9 (C5a), 151.6 (C6a), 140.6 (C4a), 138.9 (C10a), 135.9 (ArCH), 135.8 (C11^{IIa}), 132.6 (C11^{IIb}), 129.5 (C1), 129.1 (C8), 128.8 (ArCH), 128.1 (ArCCl), 127.7 (ArCH), 126.5 (C11a), 125.9 (C2), 122.8 (ArCCl), 122.2 (C10), 121.6 (C3), 119.5 (C4), 119.2 (C11^{IIIb}), 117.1 (C11^{IIIa}), 60.5 (C10b), 58.7 (C10b^{II}), 49.7 (CH$_2$N), 45.9 (C11), 39.9 (C11^{Ib}), 37.3 (C11^{Ia}), 35.6 (C10b^{I}); HR MS [ES$^+$]: m/z calcd. for C$_{30}$H$_{28}$^{79}Br^{35}Cl$_2$N$_2$O 581.0757, found 581.0756 [M+H]$^+$.

3.3.17. 3a-(4-Allyl-1-benzyl-7-bromo-2-(iodomethyl)-1,2,3,4-tetrahydroquinolin-4-yl)-5,7-dichloro-3,3a-dihydro-2H-furo[2,3-b]indole 45

To a solution of **35** (18 mg, 0.031 mmol) in base-washed CDCl$_3$ (1 mL) was added NIS (22 mg, 0.098 mmol) and the mixture was stirred at room temperature for 3 h. A saturated solution of Na$_2$S$_2$O$_3$ (aq.) (2 mL) was added and the mixture was stirred for a further 10 min before being extracted with DCM (3 × 2 mL). The combined organic extracts were dried (MgSO$_4$), filtered and concentrated in vacuo. The crude product was purified by column chromatography (Et$_3$N washed silica, 5–15% EtOAc/hexanes), giving **45** as a white solid (11 mg, 50%). I.R. (KBr) v_{max} 3320, 3074, 2921, 1607, 1548, 1486, 1420, 1255, 1206, 911, 847, 798, 729 cm^{-1}; ^{1}H NMR (500 MHz, CDCl$_3$) δ 7.40 (d, J = 2.0 Hz, 1H, C6-H), 7.38–7.27 (m, 6H, C4-H, 5 × ArC-H), 7.19 (d, J = 8.3 Hz, 1H, C5′-H), 6.93 (dd, J = 8.3, 2.0 Hz, 1H, C6′-H), 6.82 (d, J = 1.9 Hz, 1H, C8′-H), 5.70 (dddd, J = 17.8, 10.1, 7.9, 5.2 Hz, 1H, C4′^{II}-H), 5.28–5.17 (m, 2H, C4′^{III}-H$_2$), 4.64 (d, J = 17.9 Hz, 1H, CH$_2$N), 4.46 (d, J = 17.8 Hz, 1H, CH$_2$N), 4.31 (t, J = 9.1 Hz, 1H, C2-H$_2$), 3.51–3.42 (m, 1H, C2′-H), 3.26 (dd, J = 14.4, 5.2 Hz, 1H, C4′^{I}-H$_2$), 3.07 (ddd, J = 10.0, 8.5, 6.3 Hz, 1H, C2-H$_2$), 3.01 (dd, J = 9.7, 2.7 Hz, 1H, CH$_2$I), 2.54 (q, J = 7.2 Hz, 3H, 1 × C3-H$_2$, 1 × C4′^{I}-H$_2$, 1 × CH$_2$I), 2.09 (dt, J = 13.5, 10.0 Hz, 1H, C3-H$_2$), 1.58 (dd, J = 8.9, 6.7 Hz, 2H, C3′-H$_2$); ^{13}C NMR (126 MHz, CDCl$_3$) δ 193.6 (C8a), 153.8 (C7a), 149.4 (C8a′), 140.3 (C3b), 139.0 (ArC), 132.2 (C4′^{II}), 129.5 (C6), 129.1 (C5′), 128.7 (ArCH), 128.4 (ArCCl), 127.2 (ArCH), 126.5 (ArCH), 124.3 (ArCCl), 123.6 (C4), 123.5 (C7′), 120.9 (C4a′), 120.5 (C4′^{III}), 120.5 (C6′), 118.3 (C8′), 79.9 (C2), 64.9 (C3a), 57.7 (CH$_2$N), 56.8 (C2′), 45.0 (C4′), 38.5 (C4′^{I}), 37.8 (C3′), 27.5 (C3), 13.3 (CH$_2$I); HR MS [ES$^+$]: m/z calcd. for C$_{30}$H$_{26}$^{79}Br^{35}Cl$_2$IN$_2$ONa 728.9548, found 728.9554 [M+H]$^+$.

4. Conclusions

The natural product perophoramidine (**1**) continues to challenge synthetic organic chemists. This report describes how the required presence of the two chlorines and one bromine in **1** forced us into a change in synthetic approach compared to our previous reports on dehaloperophoramidine (**2**). The optimization of an NCS-mediated intramolecular C–N bond-forming reaction at the indole 2-position was achieved and led to a suitably halogen-substituted indoloquinoline core structure. Two attempts to progress further towards the structure of perophoramidine (**1**) are also described. In one of these approaches, the presence of the halogens blocked a previously observed undesired reaction pathway. However, an alternative reaction pathway occurred, leading to a major rearrangement of the core structure of the molecule and delivering an interesting furo[2,3-b]indole-containing structure. Throughout this work, small-molecule X-ray crystallographic analysis has proved essential.

Supplementary Materials: Schemes S1–S10, Tables S1–S3, Figures S1–S5 and additional experimental details. CIF files for X-ray structure of **27** and **41**.

Author Contributions: Conceptualization, N.J.W.; methodology, C.A.J., D.B.C., T.L., A.M.Z.S. and N.J.W.; investigation, C.A.J., D.B.C., T.L., A.M.Z.S. and N.J.W.; data curation, D.B.C., T.L. and A.M.Z.S.; writing—original draft, N.J.W.; writing—review and editing, D.B.C. and N.J.W.; supervision, N.J.W.; project administration, N.J.W.; funding acquisition, N.J.W. All authors have read and agreed to the published version of the manuscript.

Funding: This research was funded by EPSRC with a DTA studentship for C.A.J. And The APC was funded by University of St Andrews.

Institutional Review Board Statement: Not applicable.

Informed Consent Statement: Not applicable.

Data Availability Statement: Data is contained within the article or supplementary material.

Acknowledgments: We would like to thank and acknowledge the important contribution made to the X-ray crystallographic analysis component of work in this manuscript and other projects in our group by David Cordes, a long term colleague of Alex Slawin.

Conflicts of Interest: The authors declare no conflict of interest.

Sample Availability: Samples of the compounds are not available from the authors.

References

1. Verbitski, S.M.; Mayne, C.L.; Davis, R.A.; Concepcion, G.P.; Ireland, C.M. Isolation, structure determination, and biological activity of a novel alkaloid, perophoramidine, from the Philippine ascidian *Perophora namei*. *J. Org. Chem.* **2002**, *67*, 7124–7126. [CrossRef]
2. Siengalewicz, P.; Gaich, T.; Mulzer, J. It all began with an error: The nomofungin/communesin story. *Angew. Chem. Int. Ed.* **2008**, *47*, 8170–8176. [CrossRef]
3. Robinson, R.; Teuber, H.J. Reactions with nitrosodisulfonate. IV. Calycanthine and calycanthidine. *Chem. Ind.* **1954**, *46*, 783–784.
4. Woodward, R.B.; Yang, N.C.; Katz, T.J.; Clark, V.M.; Harley-Mason, J.; Ingleby, R.F.; Shepard, N. Calycanthine: The structure of the alkaloid and its degradation product, calycanine. *Proc. Chem. Soc.* **1960**, 76–78.
5. Hendrickson, B.; Rees, R.; Goschke, R.R. Total synthesis of the calycanthaceous alkaloids. Chimonanthine. *Proc. Chem. Soc.* **1962**, 383–384.
6. Hamor, T.A.; Robertson, J.M.; Shrivastava, H.N.; Silverton, J.V. The structure of calycanthine. *Proc. Chem. Soc.* **1960**, 78–80.
7. Hamor, T.A.; Robertson, J.M. The structure of calycanthine. X-ray analysis of the dihydrobromide dehydrate. *J. Chem. Soc.* **1962**, 194–205. [CrossRef]
8. Grant, I.J.; Hamor, T.A.; Robertson, J.M.; Sim, G.A. Structure of chimonanthine. *Proc. Chem. Soc.* **1962**, 148–149.
9. Grant, I.J.; Hamor, T.A.; Robertson, J.M.; Sim, G.A. The structure of chimonanthine. X-ray analysis of chimonanthine dihydrobromide. *J. Chem. Soc.* **1965**, 5678–5696. [CrossRef]
10. Verotta, L.; Pilati, T.; Tato, M.; Elisabetsky, E.; Amador, T.A.; Nunes, D.S. Pyrrolidinoindoline Alkaloids from *Psychotria colorata*. *J. Nat. Prod.* **1998**, *61*, 392–396. [CrossRef]
11. Fuchs, J.R.; Funk, R.L. Total Synthesis of (±)-Perophoramidine. *J. Am. Chem. Soc.* **2004**, *126*, 5068–5069. [CrossRef] [PubMed]
12. Wu, H.; Xue, F.; Xiao, X.; Qin, Y. Total Synthesis of (+)-Perophoramidine and Determination of the Absolute Configuration. *J. Am. Chem. Soc.* **2010**, *132*, 14052–14053. [CrossRef]
13. Zhang, H.; Hong, L.; Kang, H.; Wang, R. Construction of Vicinal All-Carbon Quaternary Stereocenters by Catalytic Asymmetric Alkylation Reaction of 3-Bromooxindoles with 3-Substituted Indoles: Total Synthesis of (+)-Perophoramidine. *J. Am. Chem. Soc.* **2013**, *135*, 14098–14101. [CrossRef]
14. Han, S.-J.; Vogt, F.; May, J.A.; Krishnan, S.; Gatti, M.; Virgil, S.C.; Stoltz, B.M. Evolution of a Unified, Stereodivergent Approach to the Synthesis of Communesin F and Perophoramidine. *J. Org. Chem.* **2015**, *80*, 528–547. [CrossRef]
15. Han, S.-J.; Vogt, F.; Krishnan, S.; May, J.A.; Gatti, M.; Virgil, S.C.; Stoltz, B.M. A Diastereodivergent Synthetic Strategy for the Syntheses of Communesin F and Perophoramidine. *Org. Lett.* **2014**, *16*, 3316–3319. [CrossRef]
16. Trost, B.M.; Osipov, M.; Kruger, S.; Zhang, Y. A catalytic asymmetric total synthesis of (-)-perophoramidine. *Chem. Sci.* **2015**, *6*, 349–353. [CrossRef]
17. Artman, G.D., III; Weinreb, S.M. An Approach to the Total Synthesis of the Marine Ascidian Metabolite Perophoramidine via a Halogen-Selective Tandem Heck/Carbonylation Strategy. *Org. Lett.* **2003**, *5*, 1523–1526. [CrossRef] [PubMed]
18. Wu, L.; Zhang, Q.-R.; Huang, J.-R.; Li, Y.; Su, F.; Dong, L. The application of Morita-Baylis-Hillman reaction: Synthetic studies on perophoramidine. *Tetrahedron* **2017**, *73*, 3966–3972. [CrossRef]
19. Sabahi, A.; Novikov, A.; Rainier, J.D. 2-Thioindoles as precursors to spiro-fused indolines: Synthesis of (±)-dehaloperophoramidine. *Angew. Chem. Int. Ed.* **2006**, *45*, 4317–4320. [CrossRef]
20. Ishida, T.; Ikota, H.; Kurahashi, K.; Tsukano, C.; Takemoto, Y. Dearomatizing Conjugate Addition to Quinolinyl Amidines for the Synthesis of Dehaloperophoramidine through Tandem Arylation and Allylation. *Angew. Chem. Int. Ed.* **2013**, *52*, 10204–10207. [CrossRef] [PubMed]
21. Wilkie, R.P.; Neal, A.R.; Johnston, C.A.; Voute, N.; Lancefield, C.S.; Stell, M.D.; Medda, F.; Makiyi, E.F.; Turner, E.M.; Ojo, S.O.; et al. Total synthesis of dehaloperophoramidine using a highly diastereoselective Hosomi-Sakurai reaction. *Chem. Commun.* **2016**, *52*, 10747–10750. [CrossRef]
22. Hoang, A.; Popov, K.; Somfai, P. An Efficient Synthesis of (±)-Dehaloperophoramidine. *J. Org. Chem.* **2017**, *82*, 2171–2176. [CrossRef] [PubMed]
23. Voute, N.; Philp, D.; Slawin, A.M.Z.; Westwood, N.J. Studies on the Claisen rearrangements in the indolo[2,3-b]quinoline system. *Org. Biomol. Chem.* **2010**, *8*, 442–450. [CrossRef] [PubMed]
24. Malgesini, B.; Forte, B.; Borghi, D.; Quartieri, F.; Gennari, C.; Papeo, G. A Straightforward Total Synthesis of (-)-Chaetominine. *Chem. Eur. J.* **2009**, *15*, 7922–7929. [CrossRef]
25. Arcadi, A.; Bianchi, G.; Inesi, A.; Marinelli, F.; Rossi, L. Electrochemical-mediated cyclization of 2-alkynylanilines: A clean and safe synthesis of indole derivatives. *Eur. J. Org. Chem.* **2008**, *5*, 783–787. [CrossRef]
26. Alfonsi, M.; Arcadi, A.; Aschi, M.; Bianchi, G.; Marinelli, F. Gold-Catalyzed Reactions of 2-Alkynyl-phenylamines with α,β-Enones. *J. Org. Chem.* **2005**, *70*, 2265–2273. [CrossRef]
27. Bartoli, G.; Palmieri, G.; Bosco, M.; Dalpozzo, R. The reaction of vinyl Grignard reagents with 2-substituted nitroarenes: A new approach to the synthesis of 7-substituted indoles. *Tetrahedron Lett.* **1989**, *30*, 2129–2132. [CrossRef]

28. Teng, X.; Degterev, A.; Jagtap, P.; Xing, X.; Choi, S.; Denu, R.; Yuan, J.; Cuny, G.D. Structure-activity relationship study of novel necroptosis inhibitors. *Bioorg. Med. Chem. Lett.* **2005**, *15*, 5039–5044. [CrossRef] [PubMed]
29. Vilsmeier, A.; Haack, A. Über die einwirkung von halogenphosphor auf alkyl-formanilide. Eine neue methode zur darstellung sekundärer und tertiärer p-alkylamino-benzaldehyde. *Eur. J. Org. Chem.* **1927**, *60*, 119–122. [CrossRef]
30. Flatt, A.K.; Yao, Y.; Maya, F.; Tour, J.M. Orthogonally Functionalized Oligomers for Controlled Self-Assembly. *J. Org. Chem.* **2004**, *69*, 1752–1755. [CrossRef]
31. Sapountzis, I.; Knochel, P. General preparation of functionalized o-nitroarylmagnesium halides through an iodine-magnesium exchange. *Angew. Chem. Int. Ed.* **2002**, *41*, 1610–1611. [CrossRef]
32. Sapountzis, I.; Dube, H.; Lewis, R.; Gommermann, N.; Knochel, P. Synthesis of Functionalized Nitroarylmagnesium Halides via an Iodine-Magnesium Exchange. *J. Org. Chem.* **2005**, *70*, 2445–2454. [CrossRef] [PubMed]
33. Griffith, W.P.; Ley, S.V.; Whitcombe, G.P.; White, A.D. Preparation and use of tetra-n-butylammonium per-ruthenate (TBAP reagent) and tetra-n-propylammonium per-ruthenate (TPAP reagent) as new catalytic oxidants for alcohols. *Chem. Commun.* **1987**, *21*, 1625–1627. [CrossRef]
34. Boros, E.E.; Thompson, J.B.; Katamreddy, S.R.; Carpenter, A.J. Facile Reductive Amination of Aldehydes with Electron-Deficient Anilines by Acyloxyborohydrides in TFA: Application to a Diazaindoline Scale-Up. *J. Org. Chem.* **2009**, *74*, 3587–3590. [CrossRef]
35. Mangette, J.E.; Chen, X.; Krishnamoorthy, R.; Samuel, V.A.; Csakai, A.J.; Camara, F.; Paquette, W.D.; Wang, H.-J.; Takahashi, H.; Fleck, R.; et al. 2-Trifluoroacetyl aminoindoles as useful intermediates for the preparation of 2-acylamino indoles. *Tetrahedron Lett.* **2011**, *52*, 1292–1295. [CrossRef]
36. Newhouse, T.; Lewis, C.A.; Eastman, K.J.; Baran, P.S. Scalable Total Syntheses of N-Linked Tryptamine Dimers by Direct Indole-Aniline Coupling: Psychotrimine and Kapakahines B and F. *J. Am. Chem. Soc.* **2010**, *133*, 7119–7137. [CrossRef] [PubMed]
37. Bergman, J.; Engqvist, R.; Stalhandske, C.; Wallberg, H. Studies of the reactions between indole-2,3-diones (isatins) and 2-aminobenzylamine. *Tetrahedron* **2003**, *59*, 1033–1048. [CrossRef]
38. Coste, A.; Karthikeyan, G.; Couty, F.; Evano, G. Second-generation, biomimetic total synthesis of chaetominine. *Synthesis* **2009**, *17*, 2927–2934. [CrossRef]
39. Ohno, M.; Spande, T.F.; Witkop, B. Cyclization of tryptophan and tryptamine derivatives to pyrrolo[2,3-b]indoles. *J. Am. Chem. Soc.* **1968**, *90*, 6521–6522. [CrossRef]
40. Li, Y.; Dolphin, D.; Patrick, B.O. Synthesis of a BF_2 complex of indol-2-yl-isoindol-1-ylidene-amine: A fully conjugated azadipyrromethene. *Tetrahedron Lett.* **2010**, *51*, 811–814. [CrossRef]
41. Seo, J.H.; Artman, G.D., III; Weinreb, S.M. Synthetic Studies on Perophoramidine and the Communesins: Construction of the Vicinal Quaternary Stereocenters. *J. Org. Chem.* **2006**, *71*, 8891–8900. [CrossRef] [PubMed]
42. Voute, N.; Neal, A.R.; Medda, F.; Johnston, C.A.; Slawin, A.M.Z.; Westwood, N.J. From one to two quaternary centers: Ester or nitrile α-alkylation applied to bioactive alkaloids. *Tetrahedron* **2018**, *74*, 7399–7407. [CrossRef]
43. Bowden, K.; Heilbron, I.M.; Jones, E.R.H. 13. Researches on acetylenic compounds. Part I. The preparation of acetylenic ketones by oxidation of acetylenic carbinols and glycols. *J. Chem. Soc.* **1946**, 39–45. [CrossRef]
44. Johnston, C.A.; Wilkie, R.P.; Krauss, H.; Neal, A.R.; Slawin, A.M.Z.; Lebl, T.; Westwood, N.J. Polycyclic ethers and an unexpected dearomatisation reaction during studies towards the bioactive alkaloid, perophoramidine. *Tetrahedron* **2018**, *74*, 3339–3347. [CrossRef]
45. Freeman-Cook, K.D.; Halcomb, R.L. A Symmetry-Based Formal Synthesis of Zaragozic Acid A. *J. Org. Chem.* **2000**, *65*, 6153–6159. [CrossRef] [PubMed]
46. *CrystalClear-SM Expert v2.0rc13*; Rigaku Americas: The Woodlands, TX, USA; Rigaku Corporation: Tokyo, Japan, 2009.
47. Beurskens, P.T.; Beurskens, G.; de Gelder, R.; Garcia-Granda, S.; Gould, R.O.; Israel, R.; Smits, J.M.M. *DIRDIF-99*; Crystallography Laboratory, University of Nijmegen: Nijmegen, The Netherlands, 1999.
48. Sheldrick, G.M. Crystal structure refinement with SHELXL. *Acta Crystallogr. Sect. C* **2015**, *71*, 3–8. [CrossRef]
49. *CrystalStructure v4.3.0*; Rigaku Americas: The Woodlands, TX, USA; Rigaku Corporation: Tokyo, Japan, 2018.

Article

Isothiourea-Catalyzed Enantioselective α-Alkylation of Esters via 1,6-Conjugate Addition to *para*-Quinone Methides

Jude N. Arokianathar [1], Will C. Hartley [1], Calum McLaughlin [1], Mark D. Greenhalgh [1,2], Darren Stead [3], Sean Ng [4], Alexandra M. Z. Slawin [1] and Andrew D. Smith [1,*]

[1] EaStCHEM, School of Chemistry, University of St Andrews, North Haugh, St Andrews KY16 9ST, UK; jude1991@hotmail.co.uk (J.N.A.); whartley@iciq.es (W.C.H.); mclaughl@uni-muenster.de (C.M.); Mark.Greenhalgh@warwick.ac.uk (M.D.G.); amzs@st-andrews.ac.uk (A.M.Z.S.)
[2] Department of Chemistry, University of Warwick, Coventry CV4 7AL, UK
[3] AstraZeneca, Oncology R&D, Research & Early Development, Darwin Building, 310, Cambridge Science Park, Milton Road, Cambridge CB4 0WG, UK; Darren.Stead@astrazeneca.com
[4] Syngenta, Jealott's Hill International Research Centre, Bracknell RG42 6EY, UK; sean.ng@syngenta.com
* Correspondence: ads10@st-andrews.ac.uk

Abstract: The isothiourea-catalyzed enantioselective 1,6-conjugate addition of *para*-nitrophenyl esters to 2,6-disubstituted *para*-quinone methides is reported. *para*-Nitrophenoxide, generated in situ from initial N-acylation of the isothiourea by the *para*-nitrophenyl ester, is proposed to facilitate catalyst turnover in this transformation. A range of *para*-nitrophenyl ester products can be isolated, or derivatized in situ by addition of benzylamine to give amides at up to 99% yield. Although low diastereocontrol is observed, the diastereoisomeric ester products are separable and formed with high enantiocontrol (up to 94:6 er).

Keywords: isothiourea; ammonium enolate; aryloxide; quinone methide; ester functionalization; 1,6-conjugate addition

1. Introduction

Quinone methides (QMs) are electrophilic compounds composed of a cyclohexadiene core bearing a carbonyl either *ortho* or *para* to an exocyclic alkylidene unit [1,2]. Due to their electrophilicity [3–6], QMs have been used in a variety of biological and medicinal processes [1,2,7], are present within natural products and pharmaceuticals [1,2,8–11], and have been applied as electrophiles in a variety of synthetic reactions [1,2,12–18]. While *ortho*-QMs have been used extensively in enantioselective catalysis [19], particularly as components in formal [4+2] cycloaddition reactions, the use of *para*-QMs has only recently received increased attention [19–23]. The majority of enantioselective organocatalytic methods that involve *para*-QMs have utilized Brønsted acid [24–33] or hydrogen bonding catalysts [34–38], with only a relatively small number of examples using Lewis base catalysis [39–51].

C(1)-Ammonium enolate intermediates [52–55], generated by the reaction of a tertiary amine Lewis base catalyst with a ketene, anhydride or acyl imidazole [56], have found widespread application for the synthesis of heterocyclic scaffolds in high yield and with excellent enantiocontrol. Traditionally these approaches have been limited by the requirement for the electrophilic reaction partner to contain a latent nucleophilic site to facilitate catalyst turnover. This conceptual obstacle has resulted in catalysis via C(1)-ammonium enolates being mostly applied for formal cycloaddition reactions. More recently, aryl esters have emerged as alternative C(1)-ammonium enolate precursors [55,57,58]. Significantly, following acylation of the tertiary amine catalyst by the aryl ester, a nucleophilic aryloxide is liberated, which may be exploited again in the catalytic cycle to facilitate ammonium enolate formation and catalyst turnover (Scheme 1a) [55,59]. This strategy offers a potentially general solution to allow the expansion of electrophile scope within catalytic processes using C(1)-ammonium enolate intermediates. In 2014, we applied this concept for the isothiourea-catalyzed [2,3]-rearrangement of allylic ammonium ylides (Scheme 1b) [60–64]. More recently, this approach has been used by Snaddon (Scheme 1c) [65–72], Hartwig (Scheme 1d) [73] and Gong [74,75] for co-operative isothiourea/transition metal-catalyzed α-functionalization of pentafluorophenyl esters. In both cases, an isothiourea-derived C(1)-ammonium enolate is intercepted by an electrophilic transition metal complex to affect an allylation or benzylation reaction. We have expanded the scope of electrophiles applicable within this catalyst turnover strategy to include iminium ions generated under either photoredox conditions or Brønsted acid catalysis, as well as bis-sulfone Michael acceptors, and pyridinium salts (Scheme 1e) [76–79]. Recently, Waser also reported an elegant example of this turnover strategy for the enantioselective α-chlorination of pentafluorophenyl esters [80], while Zheng and co-workers reported a related approach using diphenyl methanol as an external turnover reagent for the fluorination of carboxylic acids [81,82]. A significant challenge within this area is the identification of electrophilic reaction partners that react with the catalytically-generated C(1)-ammonium enolate, but are compatible with the nucleophilic tertiary amine catalyst and aryloxide, which is essential for catalyst turnover. Building upon this conceptual platform, it was envisaged that *para*-QMs may be suitable electrophiles to apply in formal 1,6-conjugate additions.

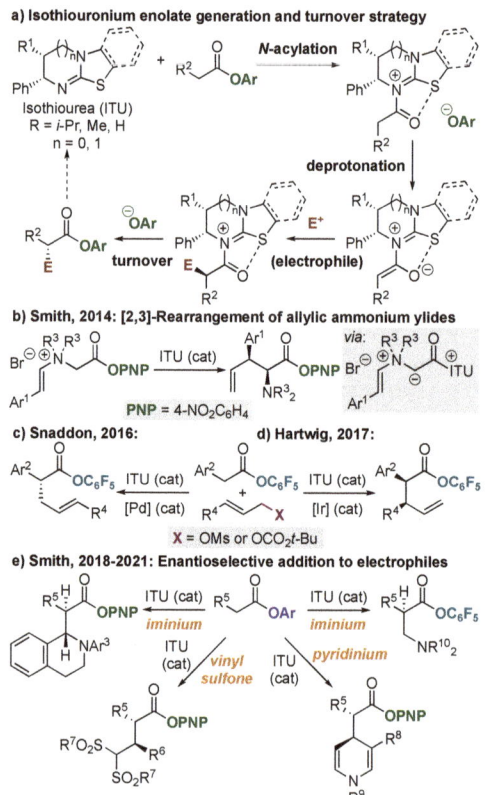

Scheme 1. Isothiourea-catalyzed enantioselective processes using aryloxide-facilitated catalyst turnover.

2. Results

2.1. Reaction Optimization

Initial studies focused on the isothiourea-catalyzed 1,6-conjugate addition of *para*-nitrophenyl (PNP) ester **1** to 2,6-di-*tert*-butyl *para*-QM **5** (Table 1). Benzylamine was added at the end of the reaction to convert the PNP ester product to the corresponding amide. Based on our previous experience, the amide was expected to be more stable to chromatographic purification. Using tetramisole HCl **6** (20 mol%) as the catalyst, *i*-Pr$_2$NEt as the base and MeCN as the solvent gave a 55:45 ratio of diastereoisomeric amide products **7** and **8** in high yield (82%) and with excellent enantioselectivity (**7**: 97:3 er; **8**: 94:6 er) (Entry 1). Chromatographic separation of the diastereoisomers was not possible, however, the enantioenrichment of both **7** and **8** could be reliably determined by chiral stationary phase (CSP)-HPLC analysis of the mixture. A control reaction in the absence of the catalyst showed no conversion (Entry 2). The use of six alternative solvents was investigated (PhMe, CH$_2$Cl$_2$, CHCl$_3$, THF, 1,4-dioxane and DMF), (see the Supporting Information for details) however, only the use of DMF provided any conversion to the product, indicating that solvent polarity may be significant for the success of this transformation. A control reaction in DMF in the absence of the catalyst, however, also led to comparable conversion to the product, consistent with the operation of a competitive Brønsted base-promoted reaction (see the Supporting Information for details). Taking MeCN as the optimal solvent, the use of eight different organic and inorganic bases was investigated (see the Supporting Information for details). Of those tested, Et$_3$N provided an improved yield of 98%, whilst maintaining comparable diastereo- and enantioselectivity (Entry 3). The use of alternative aryl esters was next probed, with pentafluorophenyl ester **2** and bis(trifluoromethyl)phenyl ester **3** giving amide products **7** and **8** in high yield, but with lower enantioselectivity than when using PNP ester **1** (Entries 4 and 5). The use of 2,4,6-trichlorophenyl ester **4** resulted in only 31% yield (Entry 6), which is consistent with previous studies in this field [65,72,76,79], and most likely reflects the increased steric hindrance of the aryloxide attenuating its nucleophilicity. Finally, using PNP ester **1**, the catalyst loading could be reduced to 5 mol% with only a small drop in stereoselectivity (Entry 7), while heating the reaction to 40 °C provided a slight improvement in yield (Entry 8). The reaction could also be performed in the absence of a base (Entry 9), however, slightly lower yield was obtained and therefore, during investigation of the substrate scope, Et$_3$N was routinely used as an auxiliary base.

2.2. Reaction Scope and Limitations

Due to the low diastereoselectivity observed using 2,6-di-*tert*-butyl *para*-QM **5**, the alternative use of 2,6-disubstituted *para*-QMs were investigated (Scheme 2). 2,6-Dimethyl, dibromo and diphenyl *para*-QMs **9–11** bearing a phenyl substituent at the exocyclic olefin were applied under optimized conditions. In all cases significantly lower conversion was observed (≤43%), and the amide products **14–16** were difficult to isolate. Based on analysis of the crude reaction mixture by ^{1}H NMR spectroscopy, the 2,6-dimethyl and dibromo-substituted analogues **14** and **15** were obtained with marginally improved diastereoselectivity (~70:30 dr), indicating that alternative substituents in these positions could prove beneficial if the products were isolable. Next, variation of the exocyclic substituent was probed. Incorporation of a methyl group at this position provided no improvement in dr, but the amide product **17** was isolated in a 50% yield and with moderate enantioenrichment for both diastereoisomers. Finally, incorporation of a 2-naphthyl substituent at this position was well tolerated, with amide **18** obtained in an 83% yield, 70:30 dr and excellent enantiocontrol (94:6 er) for the major diastereoisomer.

Table 1. Reaction optimization.

Entry	6 (mol%)	Aryl Ester	Base	Yield (%)	dr (7:8)	er (7)	er (8)
1	20	1	i-Pr$_2$NEt	82	55:45	97:3	94:6
2	0	1	i-Pr$_2$NEt	0	-	-	-
3	20	1	Et$_3$N	98	60:40	97:3	94:6
4	20	2	Et$_3$N	99	65:35	81:19	94:6
5	20	3	Et$_3$N	88	60:40	94:6	91:9
6	20	4	Et$_3$N	31	55:45	84:16	71:29
7	5	1	Et$_3$N	95	55:45	93:7	88:12
8 [a]	5	1	Et$_3$N	99	60:40	92:8	85:15
9 [a]	5	1	none	90	60:40	93:7	89:11

All reactions were carried out on a 0.25 mmol scale; isolated yields are a mixture of diastereoisomers **7** and **8**; dr was determined by ^{1}H NMR spectroscopic analysis of the crude reaction mixture; er was determined by CSP-HPLC analysis: **7** (2S,1'R:2R,1'S) and **8** (2S,1'S:2R,1'R). [a] Reaction performed at 40 °C.

Although structural variation of the *para*-QM provided marginal improvements in dr, the use of 2,6-di-*tert*-butyl *para*-QM **5** was considered most convenient for further investigations due to its stability and ease of synthesis, and the higher yields of product obtained from catalysis. To investigate if the dr obtained in these reactions was a manifestation of a kinetic or thermodynamic preference, isolation of diastereoisomeric PNP esters **19** and **20** and resubjection to catalysis conditions was attempted. Although the diastereoisomeric amides **7** and **8** had proved difficult to separate, the corresponding PNP esters **19** and **20** were chromatographically separable and displayed high stability. Epimerization studies were conducted using each diastereoisomer through sequential treatment with i-Pr$_2$NEt, (S)-TM HCl **6**, *para*-nitrophenoxide and benzylamine (Scheme 3). These experiments were followed by in situ ^{1}H NMR spectroscopic analysis and revealed no epimerization in either case. This indicates the dr obtained in the catalytic reaction most likely reflects the inherent diastereoselectivity of the transformation. Following separation of the diastereoisomers, the absolute configuration of the major diastereoisomer could also be confirmed as (2S,1'R) by single crystal X-ray crystallographic analysis [83,84]. Based on literature precedent, the (S)-configuration at C(2) was expected to be generated under catalyst-control, and therefore the absolute configuration of the minor diastereoisomer was predicted to be (2S,1'S).

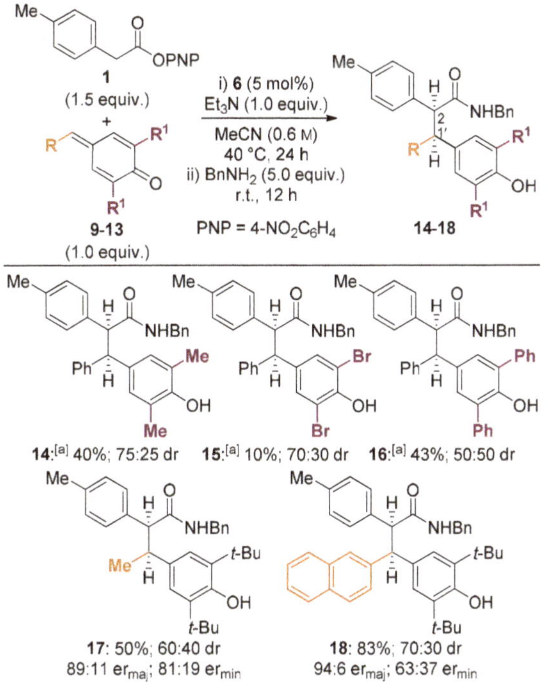

Scheme 2. Scope: Variation of *para*-quinone methide—0.25 mmol scale; only the structure of the major diastereoisomer is shown; isolated yield is a mixture of diastereoisomers; dr determined by ¹H NMR spectroscopic analysis of the crude reaction product mixture; er determined by CSP-HPLC analysis. [a] Product not isolable: conversion based on ¹H NMR analysis of the crude reaction product mixture; ers could not be determined.

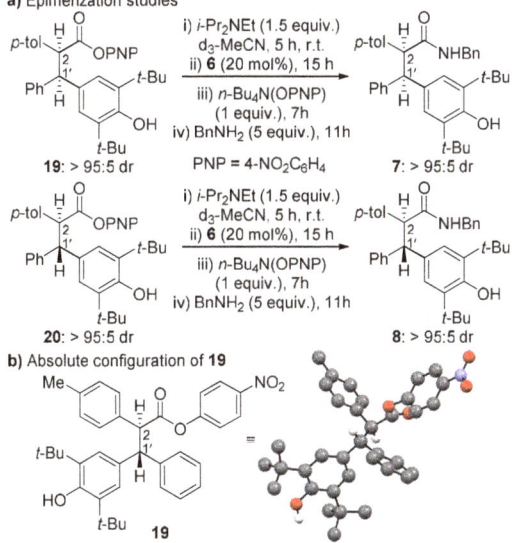

Scheme 3. Control studies and confirmation of absolute configuration of products. The majority of hydrogen atoms are omitted for clarity within X-ray crystal structure representation of **19**.

Having established the stability of the PNP ester products and demonstrated the potential to separate the diastereoisomers by column chromatography, the scope of the catalytic transformation was investigated through variation of the PNP ester substrate. In each case, the PNP ester product diastereoisomers were at least partially separable, enabling unambiguous characterization. To test the applicability of the procedure, p-tolyl-substituted PNP ester product **19** was prepared on a larger scale (1.25 mmol) (Scheme 4). A combined 85% yield of both diastereoisomers was obtained, with comparable stereoselectivity to that observed when the reaction was conducted on an analytical scale (Table 1, Entry 7). The generality of the procedure was further probed using five electronically- and sterically-differentiated PNP esters. Introduction of an electron-donating 4-methoxy substituent was well tolerated, with PNP ester **26** obtained in quantitative yield, 60:40 dr and with high enantioselectivity for both diastereoisomers. Under the optimized conditions, the introduction of an electron-withdrawing 4-trifluoromethyl group resulted in low enantioselectivity (**27**: 63:37 er), which was attributed to a competitive Brønsted base-promoted background reaction. Consistent with this hypothesis, repeating the reaction in the absence of Et$_3$N, and using the free base of the isothiourea catalyst **6**, provided PNP ester product **27** in 94% yield and significantly improved enantioselectivity (88:12 er$_{maj}$; 79:21 er$_{min}$). A similar effect was observed when using a 2-naphthyl-substituted PNP ester, with optimal enantioselectivity obtained in the absence of an auxiliary base (**28**: 91:9 er$_{maj}$; 85:15 er$_{min}$). Introduction of a sterically-imposing 1-naphthyl or a heteroaromatic thienyl substituent was also tolerated, with **29** and **30** obtained in excellent yield and with high enantioselectivity.

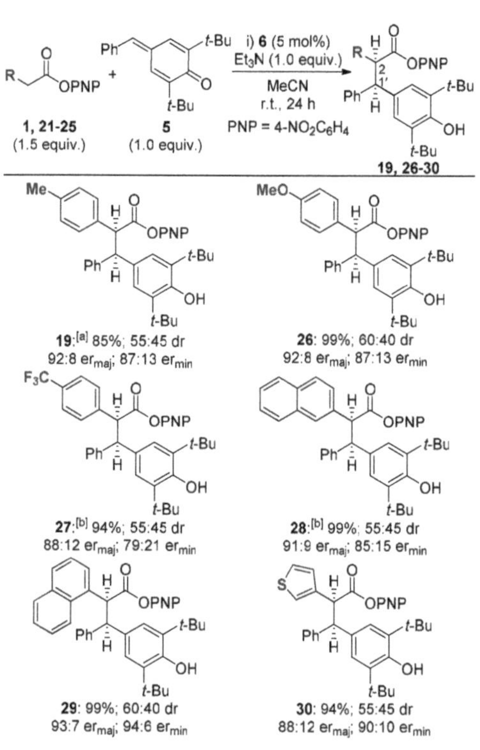

Scheme 4. Scope: Variation of *para*-nitrophenyl ester—0.25 mmol scale; only the structure of the major diastereoisomer is shown; isolated yields are given for the combination of diastereoisomers; dr determined by ^{1}H NMR spectroscopic analysis of the crude reaction product mixture; er determined by CSP-HPLC analysis. [a] 1.25 mmol scale. [b] Conducted at r.t., in the absence of Et$_3$N, and using free base of **6**.

2.3. Proposed Mechanism

The mechanism of this transformation is proposed to begin with *N*-acylation of the free base isothiourea catalyst **6** by PNP ester **31** to generate the corresponding acyl ammonium *para*-nitrophenoxide ion pair **32** (Scheme 5). Subsequent deprotonation leads to (*Z*)-ammonium enolate **33**. Based on previous mechanistic studies [78,85], and the catalytic activity observed in the absence of an auxiliary base (Table 1, Entry 9), deprotonation is likely to be affected by the *para*-nitrophenoxide counterion. 1,6-Conjugate addition of ammonium enolate **33** to *para*-QM electrophile **34**, followed by protonation, gives acyl ammonium intermediate **35**. Finally, regeneration of catalyst **6**, and concurrent release of product **37**, is proposed to be facilitated by *para*-nitrophenoxide [59–79,86,87]. Although not essential for reactivity, the addition of Et$_3$N as an auxiliary base may be beneficial as a proton shuttle, and to maintain the isothiourea catalyst in its non-protonated form **6** [77,86]. The enantioselectivity of the transformation indicates the C–C bond forming event takes place on the *Si*-face of the ammonium enolate. This selectivity can be rationalized through preferential formation of the (*Z*)-ammonium enolate [76–79,85], which is conformationally-restricted by an intramolecular 1,5-O···S interaction [61,88–104] and results in the phenyl stereodirecting group of the catalyst blocking the enolate *Re*-face. The observed poor diastereoselectivity can be tentatively rationalized by a simple stereochemical model that assumes a favored, open pre-transition state assembly where steric interactions are minimized about the forming C–C bond. Minimal differentiation between the aryl- and quinone substituents of the *para*-QM quinone leads to the two transition state assemblies **38** and **39** that give the major and minor diastereoisomers, respectively.

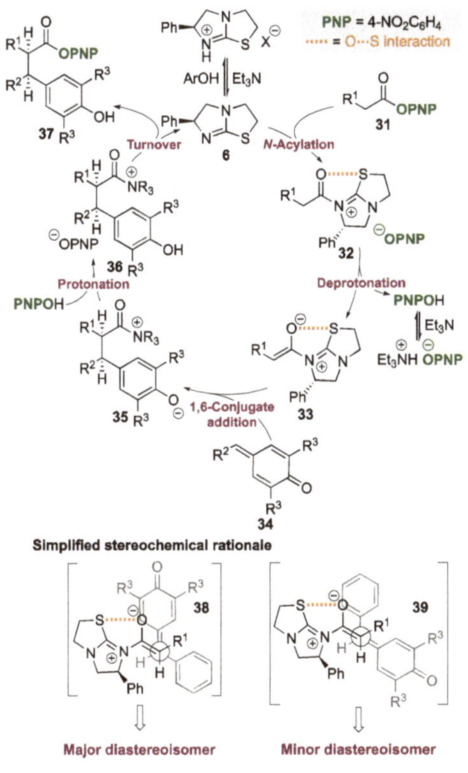

Scheme 5. Proposed mechanism (only the pathway for the formation of the major diastereoisomer is shown) and transition state assemblies. ArOH corresponds to either PNPOH or the reaction product.

3. Materials and Methods

3.1. General Procedure for the Enantioselective 1,6-Addition

In a flame-dried vial, the requisite *para*-quinone methide (1.0 equiv.), aryl ester (1.5 equiv.), (S)-TM HCl (5 mol%), Et$_3$N (1.0 equiv.) and anhydrous MeCN (0.6 M) was added and stirred at r.t. for 24 h. The reaction was then quenched with benzylamine (5.0 equiv.) and stirred at r.t. for a further 12 h before being concentrated in vacuo. The residue was diluted with EtOAc (20 mL) and washed successively with 10% citric acid (20 mL × 1), aqueous NaOH (20 mL × 3) and brine (20 mL × 1). The organic layer was extracted, dried over MgSO$_4$ and the filtrate was concentrated in vacuo. The crude material was purified by flash silica column chromatography to give the desired product.

3.2. Representative Synthesis and Characterization of Compounds 7 and 8 (Entry 8)

Following the general procedure above, 4-benzylidene-2,6-di-*tert*-butylcyclohexa-2,5-dien-1-one **5** (74 mg, 0.25 mmol), 4′-nitrophenyl 2-(*p*-tolyl)acetate **1** (102 mg, 0.375 mmol), (S)-TM HCl **6** (3 mg, 5 mol%) and Et$_3$N (35 µL, 0.25 mmol) were dissolved in anhydrous MeCN (0.42 mL). The reaction mixture was stirred at 40 °C for 24 h before being quenched with benzylamine (137 µL, 1.25 mmol) at r.t. to give a crude mixture containing the title compound in 60:40 dr. The mixture was purified by flash silica column chromatography (petroleum ether/EtOAc, 85:15) to afford diastereoisomers **7** and **8** (60:40 dr) (132 mg, 99%) as a pale yellow solid.

mp 126–128 °C; $[\alpha]_D^{20}$ +10.0 (c 1.0, CHCl$_3$); IR ν_{max} (film)/cm^{-1} 3638 (O–H) 3304 (N–H), 2955 (C–H), 1645 (C=O); HRMS (ESI$^+$) C$_{37}$H$_{43}$NO$_2$ ([M + H]$^+$), found 534.3359, requires 534.3367 (−1.4 ppm).

Data for major diastereoisomer (**7**): Chiral HPLC analysis, Chiralpak AD-H (10% *i*-PrOH/hexane, flow rate 1.5 mLmin^{-1}, 211 nm, 40 °C), t$_R$ 8.5 min and 29.3 min, 92:8 er; ^{1}H NMR (500 MHz, CDCl$_3$) δ$_H$: 1.42 (18H, s, (C(3′′′′′)C(CH$_3$)$_3$, C(5′′′′′)C(CH$_3$)$_3$), 2.24 (3H, s, C(4′′)CH$_3$), 3.90–4.05 (2H, m, C(2)H, CH$_A$Ph), 4.44 (1H, dd, J 15.0, 6.8, CH$_B$Ph), 4.82 (1H, d, J 11.7, C(1′)H), 5.14 (1H, s, OH), 5.55 (1H, t, J 5.6, NH), 6.71–6.76 (2H, m, Ar), 6.96–7.05 (2H, m, Ar), 7.09–7.15 (2H, m, Ar), 7.15–7.22 (4H, m, Ar), 7.23–7.30 (4H, m, C(4′′′)H, C(2′′′′)H, C(6′′′′)H, Ar), 7.34 (1H, t, J 7.6, Ar), 7.43–7.51 (1H, m, Ar); ^{13}C{^{1}H} NMR (126 MHz, CDCl$_3$) δ$_C$: 21.0 (C(4′′)CH$_3$), 30.4 (C(3′′′′′)C(CH$_3$)$_3$, C(5′′′′′)C(CH$_3$)$_3$), 34.4 (C(3′′′′′)C(CH$_3$)$_3$, C(5′′′′′)C(CH$_3$)$_3$), 43.6 (CH$_2$Ph), 54.1 (C(1′)), 59.5 (C(2)), 124.7 (C(2′′′′), C(6′′′′)), 125.8 (C(4′′′)), 128.2 (Ar), 127.2 (Ar), 127.3 (Ar), 128.0 (Ar), 128.4 (Ar), 128.5 (Ar), 128.6 (Ar), 129.0 (Ar), 133.8 (C(1′′′)), 135.2 (C(1′′)), 135.6 (C(3′′′′′), C(5′′′′′)), 136.4 (C(4′′)), 137.9 (*i*-Ph), 142.4 (C(1′′′′)), 152.4 (C(4′′′′′)), 172.1 (C(1)).

Selected data for minor diastereoisomer (**8**): Chiral HPLC analysis, Chiralpak AD-H (10% *i*-PrOH/hexane, flow rate 1.5 mLmin^{-1}, 211 nm, 40 °C), t$_R$ 3.8 min and 18.6 min, 85:15 er; ^{1}H NMR (500 MHz, CDCl$_3$) δ$_H$: 1.27 (18H, s, (C(3′′′′′)C(CH$_3$)$_3$, C(5′′′′′)C(CH$_3$)$_3$), 2.27 (3H, s, C(4′′)CH$_3$), 3.90–4.05 (2H, m, C(2)H, CH$_A$Ph), 4.50 (1H, dd, J 15.0, 6.8, CH$_B$Ph), 4.69 (1H, d, J 11.7, C(1′)H), 4.89 (1H, s, OH), 5.69 (1H, t, J 5.6, NH), 6.81–6.84 (2H, m, Ar); ^{13}C{^{1}H} NMR (126 MHz, CDCl$_3$) δ$_C$: 21.0 (C(4′′)CH$_3$), 30.2 (C(3′′′′′)C(CH$_3$)$_3$, C(5′′′′′)C(CH$_3$)$_3$), 34.2 (C(3′′′′′)C(CH$_3$)$_3$, C(5′′′′′)C(CH$_3$)$_3$), 43.4 (CH$_2$Ph), 54.7 (C(1′)), 59.5 (C(2)), 125.3 (C(2′′′′), C(6′′′′)), 126.3 (C(4′′′)), 127.5 (Ar), 128.2 (Ar), 128.4 (Ar), 128.8 (Ar), 132.1 (C(1′′′)), 134.9 (C(3′′′′′), C(5′′′′′)), 135.5 (C(1′′)), 136.4 (C(4′′)), 138.2 (*i*-Ph), 143.6 (C(1′′′′)), 151.7 (C(4′′′′′)), 172.0 (C(1)).

4. Conclusions

An isothiourea-catalyzed enantioselective 1,6-conjugate addition of *para*-nitrophenyl (PNP) esters to *para*-quinone methides (QMs) has been developed. Variation of the arylacetic ester and *para*-QM substrates has provided a range of functionalized products in generally excellent yields and high enantiocontrol (up to 94:6 er). An inherent limitation of the method is that the products were routinely obtained in ~60:40 dr. This diastereoselectivity was shown to arise from kinetic control, but was relatively insensitive to changes in reaction conditions and structural variation of the substrates. Although the dr could not be improved, the diastereoisomeric PNP ester products could be separated by column chromatography. The success of this catalytic methodology is proposed to rely upon the *para*-nitrophenoxide, expelled during *N*-acylation of the catalyst, to facilitate catalyst turnover and release the product. Current work in our laboratory is focused on further applications of using in situ-generated aryloxides to promote catalyst turnover in Lewis base catalysis.

Supplementary Materials: Full experimental procedures, characterization data, NMR spectra and HPLC chromatograms for all new compounds, as well as crystallographic data for product **19** (CCDC 1992504) are available online.

Author Contributions: Conceptualization, A.D.S. and J.N.A.; investigation, J.N.A.; W.C.H.; C.M.; X-ray crystallographic analysis, A.M.Z.S.; writing—original draft preparation, M.D.G.; writing—review and editing, A.D.S. and all authors; supervision, D.S., S.N. and A.D.S.; funding acquisition, A.D.S. All authors have read and agreed to the published version of the manuscript.

Funding: We thank the ERC under the European Union's Seventh Framework Programme (FP7/2007-2013)/E.R.C. grant agreement n° 279850, AstraZeneca and EPSRC (EP/M506631/1 (J.N.A.)), Syngenta and the EPSRC Centre for Doctoral Training in Critical Resource Catalysis (CRITICAT, EP/L016419/1 (W.C.H.)), and EPSRC (EP/M508214/1 (C.M.)) for funding. A.D.S. thanks the Royal Society for a Wolfson Research Merit Award. We thank the EPSRC UK National Mass Spectrometry Facility at Swansea University.

Institutional Review Board Statement: Not applicable.

Informed Consent Statement: Not applicable.

Data Availability Statement: The research data underpinning this publication can be found at DOI: 10.17630/f6cf6c80-483d-4f16-bd79-80e1537513b2.

Conflicts of Interest: The authors declare no conflict of interest.

Sample Availability: Samples of the compounds are available from the authors on request.

References and Notes

1. Rokita, S.E. *Quinone Methides*; Wiley: Hoboken, NJ, USA, 2009.
2. Toteva, M.M.; Richard, J.P. The generation and reactions of quinone methides. *Adv. Phys. Org. Chem.* **2011**, *45*, 39–91. [PubMed]
3. Turner, A.B. Quinone methides. *Q. Rev. Chem. Soc.* **1964**, *18*, 347–360. [CrossRef]
4. Wagner, H.-U.R. Gompper. In *The Chemistry of Quinonoid Compounds*; Wiley: London, UK, 1974; pp. 1145–1178.
5. Richter, D.; Hampel, N.; Singer, T.; Ofial, A.R.; Mayr, H. Synthesis and Characterization of Novel Quinone Methides: Reference Electrophiles for the Construction of Nucleophilicity Scales. *Eur. J. Org. Chem.* **2009**, 3203–3211. [CrossRef]
6. Singh, M.S. *Reactive Intermediates in Organic Chemistry: Structure and Mechanism*; Wiley-VCH: Weinheim, Germany, 2014.
7. Freccero, M. Quinone methides as alkylating and cross-linking agents. *Mini Rev. Org. Chem.* **2004**, *1*, 403–415. [CrossRef]
8. Peters, M.G. Chemical modifications of biopolymers by quinones and quinone methides. *Angew. Chem. Int. Ed.* **1989**, *28*, 555–570, *Angew. Chem.* **1989**, *101*, 572–587. [CrossRef]
9. Martin, H.J.; Magauer, T.; Mulzer, J. In Pursuit of a Competitive Target: Total Synthesis of the Antibiotic Kendomycin. *Angew. Chem. Int. Ed.* **2010**, *49*, 5614–5626, *Angew. Chem.* **2010**, *122*, 5746–5758. [CrossRef]
10. Jansen, R.; Gerth, K.; Steinmetz, H.; Reinecke, S.; Kessler, W.; Kirschning, A.; Müller, R. Elansolid A3, a Unique p-Quinone Methide Antibiotic from *Chitinophaga sancti*. *Chem. Eur. J.* **2011**, *17*, 7739–7744. [CrossRef]
11. Dehn, R.; Katsuyama, Y.; Weber, A.; Gerth, K.; Jansen, R.; Steinmetz, H.; Höfle, G.; Müller, R.; Kirschning, A. Molecular Basis of Elansolid Biosynthesis: Evidence for an Unprecedented Quinone Methide Initiated Intramolecular Diels–Alder Cycloaddition/Macrolactonization. *Angew. Chem. Int. Ed.* **2011**, *50*, 3882–3887. *Angew. Chem.* **2011**, *123*, 3968–3973. [CrossRef]
12. Gai, K.; Fang, X.; Li, X.; Xu, J.; Wu, X.; Lin, A.; Yao, H. Synthesis of spiro[2.5]octa-4,7-dien-6-one with consecutive quaternary centers via 1,6-conjugate addition induced dearomatization of para-quinone methides. *Chem. Commun.* **2015**, *51*, 15831–15834. [CrossRef]
13. López, A.; Parra, A.; Jarava-Barrera, C.; Tortosa, M. Copper-catalyzed silylation of p-quinone methides: New entry to dibenzylic silanes. *Chem. Commun.* **2015**, *51*, 17684–17687. [CrossRef]
14. Ramanjaneyulu, B.T.; Mahesh, S.; Anand, R.V. Bis(amino)cyclopropenylidene-Catalyzed 1,6-Conjugate Addition of Aromatic Aldehydes to para-Quinone Methides: Expedient Access to α,α′-Diarylated Ketones. *Org. Lett.* **2015**, *17*, 3952–3955. [CrossRef]
15. Yuan, Z.; Fang, X.; Li, X.; Wu, J.; Yao, H.; Lin, A. 1,6-Conjugated Addition-Mediated [2+1] Annulation: Approach to Spiro[2.5]octa-4,7-dien-6-one. *J. Org. Chem.* **2015**, *80*, 11123–11130. [CrossRef] [PubMed]

16. Reddy, V.; Anand, R.V. Expedient Access to Unsymmetrical Diarylindolylmethanes through Palladium-Catalyzed Domino Electrophilic Cyclization–Extended Conjugate Addition Approach. *Org. Lett.* **2015**, *17*, 3390–3393. [CrossRef]
17. Roiser, L.; Zielke, K.; Waser, M. Enantioselective Spirocyclopropanation of para-Quinone Methides using Ammonium Ylides. *Org. Lett.* **2017**, *19*, 2338–2341. [CrossRef]
18. Roiser, L.; Zielke, K.; Waser, M. Formal (4+1) Cyclization of Ammonium Ylides with Vinylogous para-Quinone Methides. *Synthesis* **2018**, *50*, 4047–4054.
19. Caruana, L.; Fochi, M.; Bernardi, L. The Emergence of Quinone Methides in Asymmetric Organocatalysis. *Molecules* **2015**, *20*, 11733–11764. [CrossRef]
20. Parra, A.; Tortosa, M. para-Quinone Methide: A New Player in Asymmetric Catalysis. *ChemCatChem* **2015**, *7*, 1524–1526. [CrossRef]
21. Chauhan, P.; Kaya, U.; Enders, D. Advances in Organocatalytic 1,6-Addition Reactions: Enantioselective Construction of Remote Stereogenic Centers. *Adv. Synth. Catal.* **2017**, *359*, 888–912. [CrossRef]
22. Li, W.; Xu, X.; Zhang, P.; Li, P. Recent Advances in the Catalytic Enantioselective Reactions of para-Quinone Methides. *Chem. Asian J.* **2018**, *13*, 2350–2359. [CrossRef]
23. Wang, J.-Y.; Hao, W.-J.; Tu, S.-J.; Jiang, B. Recent developments in 1,6-addition reactions of para-quinone methides (p-QMs). *Org. Chem. Front.* **2020**, *7*, 1743–1778. [CrossRef]
24. Wang, Z.; Wong, Y.F.; Sun, J. Catalytic Asymmetric 1,6-Conjugate Addition of para-Quinone Methides: Formation of All-Carbon Quaternary Stereocenters. *Angew. Chem. Int. Ed.* **2015**, *54*, 13711–13714. *Angew. Chem.* **2015**, *127*, 13915–13918. [CrossRef]
25. Dong, N.; Zhang, Z.-P.; Xue, X.-S.; Li, X.; Cheng, J.-P. Phosphoric Acid Catalyzed Asymmetric 1,6-Conjugate Addition of Thioacetic Acid to para-Quinone Methides. *Angew. Chem. Int. Ed.* **2016**, *55*, 1460–1464. *Angew. Chem.* **2016**, *128*, 1482–1486. [CrossRef]
26. Wong, Y.F.; Wang, Z.; Sun, J. Chiral phosphoric acid catalyzed asymmetric addition of naphthols to para-quinone methides. *Org. Biomol. Chem.* **2016**, *14*, 5751–5754. [CrossRef]
27. Chen, M.; Sun, J. How understanding the role of an additive can lead to an improved synthetic protocol without an additive: Organocatalytic synthesis of chiral diarylmethyl alkynes. *Angew. Chem.* **2017**, *129*, 12128–12132. *Angew. Chem. Int. Ed.* **2017**, *56*, 11966–11970. [CrossRef]
28. Yan, J.; Chen, M.; Sung, H.H.-Y.; Williams, I.D.; Sun, J. An Organocatalytic Asymmetric Synthesis of Chiral β,β-Diaryl-α-amino Acids via Addition of Azlactones to In Situ Generated para-Quinone Methides. *Chem. Asian J.* **2018**, *13*, 2440–2444. [CrossRef]
29. Rahman, A.; Zhou, Q.; Lin, X. Asymmetric organocatalytic synthesis of chiral 3,3-disubstituted oxindoles via a 1,6-conjugate addition reaction. *Org. Biomol. Chem.* **2018**, *16*, 5301–5309. [CrossRef]
30. Li, W.; Xu, X.; Liu, Y.; Gao, H.; Cheng, Y.; Li, P. Enantioselective Organocatalytic 1,6-Addition of Azlactones to para-Quinone Methides: An Access to α,α-Disubstituted and β,β-Diaryl-α-amino acid Esters. *Org. Lett.* **2018**, *20*, 1142–1145. [CrossRef] [PubMed]
31. Wang, J.-R.; Jiang, X.-L.; Hang, Q.-Q.; Zhang, S.; Mei, G.-J.; Shi, F. Catalytic Asymmetric Conjugate Addition of Indoles to para-Quinone Methide Derivatives. *J. Org. Chem.* **2019**, *84*, 7829–7839. [CrossRef]
32. Wang, Z.; Zhu, Y.; Pan, X.; Wang, G.; Liu, L. Synthesis of Chiral Triarylmethanes Bearing All-Carbon Quaternary Stereocenters: Catalytic Asymmetric Oxidative Cross-Coupling of 2,2-Diarylacetonitriles and (Hetero)arenes. *Angew. Chem.* **2020**, *132*, 3077–3081, *Angew. Chem.* **2020**, *59*, 3053–3057. [CrossRef]
33. Niu, J.-P.; Nie, J.; Li, S.; Ma, J.-A. Organocatalytic asymmetric synthesis of β,β-diaryl ketones via one-pot tandem dehydration/1,6-addition/decarboxylation transformation of β-keto acids and 4-hydroxybenzyl alcohols. *Chem. Commun.* **2020**, *56*, 8687–8690. [CrossRef] [PubMed]
34. Zhao, K.; Zhi, Y.; Wang, A.; Enders, D. Asymmetric Organocatalytic Synthesis of 3-Diarylmethine-Substituted Oxindoles Bearing a Quaternary Stereocenter via 1,6-Conjugate Addition to para-Quinone Methides. *ACS Catal.* **2016**, *6*, 657–660. [CrossRef]
35. Li, X.; Xu, X.; Wei, W.; Lin, A.; Yao, H. Organocatalyzed Asymmetric 1,6-Conjugate Addition of para-Quinone Methides with Dicyanoolefins. *Org. Lett.* **2016**, *18*, 428–431. [CrossRef]
36. Deng, Y.-H.; Zhang, X.-Z.; Yu, K.-Y.; Yan, X.; Du, J.-Y.; Huang, H.; Fan, C.-A. Bifunctional tertiary amine-squaramide catalyzed asymmetric catalytic 1,6-conjugate addition/aromatization of para-quinone methides with oxindoles. *Chem. Commun.* **2016**, *52*, 4183–4186. [CrossRef] [PubMed]
37. Toràn, R.; Vila, C.; Sanz-Marco, A.; Muñoz, M.C.; Pedro, J.R.; Blay, G. Organocatalytic Enantioselective 1,6-aza-Michael Addition of Isoxazolin-5-ones to p-Quinone Methides. *Eur. J. Org. Chem.* **2020**, *2020*, 627–630. [CrossRef]
38. Wang, L.; Yang, F.; Xua, X.; Jiang, J. Organocatalytic 1,6-hydrophosphination of para-quinone methides: Enantioselective access to chiral 3-phosphoxindoles bearing phosphorus-substituted quaternary carbon stereocenters. *Org. Chem. Front.* **2021**, *8*, 2002–2008. [CrossRef]
39. Caruana, L.; Kniep, F.; Johansen, T.K.; Poulsen, H.P.; Jørgensen, K.A. A New Organocatalytic Concept for Asymmetric α-Alkylation of Aldehydes. *J. Am. Chem. Soc.* **2014**, *136*, 15929–15932. [CrossRef]
40. Li, S.; Liu, Y.; Huang, B.; Zhou, T.; Tao, H.; Xiao, Y.; Liu, L.; Zhang, J. Phosphine-Catalyzed Asymmetric Intermolecular Cross-Vinylogous Rauhut–Currier Reactions of Vinyl Ketones with para-Quinone Methides. *ACS Catal.* **2017**, *7*, 2805–2809. [CrossRef]
41. Kang, T.-C.; Wu, L.-P.; Yu, Q.-W.; Wu, X.-Y. Enantioselective Rauhut–Currier-Type 1,6-Conjugate Addition of Methyl Vinyl Ketone to para-Quinone Methides. *Chem. Eur. J.* **2017**, *23*, 6509–6513. [CrossRef]

42. Wang, D.; Song, Z.-F.; Wang, W.-J.; Xu, T. Highly Regio- and Enantioselective Dienylation of p-Quinone Methides Enabled by an Organocatalyzed Isomerization/Addition Cascade of Allenoates. *Org. Lett.* **2019**, *21*, 3963–3967. [CrossRef]
43. Li, W.; Yuan, H.; Liu, Z.; Zhang, Z.; Cheng, Y.; Li, P. NHC-Catalyzed Enantioselective [4+3] Cycloaddition of Ortho-Hydroxyphenyl Substituted Para-Quinone Methides with Isatin-Derived Enals. *Adv. Synth. Catal.* **2018**, *360*, 2460–2464. [CrossRef]
44. Liu, Q.; Chen, X.-Y.; Rissanen, K.; Ender, D. Asymmetric synthesis of spiro-oxindole-ε-lactones through N-heterocyclic carbene catalysis. *Org. Lett.* **2018**, *20*, 3622–3626. [CrossRef]
45. Zhao, M.-X.; Xiang, J.; Zhao, Z.-Q.; Zhao, X.-L.; Shi, M. Asymmetric synthesis of dihydrocoumarins via catalytic sequential 1,6-addition/transesterification of α-isocyanoacetates with para-quinone methides. *Org. Biomol. Chem.* **2020**, *18*, 1637–1646. [CrossRef] [PubMed]
46. Roy, S.; Pradhan, S.; Kumar, K.; Chatterjee, I. Asymmetric organocatalytic double 1,6-addition: Rapid access to chiral chromans with molecular complexity. *Org. Chem. Front.* **2020**, *7*, 1388–1394. [CrossRef]
47. Chu, W.D.; Zhang, L.F.; Bao, X.; Zhao, X.H.; Zeng, C.; Du, J.Y.; Zhang, G.B.; Wang, F.X.; Ma, X.Y.; Fan, C.A. Asymmetric Catalytic 1,6-Conjugate Addition/Aromatization of para-Quinone Methides: Enantioselective Introduction of Functionalized Diarylmethine Stereogenic Centers. *Angew. Chem. Int. Ed.* **2013**, *52*, 9229–9233. *Angew. Chem.* **2013**, *125*, 9399–9403. [CrossRef] [PubMed]
48. Zhang, X.-Z.; Deng, Y.-H.; Yan, X.; Yu, K.-Y.; Wang, F.-X.; Ma, X.-Y.; Fan, C.-A. Diastereoselective and Enantioselective Synthesis of Unsymmetric β,β-Diaryl-α-Amino Acid Esters via Organocatalytic 1,6-Conjugate Addition of para-Quinone Methides. *J. Org. Chem.* **2016**, *81*, 5655–5662. [CrossRef]
49. Ge, L.; Lu, X.; Cheng, C.; Chen, J.; Cao, W.; Wu, X.; Zhao, G. Amide-Phosphonium Salt as Bifunctional Phase Transfer Catalyst for Asymmetric 1,6-Addition of Malonate Esters to para-Quinone Methides. *J. Org. Chem.* **2016**, *81*, 9315–9325. [CrossRef]
50. Santra, S.; Porey, A.; Jana, B.; Guin, J. N-Heterocyclic carbenes as chiral Brønsted base catalysts: A highly diastereo- and enantioselective 1,6-addition reaction. *Chem. Sci.* **2018**, *9*, 6446–6450. [CrossRef]
51. Eitzinger, A.; Winter, M.; Schörgenhumer, J.; Waser, M. Quaternary β 2, 2-amino acid derivatives by asymmetric addition of isoxazolidin-5-ones to para-quinone methides. *Chem. Commun.* **2020**, *56*, 579–582. [CrossRef]
52. Van, K.N.; Morrill, L.C.; Smith, A.D.; Romo, D. *Lewis Base Catalysis in Organic Synthesis*; Vedejs, E., Denmark, S.E., Eds.; Wiley-VCH: Weinheim, Germany, 2016; Chapter 13; pp. 527–653.
53. Morrill, L.C.; Smith, A.D. Organocatalytic Lewis base functionalisation of carboxylic acids, esters and anhydrides via C1-ammonium or azolium enolates. *Chem. Soc. Rev.* **2014**, *43*, 6214–6226. [CrossRef]
54. Gaunt, M.J.; Johansson, C.C.C. Recent developments in the use of catalytic asymmetric ammonium enolates in chemical synthesis. *Chem. Rev.* **2007**, *107*, 5596–5605. [CrossRef]
55. McLaughlin, C.; Smith, A.D. Generation and Reactivity of C (1)-Ammonium Enolates by Using Isothiourea Catalysis. *Chem. Eur. J.* **2021**, *27*, 1533–1555. [CrossRef] [PubMed]
56. Young, C.M.; Stark, D.G.; West, T.H.; Taylor, J.E.; Smith, A.D. Exploiting the Imidazolium Effect in Base-free Ammonium Enolate Generation: Synthetic and Mechanistic Studies. *Angew. Chem.* **2016**, *128*, 14606–14611. *Angew. Chem.* **2016**, *55*, 14394–14399. [CrossRef] [PubMed]
57. Hao, L.; Chen, X.; Chen, S.; Jiang, K.; Torres, J.; Chi, Y.R. Access to pyridines via DMAP-catalyzed activation of α-chloro acetic ester to react with unsaturated imines. *Org. Chem. Front.* **2014**, *1*, 148–150. [CrossRef]
58. Young, C.M.; Taylor, J.E.; Smith, A.D. Evaluating aryl esters as bench-stable C (1)-ammonium enolate precursors in catalytic, enantioselective Michael addition–lactonisations. *Org. Biomol. Chem.* **2019**, *17*, 4747–4752. [CrossRef] [PubMed]
59. Hartley, W.C.; O'Riordan, T.J.C.; Smith, A.D. Aryloxide-Promoted Catalyst Turnover in Lewis Base Organocatalysis. *Synthesis* **2017**, *49*, 3303–3310.
60. West, T.H.; Daniels, D.S.B.; Slawin, A.M.Z.; Smith, A.D. An isothiourea-catalyzed asymmetric [2, 3]-rearrangement of allylic ammonium ylides. *J. Am. Chem. Soc.* **2014**, *136*, 4476–4479. [CrossRef]
61. West, T.H.; Walden, D.M.; Taylor, J.E.; Brueckner, A.C.; Johnston, R.C.; Cheong, P.H.-Y.; Lloyd-Jones, G.C.; Smith, A.D. Catalytic Enantioselective [2, 3]-Rearrangements of Allylic Ammonium Ylides: A Mechanistic and Computational Study. *J. Am. Chem. Soc.* **2017**, *139*, 4366–4375. [CrossRef]
62. Spoehrle, S.S.M.; West, T.H.; Taylor, J.E.; Slawin, A.M.Z.; Smith, A.D. Tandem Palladium and Isothiourea Relay Catalysis: Enantioselective Synthesis of α-Amino Acid Derivatives via Allylic Amination and [2, 3]-Sigmatropic Rearrangement. *J. Am. Chem. Soc.* **2017**, *139*, 11895–11902. [CrossRef]
63. Kasten, K.; Slawin, A.M.Z.; Smith, A.D. Enantioselective Synthesis of β-Fluoro-β-aryl-α-aminopentenamides by Organocatalytic [2, 3]-Sigmatropic Rearrangement. *Org. Lett.* **2017**, *19*, 5182–5185. [CrossRef]
64. West, T.H.; Spoehrle, S.S.M.; Smith, A.D. Isothiourea-catalysed chemo-and enantioselective [2, 3]-sigmatropic rearrangements of N, N-diallyl allylic ammonium ylides. *Tetrahedron* **2017**, *73*, 4138–4149. [CrossRef]
65. Schwarz, K.J.; Amos, J.L.; Klein, J.C.; Do, D.T.; Snaddon, T.N. Uniting C1-ammonium enolates and transition metal electrophiles via cooperative catalysis: The direct asymmetric α-allylation of aryl acetic acid esters. *J. Am. Chem. Soc.* **2016**, *138*, 5214–5217. [CrossRef]
66. Schwarz, K.J.; Pearson, C.M.; Cintron-Rosado, G.A.; Liu, P.; Snaddon, T.N. Traversing Steric Limitations by Cooperative Lewis Base/Palladium Catalysis: An Enantioselective Synthesis of α-Branched Esters Using 2-Substituted Allyl Electrophiles. *Angew. Chem.* **2018**, *130*, 7926–7929. *Angew. Chem.* **2018**, *57*, 7800–7803. [CrossRef]

67. Scaggs, W.R.; Snaddon, T.N. Enantioselective α-Allylation of Acyclic Esters using B(pin)-Substituted Electrophiles: Independent Regulation of Stereocontrol Elements via Cooperative Pd/Lewis Base Catalysis. *Chem. Eur. J.* **2018**, *24*, 14378–14381. [CrossRef]
68. Fyfe, J.W.B.; Kabia, O.M.; Pearson, C.M.; Snaddon, T.N. Si-directed regiocontrol in asymmetric Pd-catalyzed allylic alkylations using C1-ammonium enolate nucleophiles. *Tetrahedron* **2018**, *74*, 5383–5391. [CrossRef]
69. Hutchings-Goetz, J.; Yang, C.; Snaddon, T.N. Enantioselective Syntheses of Strychnos and Chelidonium Alkaloids through Regio- and Stereocontrolled Cooperative Catalysis. *ACS Catal.* **2018**, *8*, 10537–10544. [CrossRef]
70. Pearson, C.M.; Fyfe, J.W.B.; Snaddon, T.N. A Regio- and Stereodivergent Synthesis of Homoallylic Amines by a One-Pot Cooperative-Catalysis-Based Allylic Alkylation/Hofmann Rearrangement Strategy. *Angew. Chem. Int. Ed.* **2019**, *58*, 10521–10527. *Angew. Chem.* **2019**, *131*, 10631–10637. [CrossRef]
71. Scaggs, W.R.; Scaggs, T.D.; Snaddon, T.N. An enantioselective synthesis of α-alkylated pyrroles via cooperative isothiourea/palladium catalysis. *Org. Biomol. Chem.* **2019**, *17*, 1787–1790. [CrossRef]
72. Schwarz, K.J.; Yang, C.; Fyfe, J.W.B.; Snaddon, T.N. Enantioselective α-Benzylation of Acyclic Esters Using π-Extended Electrophiles. *Angew. Chem. Int. Ed.* **2018**, *57*, 12102–12105. *Angew. Chem.* **2018**, *130*, 12278–12281. [CrossRef] [PubMed]
73. Jiang, X.; Beiger, J.J.; Hartwig, J.F. Stereodivergent allylic substitutions with aryl acetic acid esters by synergistic iridium and Lewis base catalysis. *J. Am. Chem. Soc.* **2017**, *139*, 87–90. [CrossRef] [PubMed]
74. Song, J.; Zhang, Z.J.; Gong, L.Z. Asymmetric [4+2] Annulation of C1 Ammonium Enolates with Copper-Allenylidenes. *Angew. Chem. Int. Ed.* **2017**, *56*, 5212–5216. *Angew. Chem.* **2017**, *129*, 5296–5300. [CrossRef] [PubMed]
75. Song, J.; Zhang, Z.J.; Chen, S.S.; Fan, T.; Gong, L.Z. Lewis base/copper cooperatively catalyzed asymmetric α-amination of esters with diaziridinone. *J. Am. Chem. Soc.* **2018**, *140*, 3177–3180. [CrossRef] [PubMed]
76. Arokianathar, J.N.; Frost, A.B.; Slawin, A.M.Z.; Stead, D.; Smith, A.D. Isothiourea-Catalyzed Enantioselective Addition of 4-Nitrophenyl Esters to Iminium Ions. *ACS Catal.* **2018**, *8*, 1067–1075. [CrossRef]
77. Zhao, F.; Shu, C.; Young, C.M.; Carpenter-Warren, C.; Slawin, A.M.Z.; Smith, A.D. Enantioselective Synthesis of α-Aryl-β2-Amino-Esters by Cooperative Isothiourea and Brønsted Acid Catalysis. *Angew. Chem. Int. Ed.* **2021**, *60*, 11892–11900. *Angew. Chem.* **2021**, *133*, 11999–12007. [CrossRef]
78. McLaughlin, C.; Slawin, A.M.Z.; Smith, A.D. Base-free Enantioselective C (1)-Ammonium Enolate Catalysis Exploiting Aryloxides: A Synthetic and Mechanistic Study. *Angew. Chem. Int. Ed.* **2019**, *131*, 15255–15263. *Angew. Chem.* **2019**, *58*, 15111–15119. [CrossRef]
79. McLaughlin, C.; Bitai, J.; Barber, L.; Slawin, A.M.Z.; Smith, A.D. Catalytic enantioselective synthesis of 1, 4-dihydropyridines via the addition of C (1)-ammonium enolates to pyridinium salts. *Chem. Sci.* **2021**. accepted for publication. [CrossRef]
80. Stockhammer, L.; Weinzierl, D.; Bögl, T.; Waser, M. Enantioselective α-Chlorination Reactions of in Situ Generated C1 Ammonium Enolates under Base-Free Conditions. *Org. Lett.* **2021**, *23*, 6143–6147. [CrossRef]
81. Kim, B.; Kim, Y.; Lee, S.Y. Stereodivergent Carbon–Carbon Bond Formation between Iminium and Enolate Intermediates by Synergistic Organocatalysis. *J. Am. Chem. Soc.* **2021**, *143*, 73–79.
82. Yuan, S.; Liao, C.; Zheng, W.-H. Paracyclophane-Based Isothiourea-Catalyzed Highly Enantioselective α-Fluorination of Carboxylic Acids. *Org. Lett.* **2021**, *23*, 4142–4146. [CrossRef]
83. Crystallographic data for (2S,1′R)-19 (CCDC 1992504), Cambridge Crystallographic Data Centre.
84. The relative and absolute configuration of all other products was assigned by analogy, with consistent diagnostic ^{1}H NMR signals of the proton at C(2) and C(1′) stereocentres, indicating the same major diastereoisomer in each case.
85. Morrill, L.C.; Douglas, J.; Lebl, T.; Slawin, A.M.Z.; Fox, D.J.; Smith, A.D. Isothiourea-mediated asymmetric Michael-lactonisation of trifluoromethylenones: A synthetic and mechanistic study. *Chem. Sci.* **2013**, *4*, 4146–4155. [CrossRef]
86. Matviitsuk, A.; Greenhalgh, M.D.; Antúnez, D.-J.B.; Slawin, A.M.Z.; Smith, A.D. Aryloxide-Facilitated Catalyst Turnover in Enantioselective α, β-Unsaturated Acyl Ammonium Catalysis. *Angew. Chem. Int. Ed.* **2017**, *56*, 12282–12287. *Angew. Chem.* **2017**, *129*, 12450–12455. [CrossRef]
87. Catalyst turnover could also be envisaged through intramolecular nucleophilic displacement by the pendant phenol to give a cyclobutanone intermediate via a formal [2+2]-cycloaddition, which may then be ring-opened by *para*-nitrophenoxide to give the final product. Based on the high steric congestion of this proposed cyclobutanone intermediate, the mechanism given in Scheme 5 is considered more probable.
88. Liu, P.; Yang, X.; Birman, V.B.; Houk, K.N. Origin of enantioselectivity in benzotetramisole-catalyzed dynamic kinetic resolution of azlactones. *Org. Lett.* **2012**, *14*, 3288–3291. [CrossRef] [PubMed]
89. Abbasov, M.E.; Hudson, B.M.; Tantillo, D.J.; Romo, D. Acylammonium salts as dienophiles in Diels–Alder/lactonization organocascades. *J. Am. Chem. Soc.* **2014**, *136*, 4492–4495. [CrossRef] [PubMed]
90. Robinson, E.R.T.; Walden, D.M.; Fallan, C.; Greenhalgh, M.D.; Cheong, P.H.-Y.; Smith, A.D. Non-bonding 1,5-S···O interactions govern chemo- and enantioselectivity in isothiourea-catalyzed annulations of benzazoles. *Chem. Sci.* **2016**, *7*, 6919–6927. [CrossRef] [PubMed]
91. Greenhalgh, M.D.; Smith, S.M.; Walden, D.M.; Taylor, J.E.; Brice, Z.; Robinson, E.R.T.; Fallan, C.; Cordes, D.B.; Slawin, A.M.Z.; Richardson, H.C.; et al. AC= O··· Isothiouronium Interaction Dictates Enantiodiscrimination in Acylative Kinetic Resolutions of Tertiary Heterocyclic Alcohols. *Angew. Chem. Int. Ed.* **2018**, *57*, 3200–3206. *Angew. Chem.* **2018**, *130*, 3254–3260. [CrossRef]
92. Young, C.M.; Elmi, A.; Pascoe, D.J.; Morris, R.K.; McLaughlin, C.; Woods, A.M.; Frost, A.B.; de la Houpliere, A.; Ling, K.B.; Smith, T.K.; et al. The Importance of 1, 5-Oxygen Chalcogen Interactions in Enantioselective Isochalcogenourea Catalysis. *Angew. Chem.* **2020**, *132*, 3734–3739. *Angew. Chem.* **2020**, *59*, 3705–3710. [CrossRef]

93. Pascoe, D.J.; Ling, K.B.; Cockroft, S.L. The origin of chalcogen-bonding interactions. *J. Am. Chem. Soc.* **2017**, *139*, 15160–15167. [CrossRef]
94. Nagao, Y.; Miyamoto, S.; Miyamoto, M.; Takeshige, H.; Hayashi, K.; Sano, S.; Shiro, M.; Yamaguchi, K.; Sei, Y. Highly Stereoselective Asymmetric Pummerer Reactions That Incorporate Intermolecular and Intramolecular Nonbonded S···O Interactions. *J. Am. Chem. Soc.* **2006**, *128*, 9722–9729. [CrossRef]
95. Beno, B.R.; Yeung, K.-S.; Bartberger, M.D.; Pennington, L.D.; Meanwell, N.A. A survey of the role of noncovalent sulfur interactions in drug design. *J. Med. Chem.* **2015**, *58*, 4383–4438. [CrossRef]
96. Breugst, M.; Koenig, J.J. σ-Hole Interactions in Catalysis. *Eur. J. Org. Chem.* **2020**, *34*, 5473–5487. [CrossRef]
97. Bleiholder, C.; Gleiter, R.; Werz, D.B.; Köppel, H. Theoretical Investigations on Heteronuclear Chalcogen—Chalcogen Interactions: On the Nature of Weak Bonds between Chalcogen Centers. *Inorg. Chem.* **2007**, *46*, 2249–2260. [CrossRef]
98. Gleiter, R.; Haberhauer, G.; Werz, D.B.; Rominger, F. From noncovalent chalcogen–chalcogen interactions to supramolecular aggregates: Experiments and calculations. *Bleiholder. Chem. Rev.* **2018**, *118*, 2010–2041. [CrossRef] [PubMed]
99. Benz, S.; López-Andarias, J.; Mareda, J.; Sakai, N.; Matile, S. Catalysis with chalcogen bonds. *Angew. Chem.* **2017**, *129*, 830–833. *Angew. Chem.* **2017**, *56*, 812–815. [CrossRef]
100. Wonner, P.; Vogel, L.; Düser, M.; Gomes, L.; Kniep, F.; Mallick, B.; Werz, D.B.; Huber, S.M. Carbon–Halogen Bond Activation by Selenium-Based Chalcogen Bonding. *Angew. Chem. Int. Ed.* **2017**, *56*, 12009–12012. *Angew. Chem.* **2017**, *129*, 12172–12176. [CrossRef] [PubMed]
101. Wonner, P.; Vogel, L.; Kniep, F.; Huber, S.M. Catalytic Carbon–Chlorine Bond Activation by Selenium-Based Chalcogen Bond Donors. *Chem. Eur. J.* **2017**, *23*, 16972–16975. [CrossRef]
102. Wonner, P.; Dreger, A.; Vogel, L.; Engelage, E.; Huber, S.M. Chalcogen Bonding Catalysis of a Nitro-Michael Reaction. *Angew. Chem. Int. Ed.* **2019**, *58*, 16923–16927. *Angew. Chem.* **2019**, *131*, 17079–17083. [CrossRef]
103. Wang, W.; Zhu, H.; Liu, S.; Zhao, Z.; Zhang, L.; Hao, J.; Wang, Y. Chalcogen–chalcogen bonding catalysis enables assembly of discrete molecules. *J. Am. Chem. Soc.* **2019**, *141*, 9175–9179. [CrossRef]
104. Wang, W.; Zhu, H.; Feng, L.; Yu, Q.; Hao, J.; Zhu, R.; Wang, Y. Dual Chalcogen–Chalcogen Bonding Catalysis. *J. Am. Chem. Soc.* **2020**, *142*, 3117–3124. [CrossRef]

Article
Azetidinium Lead Halide Ruddlesden–Popper Phases

Jiyu Tian [1,2], Eli Zysman-Colman [2,*] and Finlay D. Morrison [1,*]

1 EaStCHEM School of Chemistry, University of St Andrews, St Andrews KY16 9ST, UK; jt201@st-andrews.ac.uk
2 Organic Semiconductor Centre, EaStCHEM School of Chemistry, University of St Andrews, St Andrews KY16 9ST, UK
* Correspondence: eli.zysman-colman@st-andrews.ac.uk (E.Z.-C.); finlay.morrison@st-andrews.ac.uk (F.D.M.)

Abstract: A family of Ruddlesden–Popper ($n = 1$) layered perovskite-related phases, $Az_2PbCl_xBr_{4-x}$ with composition $0 \leq x \leq 4$ were obtained using mechanosynthesis. These compounds are isostructural with K_2NiF_4 and therefore adopt the idealised $n = 1$ Ruddlesden–Popper structure. A linear variation in unit cell volume as a function of anion average radius is observed. A tunable bandgap is achieved, ranging from 2.81 to 3.43 eV, and the bandgap varies in a second-order polynomial relationship with the halide composition.

Keywords: layered perovskite; bandgap tuning; azetidinium; Ruddlesden–Popper; structure-property relations

Citation: Tian, J.; Zysman-Colman, E.; Morrison, F.D. Azetidinium Lead Halide Ruddlesden–Popper Phases. *Molecules* **2021**, *26*, 6474. https://doi.org/10.3390/molecules26216474

Academic Editors: William T. A. Harrison, R. Alan Aitken and Paul Waddell

Received: 29 September 2021
Accepted: 21 October 2021
Published: 27 October 2021

Publisher's Note: MDPI stays neutral with regard to jurisdictional claims in published maps and institutional affiliations.

Copyright: © 2021 by the authors. Licensee MDPI, Basel, Switzerland. This article is an open access article distributed under the terms and conditions of the Creative Commons Attribution (CC BY) license (https://creativecommons.org/licenses/by/4.0/).

1. Introduction

Ruddlesden–Popper (R–P) phases are composed of layered perovskite structures with alternating layers of AMX_3 perovskite and AX rock salt along the c-axis. They are described by the general formula $A_{n+1}M_nX_{3n+1}$ (or $A'_2A''_{n-1}M_nX_{3n+1}$ in the case of two distinct A-cations), where n is a positive integer representing the number of perovskite layers that are separated by additional 'A-cation excess' rock-salt layers [1,2]. Importantly, the intergrowth rock salt layer means that the octahedra in the perovskite layers are aligned in the successive layers. In 1955, Balz and Plieth reported the first R–P phase layered structure K_2NiF_4 ($n = 1$) [3]. In 1957, Ruddlesden and Popper reported a series of layered structures in oxides, such as Sr_2TiO_4 and Ca_2TiO_4 [4]. Nowadays, the R–P phase is more commonly used to represent this type of layered perovskite structure and, increasingly, in organic–inorganic hybrid perovskites (OIHPs). Several families of layered OIHPs containing alternating layers of AMX_3 perovskite and organic cations with structures similar to R–P phases have been reported. Such examples of layered OIHPs include BA_2PbI_4 (BA = $C_4H_9NH_3^+$) [5] and PEA_2PbX_4 (PEA = $C_8H_{12}N^+$, X = Cl, Br, I) [6,7] in which the organic cations are too big to be accommodated in the cuboctahedral cavities of the 3D MX_6 framework. Without the constraint of the size of the cuboctahedral cavities, a wider range of organic A-cations would be available for layered phases. In addition, by mixing large (A') organic cations, such as those mentioned above, and small organic cations such as methylammonium (A'' = MA), organic-inorganic hybrid materials with the general formula $A'_2A''_{n-1}M_nX_{3n+1}$ can be prepared [5,8]. They show good bandgap tunability by modifying the number of layers (n) of $A''PbX_3$. Stoumpos et al. [5] reported orthorhombic crystal structures of $BA_2MA_{n-1}Pb_nX_{3n+1}$ (X = Br, I) with bandgaps changing progressively from 2.43 eV ($n = 1$) to 1.50 eV ($n = \infty$), with intermediate values of 2.17 eV ($n = 2$), 2.03 eV ($n = 3$) and 1.91 eV ($n = 4$). The thickness of the perovskite layer, n, in $(BA)_2(MA)_{n-1}Pb_nI_{3n+1}$ can be reasonably controlled by modifying the ratio of BA/MA cations in the precursor solutions. However, many so-called R–P phases reported in such compounds often do not have the required rock salt-structured interlayer between the 2D perovskite layers, resulting in an offset in the alignment of the perovskite blocks in successive layers. Such examples, therefore, do not conform to the definition of an R–P phase and are more correctly termed R–P-like

OIHPs. Such R–P-like layered OIHPs have demonstrated higher stability when exposed to light, humidity and heat stress compared to 3D perovskite analogues, which are prone to unwanted phase transition under these test conditions [9,10]. For example, Ren et al. reported an R–P-like OIHPs solar cell material with general formula $(MTEA)_2(MA)_4Pb_5I_{16}$ ($n = 5$) which achieved a power conversion efficiency up to 17.8% [11]. Their cells retained over 85% of the initial efficiency after 1000 h operation time.

Azetidinium (Az^+, $(CH_2)_3NH_3^+$) is a four-membered ring ammonium cation. In our previous study on mixed halide azetidinium lead perovskites, $AzPbBr_{3-x}X_x$ (X = Cl or I), the structure progresses from 6H to 4H to 9R perovskite polytypes with varying halide composition from Cl^- to Br^- to I^- [12]. The fact that $AzPbX_3$ (X = Cl or Br) forms a hexagonal perovskite rather than a cubic (3C) perovskite led to our study on mix-cation solid solutions of the form $AzA''PbBr_3$, $A'' = MA^+$ or FA^+ (FA^+ = formamidinium). Such systems show only partial solid solutions and phase separation of the hexagonal and cubic forms; the extent of solid solution formation also depends on the synthesis route [13]. These studies also suggest that the cation radius of Az^+ is ~310 pm, which is larger than the calculated cation radius of Az, r_{Az} = 250 pm (for comparison the reported radii for FA^+ and MA^+ are r_{FA} = 253 pm, r_{MA} = 217 pm [14], respectively). MA^+ and FA^+ are commonly used as A-site cations in OIHPs, and that adopt (pseudo-) cubic perovskite structures [15,16]. With our cation radius estimation that Az^+ is larger than MA^+ and FA^+, Az_2PbX_4 (X = Cl, Br) are found to adopt a $n = 1$ R–P phase structure. The fact that Az^+ can form a layered structure indicates that our estimation of its cation radius is more accurate than that from the computational calculation [13,14]. Furthermore, a family of mixed halide R–P phases, $Az_2PbCl_xBr_{4-x}$ with composition $0 \leq x \leq 4$ were prepared by mechanosynthesis and their structures and optical properties were analysed by powder X-ray diffraction (PXRD) and absorption spectroscopy, respectively. A linear variation in unit cell volume as a function of anion average radius is observed. The band gap was found to range from 2.81 to 3.43 eV, which varies as a second-order polynomial relationship with the halide composition.

2. Method

$PbBr_2$ (98%) and $PbCl_2$ (98%) were purchased from Alfa Aesar. Hydrobromic acid in water (48%) and AzCl (95%) were purchased from Fluorochem. All other reagents and solvents were obtained from commercial sources and used as received. AzBr were synthesised according to our previous study [17].

Preparation of $Az_2PbCl_xBr_{4-x}$ solid solutions with $0 \leq x \leq 4$ (in x = 0.67 increments) was carried out by mechanosynthesis. Appropriate molar ratios of dry AzX and PbX_2 ($AzX:PbX_2$ = 2:1, X = Cl or Br) were ground together in a Fritsch Pulverisette planetary ball mill at 600 rpm for 1 h using 60 cm^3 Teflon pots and high-wear-resistant zirconia media (nine 10 mm diameter spheres). Az_2PbBr_4 samples could also be obtained by hand grinding AzBr and $PbBr_2$ in an agate mortar and pestle for 25 min.

PXRD was carried out using a PANalytical Empyrean diffractometer with Cu $K_{\alpha 1}$ (λ = 1.5406 Å). Rietveld refinements of PXRD data using GSAS [18] were used to confirm phase formation and for the determination of lattice parameters.

Optical properties were determined from solid-state absorption spectra recorded using a Shimadzu UV-2600 spectrophotometer and bandgaps were calculated by plotting $(\alpha h\nu)^2 (cm^{-1} \cdot eV)^2$ with $h\nu$(eV) according to the Tauc method, in which α, h and ν stand for absorbance, Planck's constant and incident light frequency.

3. Results

The PXRD data for $Az_2PbCl_xBr_{4-x}$ with compositions ranging from $0 \leq x \leq 4$ were prepared by mechanosynthesis and are shown in Figure 1b. The structures of these samples were determined to be R–P $n = 1$ phase in the $I4/mmm$ space group (Figure 1a). The theoretical diffraction pattern of the tetragonal R–P phase is shown in Figure S1. Characteristic peaks of the R–P phase show systematic peak shifts to higher 2θ angle from Az_2PbBr_4 to Az_2PbCl_4, which indicate the lattice parameters decreased with more Cl content in the solid solution. The Az$^+$ cations, which are represented as solid spheres situated at the centre of electron density, form rock salt layers with the X$^-$ anions. Synthesis from solution is preferred when manufacturing devices because solutions can be easily processed into thin films by spin-coating and blade-coating methods compared to bulk powder [19]. Thus, precipitation synthesis of Az$_2$PbX$_4$ (X = Cl, Br) were also attempted (synthetic details included in the supporting information) and their PXRD data are shown in Figure S2. Although the precipitated samples contain additional phase(s) associated with additional peaks (e.g., at 6° and 11°) and have yet to be assigned to a structure. Ganguli [20] reported an empirical prediction that possible R–P phase structures are associated with a ratio of A-site and metal cation radii (r_A/r_M) in the range of 1.7 to 2.4. As discussed in our previous study [12], our estimation of the cation radius of Az$^+$ (~310 pm) differs from that calculated (250 pm) [14]. The r_{Az}/r_{Pb} calculated using our estimated radius is 2.60, while that using the literature value [14] is 2.10.

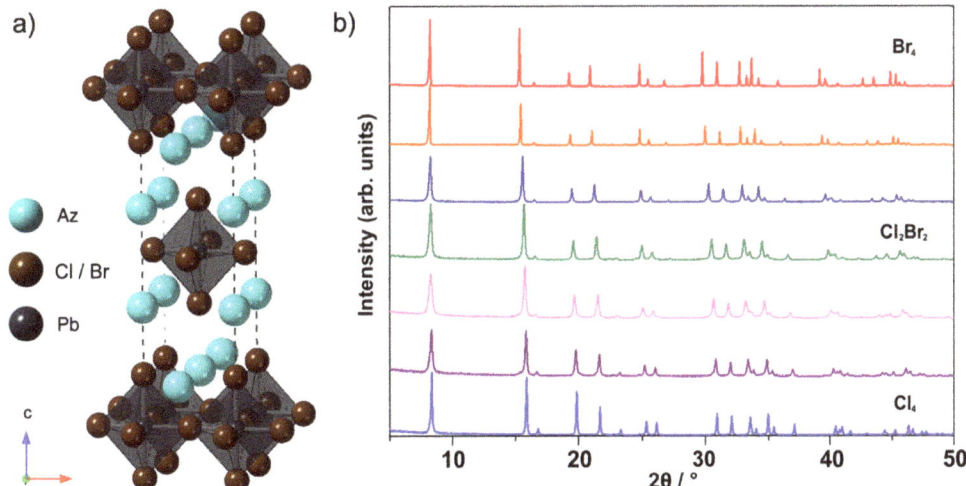

Figure 1. (a) n = 1 Ruddlesden–Popper (R–P) phase of Az$_2$PbX$_4$ (X = Cl, Br) showing alternating AzPbX$_3$ perovskite and AzX rock salt layers along the c-axis, (b) PXRD data of mix-halide layered R–P phases: Az$_2$PbCl$_x$Br$_{4-x}$ with composition $0 \leq x \leq 4$ prepared by mechanosynthesis.

Unfortunately, our attempts to synthesise single-phase Az$_2$PbI$_4$ were unsuccessful. The PXRD of mechanosynthesised Az$_2$PbI$_4$ is shown in Figure S3. In addition to the R–P phase, there are evident amounts of 9R AzPbI$_3$ phase [12,21] and the relative intensity of this phase increased with increased ball mill grinding time (1 to 3 h). PXRD of the Az$_2$PbI$_4$ sample obtained from a hand grinding synthesis showed that this method can increase the proportion of R–P phase in the samples, evidenced by the increased relative intensity of peaks associated with the R–P phase, but the presence of the 9R phase persisted across all samples. These results indicate that the 9R phase is the more stable phase compared to the R–P phase for the iodide analogue It is likely that the activation energy for the transformation of azetidinium lead iodide from a layered phase to the 9R phase is low.

For simplicity, Rietveld refinements were carried out by replacing the organic Az$^+$ cations with Mn^{2+}, as they have similar electron densities. Figure 2 shows an example of the PXRD data refinement of Az$_2$PbX$_4$ (X = Cl, Br) samples obtained from the ball mill mechanosynthesis. The refined lattice parameters of Az$_2$PbBr$_4$ are a = 5.993(6) Å and c = 21.501(1) Å, with goodness-of-fit parameters χ^2 = 10.21 and wR_p = 0.115, while those of Az$_2$PbCl$_4$ are a = 5.765(0) Å and c = 21.027(2) Å, with goodness-of-fit parameters χ^2 = 7.20 and wR_p = 0.102. The difference between the organic moieties and Mn^{2+}, which is associated with their actual atomic position and thermal motion, is one possible reason for such high χ^2 values for both refinements and may be responsible for the differences in the peak shape and intensities shown. Single crystal diffraction analysis is required for detailed structural analysis, including accurate atoms positions (particularly of the Az$^+$ cation), however, this would require preparation of sufficiently large single crystals which are challenging by this mechanosynthesis route. Nevertheless, it is clear from the rudimentary Rietveld analysis of the PXRD data that all peaks are accounted for and that the PXRD unambiguously show the formation of n = 1 R–P materials. In addition, as the peaks positions can be determined accurately the unit cell dimensions are reliable.

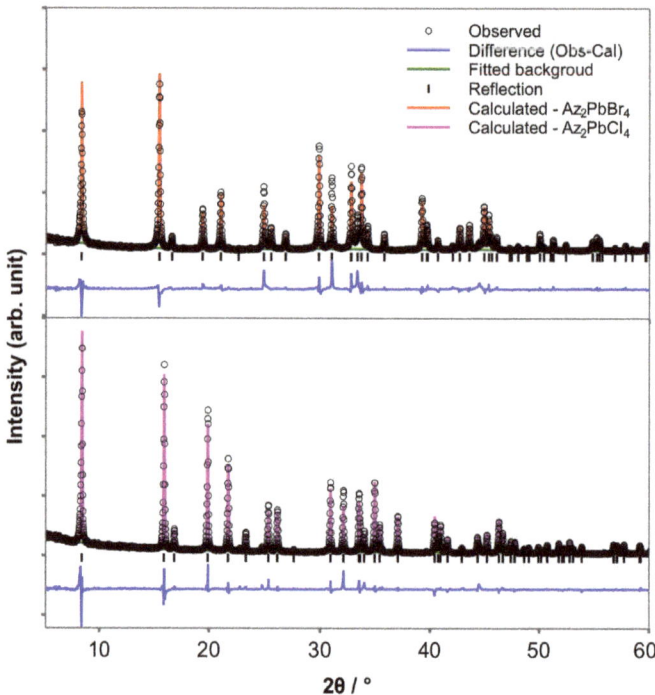

Figure 2. Rietveld refinement of PXRD data in $I4/mmm$ space group of Az$_2$PbX$_4$, X = Br (top) and Cl (bottom) obtained from mechanosynthesis with observed data (open circles), calculated data (red line for Br and magenta line for Cl), background (green lines), reflection positions (black bars) and difference plots (blue lines).

To study the mixed-halide solid solutions Az$_2$PbCl$_x$Br$_{4-x}$, the lattice parameters of each mechanosynthesised composition were determined by Rietveld refinement of PXRD data. The cell volume of these R–P phases varies linearly as a function of the average anion radius, Figure 3a (the average anion radius was calculated using r_{Br} = 196 pm and r_{Cl} = 181 pm according to Shannon [22]). This linear variation is expected in accordance with Vegard's law. The lattice parameters a and c, on the other hand, show a nonlinear relationship with the average anion radius (Figure 3b), which suggests anisotropic expansion/contraction along the a- and c-axis. The larger expansion in a is consistent with the increased X anion radius which affords a larger void for the Az$^+$ cation, resulting in less required expansion in the interlayer spacing. Based on the analysis using Mn^{2+} as a proxy for Az$^+$ we have no information regarding any orientation or dynamics of the Az$^+$ cation.

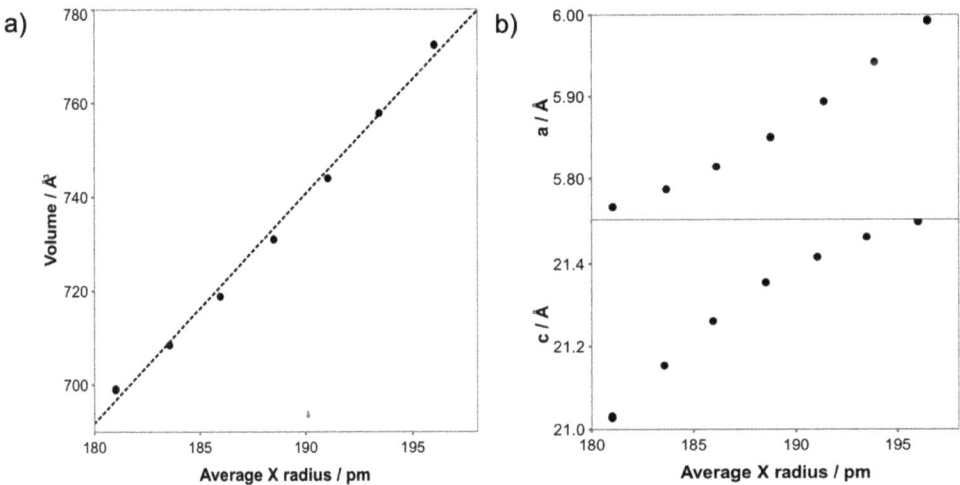

Figure 3. (a) Cell volume, (b) lattice parameters as a function of average halide anion radius for $n = 1$ R–P phases $Az_2PbCl_xBr_{4-x}$ ($0 \leq x \leq 4$) as determined from Rietveld refinement of PXRD data.

One of the benefits of mechanosynthesis is that all materials are retained during the reaction, so the overall starting composition must be retained in the post-reaction compound(s). By inference, any product(s) must have the nominal starting composition. While we do not have direct compositional analysis, the PXRD results, Figure 2, clearly show that the product formed is entirely $n = 1$ R–P phase. It has been reported that the actual composition shows a good match with the nominal composition in the mechanosynthesis of OIHPs [23,24]. Thus, the halide compositions of $Az_2PbCl_xBr_{4-x}$ are calculated according to the molar ratios of the raw materials (nominal composition).

The optical properties of $Az_2PbCl_xBr_{4-x}$ ($0 \leq x \leq 4$) solid solutions were studied by absorption spectroscopy (Figure 4a). The absorption onsets are systematically red-shifted from ca. 386 nm (Az_2PbCl_4) to ca. 457 nm (Az_2PbBr_4) with increasing average anion size (from Cl^- to Br^-). The bandgaps of Az_2PbCl_4 and Az_2PbBr_4 are calculated to be 3.43 and 2.81 eV, which are the same (within error) as the bandgap of the 6H hexagonal perovskite $AzPbCl_3$ (3.43 eV) and $AzPbBr_3$ (2.81 eV) [12]. However, unlike the linear variation in the 6H $AzPbX_3$ ($X^- = Cl^-, Br^-$), the bandgap of layered R–P Az_2PbX_4 (X = Cl, Br) shows a bowing with the average anion radius (Figure 4b). The bowing effect [25,26] simply describes the deviation of the measured band gap in continuous solid solutions from the values expected by linear interpolation of the end member values. Band gap bowing is often fitted to a second-order polynomial to account for the divergence from linearity, with a bowing parameter b as the binominal coefficient of the fitting Equation (1): [26]

$$E_g(x) = (1-x)E_{g|(x=0)} + xE_{g|(x=1)} - bx(1-x) \tag{1}$$

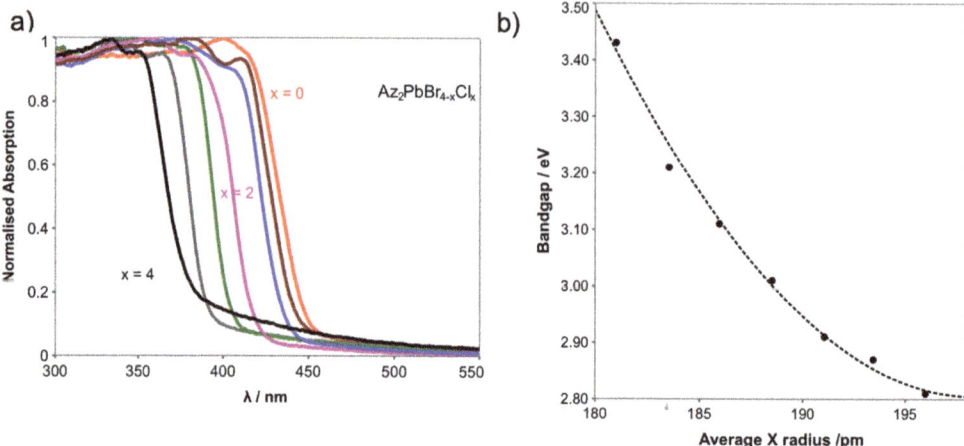

Figure 4. (a) Absorption spectra; (b) bandgap determination from the absorption spectra of samples $Az_2PbCl_xBr_{4-x}$ with composition $0 \leq x \leq 4$ plotted as a function of average halide anion radius.

The bowing parameter, b, of the mechanosynthesised mixed halide layered $Az_2PbCl_xBr_{4-x}$ ($0 \leq x \leq 4$) is 0.47 with a goodness-of-fit R^2 value of 0.995. The bowing parameter of mixed halide OIHPs are usually smaller, variously reported as 7×10^{-4} to 0.33 for $MAPbBr_{3-x}X_x$ (X = Cl or I), [27,28] compared to the bowing parameters (0.4 to 1.33) found for other mixed metal perovskite systems such as $MA_3(Sb_{1-x}Bi_x)I_9$ (0.4 for Bi rich region and 1.3 for Sb rich region) and 1.06 for $MA(Pb_{1-x}Sn_x)I_3$ [25,26,29].

4. Conclusions

$n = 1$ Ruddlesden–Popper (R–P) layered perovskite phases were successfully obtained by mechanosynthesis in the mixed halide solid solution $Az_2PbCl_xBr_{4-x}$ with composition $0 \leq x \leq 4$. Az_2PbX_4 (X = Cl, Br) was determined to be the conventional R–P $n = 1$ (K_2NiF_4) structure with a space group of $I4/mmm$. A linear variation in unit cell volume as a function of anion average radius is observed. The band gap of the R–P phases Az_2PbCl_4 and Az_2PbBr_4 are determined to be 3.43 and 2.81 eV, which is the same (within error) as the bandgap of 6H hexagonal perovskite $AzPbCl_3$ (3.43 eV) and $AzPbBr_3$ (2.81 eV) [12]. A bowing effect with a bowing parameter of 0.47 is observed in the band gap-composition relationship of R–P layered mixed halide solid solutions, compared to the linear relationship observed in the 6H hexagonal perovskite.

Supplementary Materials: The following are available online. Supporting Information data include synthetic details of precipitation synthesis of Az_2PbX_4 (X = Cl, Br) (Figures S1 and S2) and synthesis of Az_2PbI_4 (Figure S3). Also, include selected crystallographic data obtained powder X-ray diffraction of samples prepared by mechanosynthesis (Table S1).

Author Contributions: Conceptualization, J.T., E.Z.-C. and F.D.M.; methodology, J.T.; validation, J.T., E.Z.-C. and F.D.M.; formal analysis, J.T.; data curation, J.T.; writing—original draft preparation, J.T.; supervision, E.Z.-C. and F.D.M. All authors have read and agreed to the published version of the manuscript.

Funding: This research received no external funding. The publication of this work received support from the St Andrews Institutional Open Access Fund.

Institutional Review Board Statement: Not applicable.

Informed Consent Statement: Not applicable.

Data Availability Statement: The research data supporting this publication can be accessed at https://doi.org/10.17630/fd5aab9b-fced-4926-afee-5eb56e2e6a5e (accessed on 15 October 2021).

Acknowledgments: We thank the Chinese Scholarship Council for support to JT (CSC No. 201603780020).

Conflicts of Interest: The authors declare no conflict of interest.

Sample Availability: Samples of the $Az_2PbCl_xBr_{4-x}$ with composition $0 \leq x \leq 4$ are available from the authors.

References

1. Lichtenberg, F.; Herrnberger, A.; Wiedenmann, K. Synthesis, structural, magnetic and transport properties of layered perovskite-related titanates, niobates and tantalates of the type $A_nB_nO_{3n+2}$, $A'A_{k-1}B_kO_{3k+1}$ and $A_mB_{m-1}O_{3m}$. *Prog. Solid State Chem.* **2008**, *36*, 253–387. [CrossRef]
2. Aleksandrov, K.S.; Beznosikov, V.V. Hierarchies of perovskite-like crystals (Review). *Phys. Solid State* **1997**, *39*, 695–715. [CrossRef]
3. Balz, D.; Plieth, K. Die Struktur des Kaliumnickelfluorids, K_2NiF. *Z. Elektrochem. Ber. Bunsenges. Phys. Chem.* **1955**, *59*, 545–551. [CrossRef]
4. Ruddlesden, S.N.; Popper, P. New compounds of the K_2NiF_4 type. *Acta Crystallogr.* **1957**, *10*, 538–539. [CrossRef]
5. Stoumpos, C.C.; Cao, D.H.; Clark, D.J.; Young, J.; Rondinelli, J.M.; Jang, J.I.; Hupp, J.T.; Kanatzidis, M.G. Ruddlesden-Popper hybrid lead iodide perovskite 2D homologous semiconductors. *Chem. Mater.* **2016**, *28*, 2852–2867. [CrossRef]
6. Du, K.Z.; Tu, Q.; Zhang, X.; Han, Q.; Liu, J.; Zauscher, S.; Mitzi, D.B. Two-dimensional lead(II) halide-based hybrid perovskites templated by acene alkylamines: Crystal structures, optical properties, and piezoelectricity. *Inorg. Chem.* **2017**, *56*, 9291–9302. [CrossRef]
7. Mitzi, D.B. A layered solution crystal growth technique and the crystal structure of $(C_6H_5C_2H_4NH_3)_2PbCl_4$. *J. Solid State Chem.* **1999**, *145*, 694–704. [CrossRef]
8. Spanopoulos, I.; Hadar, I.; Ke, W.; Tu, Q.; Chen, M.; Tsai, H.; He, Y.; Shekhawat, G.; Dravid, V.P.; Wasielewski, M.R.; et al. Uniaxial expansion of the 2D Ruddlesden–Popper perovskite family for improved environmental stability. *J. Am. Chem. Soc.* **2019**, *141*, 5518–5534. [CrossRef]
9. Tsai, H.; Nie, W.; Blancon, J.C.; Stoumpos, C.C.; Asadpour, R.; Harutyunyan, B.; Neukirch, A.J.; Verduzco, R.; Crochet, J.J.; Tretiak, S.; et al. High-efficiency two-dimensional Ruddlesden–Popper perovskite solar cells. *Nature* **2016**, *536*, 312–317. [CrossRef] [PubMed]
10. Leng, K.; Abdelwahab, I.; Verzhbitskiy, I.; Telychko, M.; Chu, L.; Fu, W.; Chi, X.; Guo, N.; Chen, Z.; Chen, Z.; et al. Molecularly thin two-dimensional hybrid perovskites with tunable optoelectronic properties due to reversible surface relaxation. *Nat. Mater.* **2018**, *17*, 908–914. [CrossRef]
11. Ren, H.; Yu, S.; Chao, L.; Xia, Y.; Sun, Y.; Zuo, S.; Li, F.; Niu, T.; Yang, Y.; Ju, H.; et al. Efficient and stable Ruddlesden–Popper perovskite solar cell with tailored interlayer molecular interaction. *Nat. Photonics* **2020**, *14*, 154–163. [CrossRef]
12. Tian, J.; Cordes, D.B.; Slawin, A.M.Z.; Zysman-Colman, E.; Morrison, F.D. Progressive polytypism and bandgap tuning in azetidinium lead halide perovskites. *Inorg. Chem.* **2021**, *60*, 12247–12254. [CrossRef] [PubMed]
13. Tian, J.; Zysman-Colman, E.; Morrison, F.D. Compositional variation in hybrid organic-inorganic lead halide perovskites: Kinetically versus thermodynamically controlled synthesis. *Chem. Mater.* **2021**, *33*, 3650–3659. [CrossRef]
14. Kieslich, G.; Sun, S.; Cheetham, A.K.; Cheetham, T.; Gregor, K.; Shijing, S.; Anthony, K.C. Solid-state principles applied to organic-inorganic perovskites: New tricks for an old dog. *Chem. Sci.* **2014**, *5*, 4712–4715. [CrossRef]
15. Levchuk, I.; Osvet, A.; Tang, X.; Brandl, M.; Perea, J.D.; Hoegl, F.; Matt, G.J.; Hock, R.; Batentschuk, M.; Brabec, C.J. Brightly luminescent and color-tunable formamidinium lead halide perovskite $FAPbX_3$ (X = Cl, Br, I) colloidal nanocrystals. *Nano Lett.* **2017**, *17*, 2765–2770. [CrossRef] [PubMed]
16. Cao, M.; Tian, J.; Cai, Z.; Peng, L.; Yang, L.; Wei, D. Perovskite heterojunction based on $CH_3NH_3PbBr_3$ single crystal for high-sensitive self-powered photodetector. *Appl. Phys. Lett.* **2016**, *109*, 233303. [CrossRef]
17. Tian, J.; Cordes, D.B.; Quarti, C.; Beljonne, D.; Slawin, A.M.Z.; Zysman-Colman, E.; Morrison, F.D. Stable 6H organic-inorganic hybrid lead perovskite and competitive formation of 6H and 3C perovskite structure with mixed A cations. *ACS Appl. Energy Mater.* **2019**, *2*, 5427–5437. [CrossRef]
18. Larson, A.C.; Von Dreele, R.B. *General Structure Analysis System (GSAS)*; Los Alamos National Laboratory: Carlsbad, NM, USA, 2004; pp. 86–748.
19. Yu, J.C.; Kim, D.B.; Jung, E.D.; Lee, B.R.; Song, M.H. High-performance perovskite light-emitting diodes via morphological control of perovskite films. *Nanoscale* **2016**, *8*, 7036–7042. [CrossRef]
20. Ganguli, D. Cationic radius ratio and formation of K_2NiF_4-type compounds. *J. Solid State Chem.* **1979**, *30*, 353–356. [CrossRef]
21. Panetta, R.; Righini, G.; Colapietro, M.; Barba, L.; Tedeschi, D.; Polimeni, A.; Ciccioli, A.; Latini, A. Azetidinium lead iodide: Synthesis, structural and physico-chemical characterization. *J. Mater. Chem. A* **2018**, *6*, 10135–10148. [CrossRef]
22. Shannon, R.D. Revised effective ionic radii and systematic studies of interatomic distances in halides and chalcogenides. *Acta Crystallogr. Sect. A* **1976**, *32*, 751–767. [CrossRef]
23. Pal, P.; Saha, S.; Banik, A.; Sarkar, A.; Biswas, K. All-solid-state mechanochemical synthesis and post-synthetic transformation of inorganic perovskite-type halides. *Chem. Eur. J.* **2018**, *24*, 1811–1815. [CrossRef] [PubMed]

24. Saski, M.; Prochowicz, D.; Marynowski, W.; Lewiński, J. Mechanosynthesis, optical, and morphological properties of MA, FA, CsSnX$_3$ (X = I, Br) and phase-pure mixed-halide MASnI$_x$Br$_{3-x}$ perovskites. *Eur. J. Inorg. Chem.* **2019**, *2019*, 2680–2684. [CrossRef]
25. Lee, S.; Levi, R.D.; Qu, W.; Lee, S.C.; Randall, C.A. Band-gap nonlinearity in perovskite structured solid solutions. *J. Appl. Phys.* **2010**, *107*, 023523. [CrossRef]
26. Chatterjee, S.; Payne, J.; Irvine, J.T.S.; Pal, A.J. Bandgap bowing in a zero-dimensional hybrid halide perovskite derivative: Spin-orbit coupling: Versus lattice strain. *J. Mater. Chem. A* **2020**, *8*, 4416–4427. [CrossRef]
27. Noh, J.H.; Im, S.H.; Heo, J.H.; Mandal, T.N.; Seok, S. Il Chemical management for colorful, efficient, and stable inorganic-organic hybrid nanostructured solar cells. *Nano Lett.* **2013**, *13*, 1764–1769. [CrossRef]
28. Wang, W.; Su, J.; Zhang, L.; Lei, Y.; Wang, D.; Lu, D.; Bai, Y. Growth of mixed-halide perovskite single crystals. *CrystEngComm* **2018**, *20*, 1635–1643. [CrossRef]
29. Hu, Z.; Lin, Z.; Su, J.; Zhang, J.; Chang, J.; Hao, Y. A review on energy band-gap engineering for perovskite photovoltaics. *Sol. RRL* **2019**, *3*, 1900304. [CrossRef]

Article

Low-Dimensional Architectures in Isomeric *cis*-PtCl$_2${Ph$_2$PCH$_2$N(Ar)CH$_2$PPh$_2$} Complexes Using Regioselective-N(Aryl)-Group Manipulation

Peter De'Ath, Mark R. J. Elsegood, Noelia M. Sanchez-Ballester and Martin B. Smith *

Department of Chemistry, Loughborough University, Loughborough LE11 3TU, UK; P.DeAth@lboro.ac.uk (P.D.); m.r.j.elsegood@lboro.ac.uk (M.R.J.E.); n.m.sanchez-ballester@lboro.ac.uk (N.M.S.-B.)
* Correspondence: m.b.smith@lboro.ac.uk

Citation: De'Ath, P.; Elsegood, M.R.J.; Sanchez-Ballester, N.M.; Smith, M.B. Low-Dimensional Architectures in Isomeric *cis*-PtCl$_2${Ph$_2$PCH$_2$N(Ar)CH$_2$PPh$_2$} Complexes Using Regioselective-N(Aryl)-Group Manipulation. *Molecules* 2021, 26, 6809. https://doi.org/10.3390/molecules26226809

Academic Editors: William T. A. Harrison, R. Alan Aitken and Paul Waddell

Received: 28 September 2021
Accepted: 7 November 2021
Published: 11 November 2021

Publisher's Note: MDPI stays neutral with regard to jurisdictional claims in published maps and institutional affiliations.

Copyright: © 2021 by the authors. Licensee MDPI, Basel, Switzerland. This article is an open access article distributed under the terms and conditions of the Creative Commons Attribution (CC BY) license (https://creativecommons.org/licenses/by/4.0/).

Abstract: The solid-state behaviour of two series of isomeric, phenol-substituted, aminomethylphosphines, as the free ligands and bound to Pt^{II}, have been extensively studied using single crystal X-ray crystallography. In the first library, isomeric diphosphines of the type Ph$_2$PCH$_2$N(Ar)CH$_2$PPh$_2$ [**1a–e**; Ar = C$_6$H$_3$(Me)(OH)] and, in the second library, amide-functionalised, isomeric ligands Ph$_2$PCH$_2$N{CH$_2$C(O)NH(Ar)}CH$_2$PPh$_2$ [**2a–e**; Ar = C$_6$H$_3$(Me)(OH)], were synthesised by reaction of Ph$_2$PCH$_2$OH and the appropriate amine in CH$_3$OH, and isolated as colourless solids or oils in good yield. The non-methyl, substituted diphosphines Ph$_2$PCH$_2$N{CH$_2$C(O)NH(Ar)}CH$_2$PPh$_2$ [**2f**, Ar = 3-C$_6$H$_4$(OH); **2g**, Ar = 4-C$_6$H$_4$(OH)] and Ph$_2$PCH$_2$N(Ar)CH$_2$PPh$_2$ [**3**, Ar = 3-C$_6$H$_4$(OH)] were also prepared for comparative purposes. Reactions of **1a–e**, **2a–g**, or **3** with PtCl$_2$(η^4-cod) afforded the corresponding square-planar complexes **4a–e**, **5a–g**, and **6** in good to high isolated yields. All new compounds were characterised using a range of spectroscopic (^{1}H, ^{31}P{^{1}H}, FT–IR) and analytical techniques. Single crystal X-ray structures have been determined for **1a**, **1b**·CH$_3$OH, **2f**·CH$_3$OH, **2g**, **3**, **4b**·(CH$_3$)$_2$SO, **4c**·CHCl$_3$, **4d**·$\frac{1}{2}$Et$_2$O, **4e**·$\frac{1}{2}$CHCl$_3$·$\frac{1}{2}$CH$_3$OH, **5a**·$\frac{1}{2}$Et$_2$O, **5b**, **5c**·$\frac{1}{4}$H$_2$O, **5d**·Et$_2$O, and **6**·(CH$_3$)$_2$SO. The free phenolic group in **1b**·CH$_3$OH, **2f**·CH$_3$OH, **2g**, **4b**·(CH$_3$)$_2$SO, **5a**·$\frac{1}{2}$Et$_2$O, **5c**·$\frac{1}{4}$H$_2$O, and **6**·(CH$_3$)$_2$SO exhibits various intra- or intermolecular O–H···X (X = O, N, P, Cl) hydrogen contacts leading to different packing arrangements.

Keywords: amide groups; isomers; late-transition metals; P-ligands; phenols; secondary interactions; single crystal X-ray crystallography

1. Introduction

Tertiary phosphines, and their phosphine oxides, have played an important role in the study of supramolecular and self-assembly processes [1–3]. Their synthetic versatility, coupled with ease of substituent modification, has no doubt played a significant contribution over the years. Hydrogen bonding interactions are routinely encountered in supramolecular ligand systems as illustrated by the elegant studies from Breit [4], Reek [5], and others [6,7]. More recently, amongst other common types of non-covalent interactions, those based on halogen bonding [8,9] and H$^{δ+}$···H$^{δ−}$ have been reported [10].

For a number of years, we [11–16], and others [17–22], have been interested in aminomethylphosphines, readily amenable by Mannich condensation reactions. Such interest stems from the relative ease of accessing *P*-monodentate ligands based on a P–C–N linker [11,15,16,19,20,22] or *P/P*-bidentate derivatives bearing a P–C–N–C–P backbone [12–14,17–19,21]. Previously, we have shown that the N-arene group can be easily tuned with, for example, various H-bonding donor/acceptor sites based on –CO$_2$H/OH groups [12–16]. In continuation of these studies, we report here our work on the regioselective positioning of amide/hydroxy and methyl groups within a series of aminomethylphosphines, both as the free ligands and when coordinated to a square-planar Pt(II) metal centre. Our rationale for introducing an –C(O)NH– group is based on the known use of this functionality in supramolecular chemistry [23] and, furthermore, the recent interest in amide-modified phosphines for their variable coordination chemistry [24–26], binding nitroaromatics [27], and relevance to catalysis based on Pd [28]. Our choice of metal fragment in this work, "*cis*-PtCl$_2$", is based on its capability to support a relatively small bite angle diphosphine ligand in a *cis*, six-membered ring conformation, and to provide up to two "acceptor" sites for potential H-bonding [29]. For this purpose, we elected to pursue a double Mannich condensation reaction of Ph$_2$PCH$_2$OH with a series of isomeric primary amines bearing either OH/CH$_3$ groups and/or an amide spacer between the arene and P–C–N–C–P backbone (Chart 1).

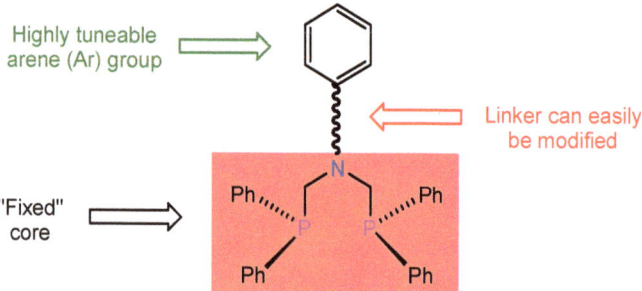

Chart 1. Potential modification sites of a Ph$_2$P–C–N(Ar)–C–PPh$_2$ backbone.

2. Results and Discussion

2.1. Ligand Synthesis

We [11–16,29], and others [17,19–22], have previously used Mannich condensations as a versatile method for the synthesis of aminomethylphosphines. Accordingly, two equivalents of Ph$_2$PCH$_2$OH were reacted with one equivalent of the amine, for 24 h at r.t. under N$_2$, yielding the desired phenol-substituted ditertiary phosphines **1a–e** and **3** (Scheme 1).

	A	B	C	D
1a	OH	H	H	CH$_3$
1b	CH$_3$	H	OH	H
1c	OH	H	CH$_3$	H
1d	H	OH	CH$_3$	H
1e	H	CH$_3$	OH	H
3	H	OH	H	H

	A	B	C	D
2a	OH	H	H	CH$_3$
2b	CH$_3$	H	OH	H
2c	OH	H	CH$_3$	H
2d	H	OH	CH$_3$	H
2e	H	CH$_3$	OH	H
2f	H	OH	H	H
2g	H	H	OH	H

Scheme 1. Synthesis of **1a–e**, **2a–g**, and **3**.

For **1a–e**, colourless solids were isolated in 38–97% yields and found to be air stable in the solid state, but oxidise rapidly in solution. Compounds **1a–e** and **3** exhibit single resonances in their $^{31}P\{^1H\}$ NMR spectra (in d^6-dmso) around $\delta(P)$ −26 ppm [12–15,29], indicating the presence of only one P^{III} environment. The ligands were also characterised by ^1H NMR, FT–IR, and elemental analysis (Table 1). In particular, the absence of an NH resonance, in the ^1H NMR spectra, confirmed that double condensation had occurred.

Table 1. Selected spectroscopic and analytical data for compounds **1a–3** [a].

Compound [a]	δ(P) [b]	δ(H) /OH (NH)	δ(H) /arom. H.	δ(H) /CH$_2$	δ(H) /CH$_2$ [d]	δ(H) /CH$_3$	ν_{OH} (ν_{NH}) [e]	Microanalysis (CHN)
1a (79)	−27.5	8.62	7.33–7.23, 6.76, 6.69–6.57		4.15 (2.4)	2.10	3398	Calc. for C$_{33}$H$_{31}$NOP$_2$, C, 76.29; H, 6.01; N, 2.70 Found, C, 76.07; H, 6.13; N, 2.78
1b (56)	−27.3	9.06	7.36–7.26, 7.15, 6.50, 6.44		3.96 (5.6)	1.74	3282	Calc. for C$_{33}$H$_{31}$NOP$_2$.2MeOH, C, 72.03; H, 6.74; N, 2.40 Found, C, 72.45; H, 6.04; N, 2.58
1c (97)	−27.5	8.77	7.44–7.22, 6.86, 6.54, 6.48		4.09 (3.4)	2.12	3389	Calc. for C$_{33}$H$_{31}$NOP$_2$, C, 76.29; H, 6.01; N, 2.70 Found, C, 75.99; H, 6.00; N, 2.76
1d (38)	−26.7	8.63	7.40–7.30, 6.55		4.02 (3.2)	1.96	3432	Calc. for C$_{33}$H$_{31}$NOP$_2$, C, 76.29; H, 6.01; N, 2.70 Found, C, 75.53; H, 6.05; N, 2.74
1e (96)	−26.4	9.06	7.49–7.33, 6.85, 6.50, 6.27		3.88 (3.6)	2.08	3387	Calc. for C$_{33}$H$_{31}$NOP$_2$.MeOH, C, 74.03; H, 6.40; N, 2.54 Found, C, 74.81; H, 5.93; N, 2.61
2a (81)	−26.0	8.15	7.77–7.19	5.06	3.62 (8.0)	1.19	-	-
2b (89)	−26.0	7.83	7.60–7.21	5.07	3.69 (3.6)	1.63	-	-
2c (88)	−26.5 [c]	9.34 (8.17)	7.71–7.19	5.27	3.61 (4.8)	1.63	-	-
2d (65)	−27.1	9.05 (8.68)	7.55–7.32, 6.95, 6.61, 6.41	3.69	3.81 (4.8)	2.04	3047 (3228)	Calc. for C$_{35}$H$_{34}$N$_2$O$_2$P$_2$, C, 72.91; H, 5.94; N, 4.86 Found, C, 72.72; H, 5.95; N, 4.88
2e (80)	−27.1	9.29 (9.08)	7.46–7.35, 7.29, 6.86, 6.15	3.73	3.82 (4.4)	2.08	3178 (3317)	Calc. for C$_{35}$H$_{34}$N$_2$O$_2$P$_2$, C, 72.91; H, 5.94; N, 4.86 Found, C, 72.71; H, 5.94; N, 4.82
2f (70)	−27.5	9.31 (9.07)	7.41–7.03, 6.94, 6.40, 6.30	3.69	3.77 (4.4)		3163 (3283)	Calc. for C$_{34}$H$_{32}$N$_2$O$_2$P$_2$, C, 72.59; H, 5.73; N, 4.98 Found, C, 72.10; H, 5.80; N, 4.95
2g (85)	−26.8	9.09 (8.78)	7.36–7.25, 6.83, 6.53	3.61	3.72 (4.4)		3300 (3257)	Calc. for C$_{34}$H$_{32}$N$_2$O$_2$P$_2$, C, 72.59; H, 5.73; N, 4.98 Found, C, 72.15; H, 5.72; N, 4.95
3 (53)	−27.6	9.12	7.38–7.31, 6.92, 6.30, 6.13		3.85		3376	Calc. for C$_{32}$H$_{29}$NOP$_2$, C, 76.03; H, 5.78; N, 2.77 Found, C, 75.67; H, 5.71; N, 2.74

[a] Isolated yields in parentheses. [b] Recorded in (CD$_3$)$_2$SO unless otherwise stated. [c] Recorded in CDCl$_3$. [d] ^2J(PH) coupling in brackets. [e] Recorded as KBr discs.

The synthesis of ditertiary phosphines, containing a flexible backbone presenting extra donor/acceptor sites with additional H-bonding capability, is described here with the opportunity to enhance solid-state packing behaviour. The precursors for the synthesis of the desired functionalised ditertiary phosphines **2a–g** were prepared using, in step (i), 1 equiv. of primary amine, *N*-carbobenzyloxyglycine (1 equiv.) and dicyclohexylcarbodiimide (DCC, 1 equiv.) in THF affording the corresponding carbamates followed by, in step (ii), treatment with Pd/C and cyclohexene in C_2H_5OH, to give the desired primary alkylamines in moderate to good yields [30,31]. Using a similar procedure to that described for **1a–e**, the amide-functionalised diphosphines **2a–e** were prepared in 65–89% yields by condensation using 1 equiv. of primary amine and two equiv. of Ph_2PCH_2OH at r.t. in CH_3OH (Scheme 1). Furthermore, the phenol-substituted phosphines **2f** and **2g** were synthesised to investigate what effect, if any, an absent methyl group on the N-arene ring displays. In the case of **2d–g**, the diphosphines were obtained as solids whereas **2a–c** were obtained as yellow oils that were sufficiently pure to be used in complexation studies. All compounds displayed a single ^{31}P NMR resonance around $\delta(P)$ −26 ppm [12–15,29] indicating the inclusion of an amide spacer has negligible effect on the ^{31}P chemical shift. Other spectroscopic and analytical data are given in Table 1.

2.2. Single Crystal X-ray Studies of **1a**, **1b·CH₃OH**, **2f·CH₃OH**, **2g**, and **3**

X-ray quality crystals of **1a**, **1b·CH₃OH**, **2f·CH₃OH**, **2g**, and **3** were obtained by slow evaporation of a methanol solution, while for **2g** diethyl ether was diffused into a deuterochloroform/methanol solution (Table 2).

Table 2. Details of the X-ray data collections and refinements for compounds **1a**, **1b·CH₃OH**, **2f·CH₃OH**, **2g**, and **3**.

Compound	1a	1b·CH₃OH	2f·CH₃OH	2g	3
Formula	$C_{33}H_{31}NOP_2$	$C_{34}H_{35}NO_2P_2$	$C_{35}H_{36}N_2O_3P_2$	$C_{34}H_{32}N_2O_2P_2$	$C_{32}H_{29}NOP_2$
M	519.53	551.57	594.60	562.55	505.50
Crystal dimensions	0.42 × 0.15 × 0.03	0.13 × 0.12 × 0.02	0.24 × 0.18 × 0.16	0.25 × 0.18 × 0.15	0.31 × 0.28 × 0.03
Crystal morphology and colour	Plate, colourless	Block, colourless	Block, colourless	Block, colourless	Plate, colourless
Crystal system	Monoclinic	Monoclinic	Triclinic	Monoclinic	Triclinic
Space group	$P2_1/n$	$P2_1/c$	$P\bar{1}$	Ia	$P\bar{1}$
a/Å	17.367(5)	10.3050(3)	12.6198(3)	11.6234(10)	10.5860(4)
b/Å	8.522(2)	32.8017(10)	16.2027(4)	21.7359(19)	10.7397(4)
c/Å	20.382(6)	8.5189(2)	17.8529(4)	11.6340(10)	13.4172(6)
α/°			64.0678(10)		73.1667(6)
β/°	114.673(4)	92.7318(16)	76.7403(14)	93.8717(14)	80.4518(7)
γ/°			75.5070(14)		63.1422(6)
V/Å³	2741.2(13)	2876.30(14)	3148.15(13)	2932.6(4)	1301.45(9)
Z	4	4	4	4	2
λ/Å	0.71073	0.71073	0.71073	0.71073	0.71073
T/K	150(2)	120(2)	120(2)	150(2)	150(2)
Density (calcd.)/Mg/m³	1.259	1.274	1.255	1.274	1.290
μ/mm⁻¹	0.185	0.183	0.176	0.182	0.193
θ range/°	2.02–26.60	3.03–27.53	3.24–25.00	1.87–28.82	1.59–30.62
Measured reflections	23,525	27,247	61,330	12,577	15,576
Independent reflections	5708	6545	11,047	6586	7814
Observed reflections ($F^2 > 2\sigma(F^2)$)	3115	5019	7559	5293	6116
R_{int}	0.124	0.058	0.095	0.039	0.027
$R[F^2 > 2\sigma(F^2)]$ [a]	0.0743	0.0799	0.0517	0.0389	0.0441
$wR2$ [all data] [b]	0.2205	0.1650	0.1220	0.0861	0.1248
Largest difference map features/eÅ⁻³	1.40, −0.49	0.46, −0.52	0.38, −0.30	0.29, −0.16	0.51, −0.21

[a] $R = \sum ||Fo| - |Fc|| / \sum |Fo|$. [b] $wR2 = [\sum[w(Fo^2 - Fc^2)^2]/\sum[w(Fo^2)^2]]^{1/2}$.

The geometry around each phosphorus atom is essentially pyramidal as would be anticipated (Figures 1–5). The P^{III} atoms are in an *anti* conformation, presumably to minimise steric repulsions between the phenyl groups. The geometry about the N(1) centre is approx. pyramidal [Σ(C–N(1)–C) angles: 337.0(3)° for **1a**; 335(2)° for **1b**·CH$_3$OH; 335.2(2)/336.6(2)° for **2f**·CH$_3$OH; 333.7(2)° for **2g**] and approximately trigonal planar for **3** [Σ(C–N–C) = 359.05(11)°]. In **1a** and **1b**·CH$_3$OH, the N-arene ring [C(3) > C(8)] is twisted by ca. 88° (**1a**) and 86° (**1b**·CH$_3$OH) [12,32] such that it is almost perpendicular to the C(1)–N(1)–C(2) plane, whereas for **3**, the twist of the C(1)–N(1)–C(2) fragment is around 9° from co-planarity with the N-arene group, apparently as a result of the intermolecular H-bonding requirements (*vide infra*).

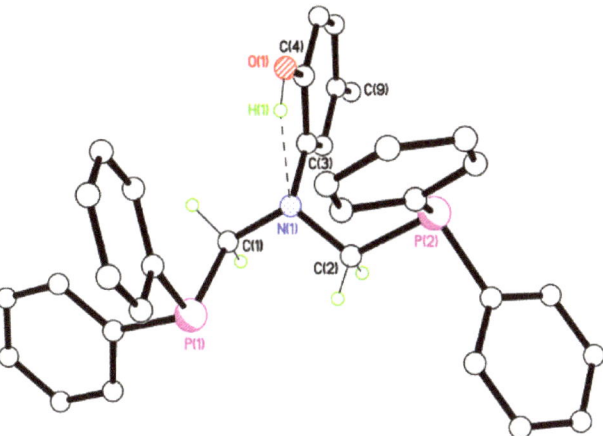

Figure 1. Molecular structure of **1a**. All hydrogens, except on C(1), C(2) and O(1), have been omitted for clarity.

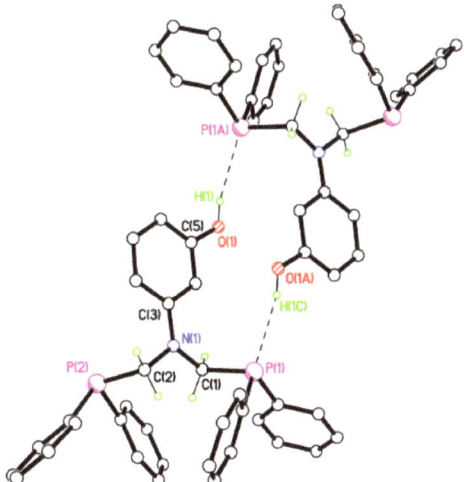

Figure 2. Molecular structure of **3** showing a dimer pair. All hydrogens, except on C(1), C(2) and O(1), have been omitted for clarity. Symmetry code: A = 1 − x, 1 − y, 1 − z.

115

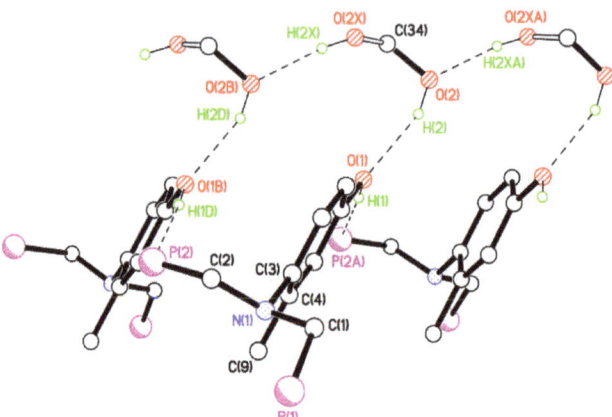

Figure 3. Crystal structure packing plot for **1b**·CH$_3$OH. Most H atoms, two Ph groups per P atom have been omitted for clarity. Symmetry code: A = x, −y + $\frac{1}{2}$, z + $\frac{1}{2}$.

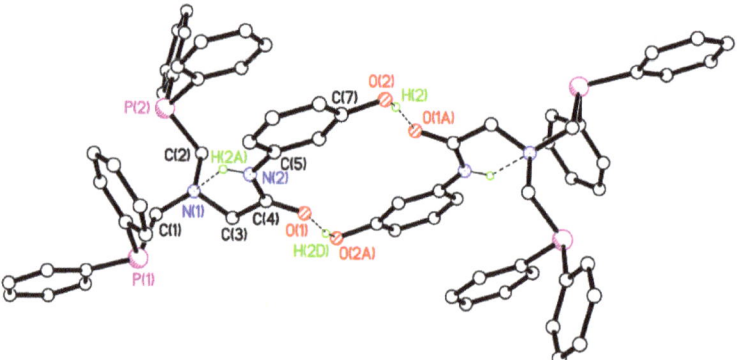

Figure 4. Dimers of **2f** forming $R^2_2(16)$ graph set motifs. Most H atoms omitted for clarity. The second unique molecule which adopts a similar, centrosymmetric motif, is not shown.

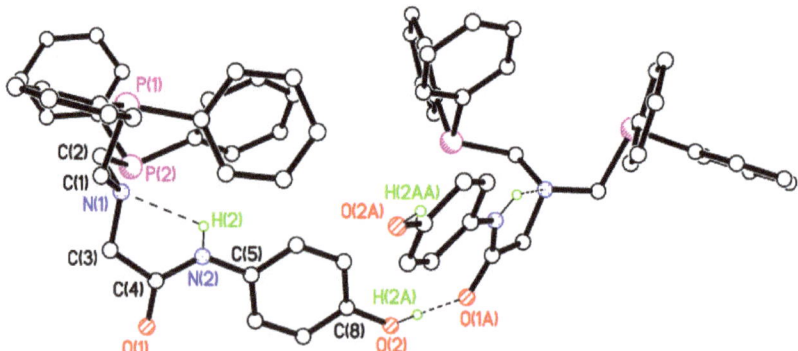

Figure 5. Intra- and intermolecular interactions in the crystal structure of **2g**. Most H atoms omitted for clarity. Symmetry operator A = x, −y + $^3/_2$, z − $\frac{1}{2}$.

2.3. Secondary Interactions in **1a**, **1b·CH$_3$OH**, **2f·CH$_3$OH**, **2g**, *and* **3**

The synthons observed in the solid state for these highly modular ligands may be dictated by various factors including the nature of the ligand, the flexibility of the P–C–N–C–P backbone, the predisposition of the OH/CH$_3$ groups about the N-arene ring, and the solvent used in the crystallisation. In order to probe the OH/CH$_3$ interplay of groups, the crystal structure of **1a**, with the –OH group in the *ortho* position with respect to the N(1) atom, is described first. Ligand **1a** crystallises with an intramolecular S(5) [33–35] H-bonded ring with d = 2.26(5) Å [denoting the hydrogen (H) to acceptor (A) distance in an H-bond D–H···A] [36] for the O–H···N interaction (Figure 1). The intramolecular H-bonding in **1a** limits the dimensionality of the packing of the diphosphine ligand. Therefore, the structure of **1a** is essentially zero-dimensional (Table 3).

Table 3. Selected data (D···A/Å, $\angle D$–H···A/°) for key inter- and intramolecular contacts for compounds **1a**, **1b·CH$_3$OH**, **2f·CH$_3$OH**, **2g**, and **3**.

	1a	1b·CH$_3$OH	2f·CH$_3$OH [a]	2g	3
O–H···N$_{intra}$	2.745(4), 119(4)				
O–H$_{MeOH}$···O$_{inter}$		2.844(8), 157			
O–H$_{MeOH}$···O$_{MeOH}$		2.781(11), 172			
O–H···P$_{inter}$		3.432(3), 173			3.4400(12), 167(2)
O–H···(O)C$_{inter}$			2.671(3), 171(3) [2.659(3), 165(3)]	2.706(4), 169(4)	
N–H···N$_{intra}$			2.695(3), 114(2) [2.714(3), 117(2)]	2.748(4), 114(3)	

[a] Values in parentheses are for the second independent molecule.

Compound **3**, where the –OH functional group is in the *meta* position with respect to the tertiary N(1) atom, aggregates in the solid state in such a way that fairly weak hydrogen bonds, O–H···P [d = 2.60(2) Å], form between symmetry-related molecules, creating dimers in which two ligands are held in an $R^2_2(16)$ H-bonding motif (Figure 2). The distance between symmetry-related nitrogen atoms is 8.257 Å. The structure of **3** shows a 0D arrangement.

Compound **1b·CH$_3$OH**, which contains the –OH group in a *para* position with respect to the N-arene, displays a similar structure to **3** with intramolecular O–H···P interactions at d = 2.60 Å. However, instead of forming dimers, there are 1D zig-zag chains in the *c* direction (Figure 3). The *para* hydroxyl oxygen acts as an acceptor for an O–H···O intermolecular H-bond from approximately alternate CH$_3$OH molecules of crystallisation with d = 2.05 Å. These CH$_3$OH molecules are 50/50 disordered with the second component H-bonding to its neighbour with d = 1.95 Å. Selected hydrogen parameters for **1b·CH$_3$OH** are listed in Table 3.

Compound **2f·CH$_3$OH** crystallises with two, similarly behaved, molecules in the asymmetric unit. A pair of H-bonded molecules, related by inversion symmetry, and with d = 1.81(3) Å for the intermolecular O–H···O interaction [1.78(3) Å for molecule 2] affords $R^2_2(16)$ ring motifs (Figure 4). The intramolecular N–H···N S(5) H-bond motif with d = 2.25(3) Å [2.26(3) Å for molecule 2] results in an intermediate twist angle of 64.23(13)° [but a rather more perpendicular 78.70(8)° for molecule 2] between planes C(1)/N(1)/C(2) and ring C(5) > C(10) [plane C(35)/N(4)/C(36) and ring C(39) > C(44) for molecule 2]. The *meta* hydroxy group in **2f** facilitates 0D dimer formation, as opposed to the chains observed in **2g** (*vida infra*).

For **2g**, molecules form H-bonded, 1D, zig-zag chains in the *c* direction via strong O–H···O interactions with d = 1.83(5) Å (Figure 5). The intramolecular N–H···N S(5), H-bond motif with d = 2.29(3) Å again results in an almost perpendicular twist angle of 82.09(15)° between planes C(1)/N(1)/C(2) and arene ring C(5) > C(10). The *para* hydroxy group promotes chain formation.

2.4. Dichloroplatinum(II) Complexes of 1a–e, 2a–g, and 3

The synthesis of P,P-chelate complexes cis-PtCl$_2$(1a–e) [4a–e], cis-PtCl$_2$(2a–g) [5a–g], and cis-PtCl$_2$(3) [6] (Chart 2) was achieved by stirring the ligands and PtCl$_2$(η^4-cod) (1:1 ratio) in CH$_2$Cl$_2$ for 1.5 h with displacement of the cod ligand. The products were isolated in good yields as colourless solids. Downfield shifts of the ^{31}P NMR resonances were observed for all complexes, with $^1J_{PtP}$ coupling constants of approx. 3400 Hz, indicative of a *cis* conformation [29]. This was further supported by two characteristic ν_{PtCl} IR vibrations in the range of 279–316 cm^{-1} (Table 4). Furthermore, compounds 4a–e, 5a–g, and 6 present ν(NH) and ν(OH) IR absorptions in the range 3050–3465 cm^{-1} and also a strong band in the region of 1653–1675 cm^{-1}, indicative of ν(C=O amide).

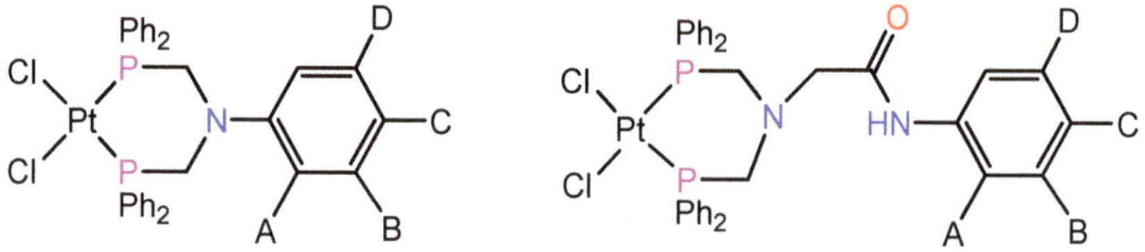

	A	B	C	D
4a	OH	H	H	CH$_3$
4b	CH$_3$	H	OH	H
4c	OH	H	CH$_3$	H
4d	H	OH	CH$_3$	H
4e	H	CH$_3$	OH	H
6	H	OH	H	H

	A	B	C	D
5a	OH	H	H	CH$_3$
5b	CH$_3$	H	OH	H
5c	OH	H	CH$_3$	H
5d	H	OH	CH$_3$	H
5e	H	CH$_3$	OH	H
5f	H	OH	H	H
5g	H	H	OH	H

Chart 2. Structures of compounds 4a–e, 5a–g, and 6.

Table 4. Selected spectroscopic and analytical data for compounds 4a–6 [a].

Compound [a]	δ(P) [b]	δ(H) /OH (NH)	δ(H) /arom. H.	δ(H) /CH$_2$	δ(H) /CH$_2$	δ(H) /CH$_3$	ν_{OH} (ν_{NH}) [f]	ν_{PtCl}	Microanalysis (CHN)
4a (98)	−9.4 [d] (3424)	9.25	7.89–7.80, 7.64–7.46, 6.68, 5.90		4.21	1.93	3314	316, 289	Calc. for C$_{33}$H$_{31}$Cl$_2$NOP$_2$Pt·CH$_2$Cl$_2$, C, 46.91; H, 3.82; N, 1.61 Found, C, 47.07; H, 3.77; N, 1.69
4b (89)	−4.9 [d] (3426)	9.22	7.96–7.53, 6.96, 6.49, 6.33		4.19	1.29	3373	315, 282	Calc. for C$_{33}$H$_{31}$Cl$_2$NOP$_2$Pt, C, 50.46; H, 3.98; N, 1.78 Found, C, 50.51; H, 4.13; N, 1.83
4c (78)	−8.6 [d] (3436)	9.42	7.89–7.84, 7.56–7.43, 6.59, 6.30, 6.05		4.16	2.06	3433	309, 290	Calc. for C$_{33}$H$_{31}$Cl$_2$NOP$_2$Pt·0.5CH$_2$Cl$_2$, C, 48.96; H, 3.87; N, 1.68 Found, C, 49.42; H, 3.96; N, 1.73
4d (98)	−11.7 [d] (3410)	8.44	7.94–7.87, 7.78–7.62, 6.86, 6.47		4.43	2.02	3421	314, 290	Calc. for C$_{33}$H$_{31}$Cl$_2$NOP$_2$Pt, C, 50.46; H, 3.98; N, 1.78 Found, C, 50.24; H, 3.98; N, 1.85
4e (81)	−7.8 [d] (3421)	9.01	7.96–7.85, 7.59–7.45, 6.75, 6.27, 6.03		4.33	2.09	3416	316, 284	Calc. for C$_{33}$H$_{31}$Cl$_2$NOP$_2$Pt, C, 50.46; H, 3.98; N, 1.78 Found, C, 50.66; H, 4.61; N, 1.70

Table 4. Cont.

Compound [a]	δ(P) [b]	δ(H) /OH (NH)	δ(H) /arom. H.	δ(H) /CH$_2$	δ(H) /CH$_2$	δ(H) /CH$_3$	ν$_{OH}$ (ν$_{NH}$) [f]	ν$_{PtCl}$	Microanalysis (CHN)
5a (89)	−9.8 [d,e] (3398)	9.45 (8.91)	7.84–7.80, 7.53–7.44, 6.69	3.49	4.05	2.22	3051 (3249)	305, 283	Calc. for C$_{35}$H$_{34}$Cl$_2$N$_2$O$_2$P$_2$Pt.0.5CH$_2$Cl$_2$, C, 48.63; H, 3.74; N, 3.15 Found, C, 49.00; H, 4.07; N, 3.13
5b (65)	−11.0 [d] (3397)	9.16 (8.61)	7.83–7.80, 7.57–7.41, 7.05, 6.48	4.03	4.03	1.80	3050 (3350)	316, 283	Calc. for C$_{35}$H$_{34}$Cl$_2$N$_2$O$_2$P$_2$Pt, C, 49.89; H, 4.07; N, 3.32 Found, C, 49.32; H, 4.17; N, 3.25
5c (73)	−9.9 [d] (3405)	9.56 (8.94)	7.85–7.77, 7.59–7.38, 6.63, 6.51	3.17	4.05	2.17	3075 (3347)	315, 290	Calc. for C$_{35}$H$_{34}$Cl$_2$N$_2$O$_2$P$_2$Pt, C, 49.89; H, 4.07; N, 3.32 Found, C, 49.28; H, 4.05; N, 2.91
5d (99)	−9.8 [c,d] (3406)	9.17 (8.90)	7.98–7.50, 6.97–6.84, 6.68, 6.73	3.20	4.66	2.13	3323 (3465)	309, 283	Calc. for C$_{35}$H$_{34}$Cl$_2$N$_2$O$_2$P$_2$Pt.0.5C$_4$H$_{10}$O, C, 50.52; H, 4.47; N, 3.19 Found, C, 50.91; H, 4.53; N, 3.61
5e (90)	−9.7 [c,d] (3406)	9.46 (9.21)	7.94–7.78, 7.54–7.42, 7.09, 6.87, 6.69	3.43	4.12	2.02	3287 (3439)	312, 286	Calc. for C$_{35}$H$_{34}$Cl$_2$N$_2$O$_2$P$_2$Pt, C, 49.89; H, 4.07; N, 3.32 Found, C, 49.77; H, 3.95; N, 3.38
5f (85)	−9.5 [c,d] (3425)	9.62 (9.36)	7.91–7.86, 7.60–7.42, 7.05, 6.83, 6.45	3.47	4.18		3053 (3312)	304, 279	Calc. for C$_{34}$H$_{32}$Cl$_2$N$_2$O$_2$P$_2$Pt, C, 49.29; H, 3.89; N, 3.38 Found, C, 48.98; H, 3.38; N, 3.37
5g (84)	−9.5 [c,d] (3405)	9.52 (9.31)	8.01–7.97, 7.70–7.61, 7.34, 6.78	3.49	4.26		3054 (3325)	311, 287	Calc. for C$_{34}$H$_{32}$Cl$_2$N$_2$O$_2$P$_2$Pt, C, 49.29; H, 3.89; N, 3.38 Found, C, 48.72; H, 3.66; N, 3.33
6 (89)	−4.0 [d] (3436)	8.45	7.45–7.05, 6.89–6.76, 6.31,		4.31		3356	311, 289	Calc. for C$_{32}$H$_{29}$Cl$_2$NOP$_2$Pt, C, 49.82; H, 3.79; N, 1.82 Found, C, 49.31; H, 3.58; N, 1.79

[a] Isolated yields in parentheses. [b] Recorded in (CD$_3$)$_2$SO unless otherwise stated. [c] Recorded in CDCl$_3$. [d] ^{1}J(PtP) coupling in parentheses. [e] Recorded in CDCl$_3$/CD$_3$OD. [f] Recorded as KBr discs.

2.5. Single Crystal X-ray Studies of Complexes 4b·(CH$_3$)$_2$SO, 4c·CHCl$_3$, 4d·½Et$_2$O, 4e·½CHCl$_3$·½CH$_3$OH, 5a·½Et$_2$O, 5b, 5c·¼H$_2$O, 5d·Et$_2$O, and 6·(CH$_3$)$_2$SO

Detailed single crystal X-ray analysis (Tables 5 and 6) of complexes 4b·(CH$_3$)$_2$SO, 4c·CHCl$_3$, 4d·½Et$_2$O, 4e·½CHCl$_3$·½CH$_3$OH, 5a·½Et$_2$O, 5b, 5c·¼H$_2$O, 5d·Et$_2$O, and 6·(CH$_3$)$_2$SO shows that the geometry about each Pt(II) centre is approximately square planar [P–Pt–P range 90.23(9)–96.52(3)°] (Tables 7 and 8). The Pt–Cl and Pt–P bond distances are consistent with literature values [29] and the conformation of the Pt–P–C–N–C–P six-membered ring in each complex is best described as a boat. The dihedral angle measured between the P$_2$C$_2$ plane and N-arene ring least-squares planes varies between 50.98(12)° [in 6·(CH$_3$)$_2$SO] and 90° (in 5d·Et$_2$O), the difference of ca. 39° may tentatively be explained by the predisposition of the –OH group about the N-arene group and subsequent H-bonding requirements. Upon metal chelation, a degree of freedom, compared with the free ligands 1a, 1b·CH$_3$OH, 2f·CH$_3$OH, 2g, and 3 has been removed, as the P–C–N–C–P backbone is locked into a specific conformation. Unfortunately, we were unable to obtain suitable X-ray quality crystals of compounds 4a and 5e–g.

Table 5. Details of the X-ray data collections and refinements for compounds 4b·(CH$_3$)$_2$SO, 4c·CHCl$_3$, 4d·$\frac{1}{2}$OEt$_2$, and 4e·$\frac{1}{2}$CHCl$_3$·$\frac{1}{2}$CH$_3$OH.

Compound	4b·(CH$_3$)$_2$SO	4c·CHCl$_3$	4d·$\frac{1}{2}$OEt$_2$	4e·$\frac{1}{2}$CHCl$_3$·$\frac{1}{2}$CH$_3$OH
Formula	C$_{35}$H$_{37}$Cl$_2$NO$_2$P$_2$PtS	C$_{34}$H$_{32}$Cl$_5$NOP$_2$Pt	C$_{35}$H$_{36}$Cl$_2$NO$_{1.5}$P$_2$Pt	C$_{34}$H$_{33.5}$Cl$_{3.5}$NO$_{1.5}$P$_2$Pt
M	863.64	904.88	822.58	861.22
Crystal dimensions	0.19 × 0.02 × 0.01	0.30 × 0.18 × 0.04	0.13 × 0.06 × 0.03	0.15 × 0.04 × 0.02
Crystal morphology and colour	Needle, colourless	Plate, colourless	Lath, colourless	Needle, colourless
Crystal system	Tetragonal	Monoclinic	Monoclinic	Monoclinic
Space group	P4$_3$	P2$_1$/n	P2$_1$/c	P2$_1$/n
a/Å	11.373(3)	11.6938(4)	15.7344(6)	21.4521(4)
b/Å		16.7052(6)	17.0714(6)	12.5164(2)
c/Å	26.773(6)	18.2242(7)	13.9632(5)	24.5837(4)
α/°				
β/°		99.7066(6)	92.0800(4)	92.2343(5)
γ/°				
V/Å^3	3463(2)	3509.1(2)	3748.2(2)	6595.78(19)
Z	4	4	4	8
λ/Å	0.71073	0.71073	0.6710	0.71073
T/K	150(2)	150(2)	150(2)	120(2)
Density (calcd.)/Mg/m^3	1.657	1.713	1.458	1.735
µ/mm^{-1}	4.391	4.500	3.425	4.666
θ range/°	1.79–26.09	1.67–31.09	1.78–31.10	2.94–27.49
Measured reflections	29,848	32,600	48,268	84,353
Independent reflections	6852	10,997	13,239	15,063
Observed reflections ($F^2 > 2\sigma(F^2)$)	5560	8926	10,918	12,905
R$_{int}$	0.110	0.043	0.039	0.049
R[$F^2 > 2\sigma(F^2)$] [a]	0.0473	0.0303	0.0266	0.0561
wR2 [all data] [b]	0.1015	0.0660	0.0646	0.1202
Largest difference map features/eÅ^{-3}	1.43, −0.91	1.29, −1.08	0.84, −0.67	1.64, −1.48

[a] $R = \Sigma ||Fo| - |Fc|| / \Sigma |Fo|$. [b] $wR2 = [\Sigma[w(Fo^2 - Fc^2)^2]/\Sigma[w(Fo^2)^2]]^{1/2}$.

Table 6. Details of the X-ray data collections and refinements for compounds 5a·$\frac{1}{2}$OEt$_2$, 5b, 5c·$\frac{1}{4}$H$_2$O, 5d·OEt$_2$, and 6·(CH$_3$)$_2$SO.

Compound	5a·$\frac{1}{2}$OEt$_2$	5b	5c·$\frac{1}{4}$H$_2$O	5d·OEt$_2$	6·(CH$_3$)$_2$SO
Formula	C$_{37}$H$_{39}$Cl$_2$N$_2$O$_{2.50}$P$_2$Pt	C$_{35}$H$_{34}$Cl$_2$N$_2$O$_2$P$_2$Pt	C$_{35}$H$_{34.5}$Cl$_2$N$_2$O$_{2.25}$P$_2$Pt	C$_{39}$H$_{44}$Cl$_2$N$_2$O$_3$P$_2$Pt	C$_{34}$H$_{35}$Cl$_2$NO$_2$P$_2$PtS
M	879.63	842.57	842.57	916.69	849.62
Crystal dimensions	0.12 × 0.06 × 0.05	0.05 × 0.02 × 0.01	0.09 × 0.05 × 0.02	0.13 × 0.12 × 0.02	0.32 × 0.11 × 0.02
Crystal morphology and colour	Block, colourless	Plate, colourless	Plate, colourless	Plate, colourless	Needle, colourless
Crystal system	Trigonal	Monoclinic	Triclinic	Orthorhombic	Monoclinic
Space group	P3$_2$	P2$_1$/n	P$\bar{1}$	Pbcm	P2$_1$/n
a/Å	24.3688(7)	18.2384(7)	8.4021(6)	10.125(6)	9.7763(4)
b/Å		8.1955(3)	10.3896(7)	19.790(11)	13.0930(5)
c/Å	10.6567(6)	23.5809(10)	21.8810(15)	18.407(10)	25.8715(10)
α/°			92.8380(10)		
β/°		111.4543(5)	97.9841(9)		95.1690(6)
γ/°			106.6253(8)		
V/Å^3	5480.5(4)	3280.5(2)	1804.5(2)	3688(4)	3298.1(2)
Z	6	4	2	4	4
λ/Å	0.7848	0.6910	0.6710	0.71073	0.71073
T/K	150(2)	120(2)	150(2)	150(2)	150(2)
Density (calcd.)/Mg/m^3	1.599	1.706	1.559	1.651	1.711
µ/mm^{-1}	5.266	4.225	3.561	4.077	4.609
θ range/°	3.69–33.17	1.19–31.01	2.71–30.94	2.01–25.00	1.58–30.64
Measured reflections	48,822	37,553	22,910	25,092	38,887

Table 6. Cont.

Compound	5a·½OEt₂	5b	5c·¼H₂O	5d·OEt₂	6·(CH₃)₂SO
Independent reflections	19,298	10,642	12,184	3357	10,104
Observed reflections ($F^2 > 2\sigma(F^2)$)	17,145	8283	10,104	2051	7753
R_{int}	0.071	0.063	0.053	0.1504	0.0572
$R[F^2 > 2\sigma(F^2)]$ [a]	0.0542	0.0363	0.0507	0.0746	0.0356
$wR2$ [all data] [b]	0.1551	0.0842	0.1341	0.2065	0.0816
Largest difference map features/eÅ⁻³	1.59, −1.71	1.40, −1.47	2.34, −3.46	2.77, −1.91	1.97, −1.50

[a] $R = \sum ||Fo| - |Fc||/\sum |Fo|$. [b] $wR2 = [\sum[w(Fo^2 - Fc^2)^2]/\sum[w(Fo^2)^2]]^{1/2}$.

Table 7. Selected bond distances and angles for dichloroplatinum(II) compounds 4b·(CH₃)₂SO, 4c·CHCl₃, 4d, and 4e·½CHCl₃·½CH₃OH.

Bond Length (Å)	4b·(CH₃)₂SO	4c·CHCl₃	4d	4e·½CHCl₃·½MeOH [a]
Pt(1)–P(1)	2.223(4)	2.2226(6)	2.2257(6)	2.2386(18) [2.2353(18)]
Pt(1)–P(2)	2.225(4)	2.2186(7)	2.2146(6)	2.2475(18) [2.2464(18)]
Pt(1)–Cl(1)	2.358(4)	2.3625(6)	2.3558(6)	2.3560(18) [2.3574(17)]
Pt(1)–Cl(2)	2.359(4)	2.3484(6)	2.3553(6)	2.3694(17) [2.3616(18)]
Bond angle (°)				
Cl(1)–Pt(1)–P(1)	87.91(14)	86.12(2)	87.63(2)	87.30(7) [86.87(7)]
Cl(1)–Pt(1)–P(2)	174.97(15)	176.08(3)	175.93(2)	176.50(7) [176.84(7)]
Cl(1)–Pt(1)–Cl(2)	88.95(13)	88.68(2)	90.43(2)	88.13(7) [87.34(7)]
Cl(2)–Pt(1)–P(2)	86.71(13)	88.94(2)	85.53(2)	88.54(7) [90.04(7)]
Cl(2)–Pt(1)–P(1)	176.37(15)	174.73(2)	177.81(2)	169.14(7) [170.73(7)]
P(1)–Pt(1)–P(2)	96.35(14)	96.30(3)	96.42(2)	96.17(7) [95.95(7)]

[a] Values in parentheses are for the second independent molecule.

Table 8. Selected bond distances and angles for dichloroplatinum(II) compounds 5a·½OEt₂, 5b, 5c·¼H₂O, 5d·OEt₂, and 6·(CH₃)₂SO.

Bond Length (Å)	5a·½OEt₂ [a]	5b	5c·¼H₂O	5d·OEt₂	6·(CH₃)₂SO
Pt(1)–P(1)	2.233(3) [2.234(3)]	2.2172(9)	2.2268(12)	2.220(3)	2.2219(9)
Pt(1)–P(2)	2.230(3) [2.229(3)]	2.2249(9)	2.2196(12)	[c]	2.2288(9)
Pt(1)–Cl(1)	2.381(3) [2.378(3)]	2.3685(9)	2.347(4) [b]	2.348(3)	2.3421(9)
Pt(1)–Cl(2)	2.361(3) [2.365(3)]	2.3425(9)	2.3638(12)	[c]	2.3618(10)
Bond angle (°)					
Cl(1)–Pt(1)–P(1)	86.31(10) [86.48(10)]	85.73(3)	92.55(11)	89.98(12)	88.81(3)
Cl(1)–Pt(1)–P(2)	177.81(12) [177.51(12)]	176.29(3)	167.1(2)	176.96(13)	173.98(3)
Cl(1)–Pt(1)–Cl(2)	90.20(13) [90.16(13)]	89.33(3)	88.17(11)	87.38(17)	88.78(4)
Cl(2)–Pt(1)–P(2)	87.75(14) [87.51(13)]	88.31(3)	87.76(4)	89.98(12)	87.17(3)
Cl(2)–Pt(1)–P(1)	175.76(12) [175.67(12)]	174.72(3)	178.70(5)	176.96(13)	174.52(4)
P(1)–Pt(1)–P(2)	90.20(13) [95.81(11)]	96.52(3)	91.30(4)	92.62(17)	94.83(3)

[a] Values in parentheses are for the second independent molecule. [b] 2-fold disorder. [c] Molecule lies on a mirror plane.

Despite the *ortho* position of the hydroxy group in **4c**·CHCl₃, molecules do not form an intramolecular S(5) O–H···N interaction as seen in **1a** (Figure 1), instead forming a bifurcated H-bond with the two coordinated chloride ligands of an adjacent molecule (Figure 6). This generates a 1D chain, and also attracts a bifurcated H-bonded chloroform molecule. There are somewhat asymmetric distances d for H(1C) to Cl(1) and Cl(2) are 2.45(4) and 2.76(4) Å, while those from H(34) to Cl(1) and Cl(2) are 2.66 and 2.86 Å, so are also asymmetric. The twist angle between planes P(1)/P(2)/C(1)/C(2) and ring C(3) > C(8) is 84.83(8)°, so is almost perpendicular. Atoms N(1) and Pt(1) lie 0.795(4) and 0.024(2) Å away from the P(1)/P(2)/C(1)/C(2) plane, respectively. The hinge angle across the P(1)–P(2) vector is 2.51(5)°. Selected hydrogen bonding geometric parameters for 4c·CHCl3 are shown in Table 9.

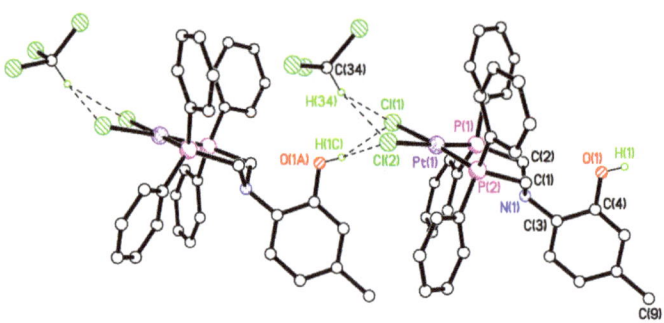

Figure 6. H-bonded packing arrangement in the crystal structure of 4c·CHCl3. Most H atoms omitted for clarity. Symmetry operator A = $x+\frac{1}{2}, -y+\frac{1}{2}, z+\frac{1}{2}$.

Table 9. Selected data ($D\cdots A$/Å, $\angle D$–$H\cdots A$/°) for key inter- and intramolecular contacts for compounds 4b·(CH3)2SO, 4c·CHCl3, 4e·½CHCl3·½CH3OH, 5a·½OEt2, 5b, 5c·¼H2O, 5d·OEt2, and 6·(CH3)2SO.

	4b·(CH3)2SO	4c·CHCl3	4e·½CHCl3·½CH3OH	5a·½OEt2 [a]	5b	5c·¼H2O	5d·OEt2	6·(CH3)2SO
O–H···(O)C_inter						3.714(14), 169		
N–H···N_intra					2.711(5), 107(4)	2.776(12), 108 [b]		
O–H_inter···O_MeOH			2.624(10), 160					
O–H_inter···ClPt		3.145(2), 145(3) 3.361(2), 133(3)	3.197(6), 160(9)			3.065(3), 161(5)		
O–H_inter···O(CH3)2SO	2.716(17), 170							1.79(2), 173(5)
N–H_inter···ClPt				3.328(12), 145(16) 3.320(11), 159(16)			3.505(15), 138(6)	
O–H···(O)C_intra				2.596(13), 157 2.610(13), 175(20)				

[a] Values in parentheses are for the second independent molecule. [b] For the major disorder component; 2.658(12), 117 for the minor component.

Compound 6·(CH3)2SO, in which the –OH group is meta to the N-arene group H-bonds to the DMSO molecule of crystallisation resulting in a 0D structure (Figure 7). The distance d for this H-bond is 1.79(2) Å. The twist angle between plane P(1)/P(2)/C(1)/C(2) and ring C(3) > C(8) is 50.98(12)°. Atoms N(1) and Pt(1) lie 0.758(4) and 0.404(2) Å away from the P(1)/P(2)/C(1)/C(2) plane, respectively, so is more chair-shaped than some of the other platinum(II) complexes reported here. The hinge angle across the P(1)–P(2) vector is 11.87(13)°.

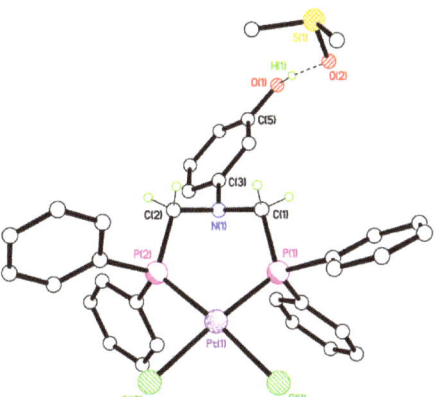

Figure 7. Crystal structure of **6**·$(CH_3)_2SO$ showing the hydroxyl group H-bonding to the $(CH_3)_2SO$ molecule of crystallisation. Most H-atoms omitted for clarity.

For **4d**·$\frac{1}{2}Et_2O$ (Figure 8) a molecule of badly disordered diethyl ether, modelled by the Platon Squeeze procedure, is not shown, but is in the vicinity of the hydroxy group and may H-bond to it resulting in a 0D structure. The twist angle between plane P(1)/P(2)/C(1)/C(2) and ring C(3) > C(8) is 67.82(7)°. Atoms N(1) and Pt(1) lie 0.797(3) and 0.2378(16) Å away from the P(1)/P(2)/C(1)/C(2) plane, respectively. The hinge angle across the P(1)–P(2) vector is 9.20(9)°.

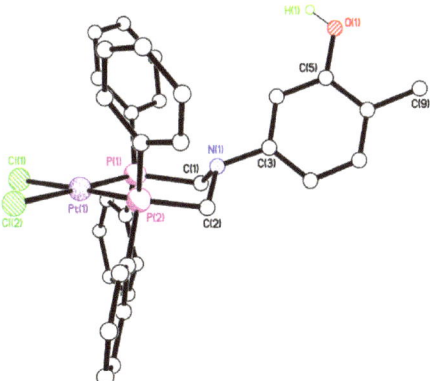

Figure 8. Crystal structure of **4d**·$\frac{1}{2}Et_2O$. Most H atoms and the disordered OEt_2 molecule omitted for clarity.

The crystal structure of **4b**·$(CH_3)_2SO$ shows the hydroxy group H-bonding to the DMSO molecule of crystallisation (Figure 9a). The distance d for this H-bond is 1.89 Å. The twist angle between plane P(1)/P(2)/C(1)/C(2) and ring C(3) > C(8) is 72.2(4)°. Atoms N(1) and Pt(1) lie 0.781(17) and 0.180(10) Å away from the P(1)/P(2)/C(1)/C(2) plane, respectively. The hinge angle across the P(1)–P(2) vector is 8.7(6)°. Molecules form 1D, weakly H-bonded, undulating chains in the c direction via the methylene H atoms on C(1) and C(2) to a single, coordinated chloride ligand in an adjacent molecule (Figure 9b). Selected hydrogen bonding parameters for **4b**·$(CH_3)_2SO$ are shown in Table 9.

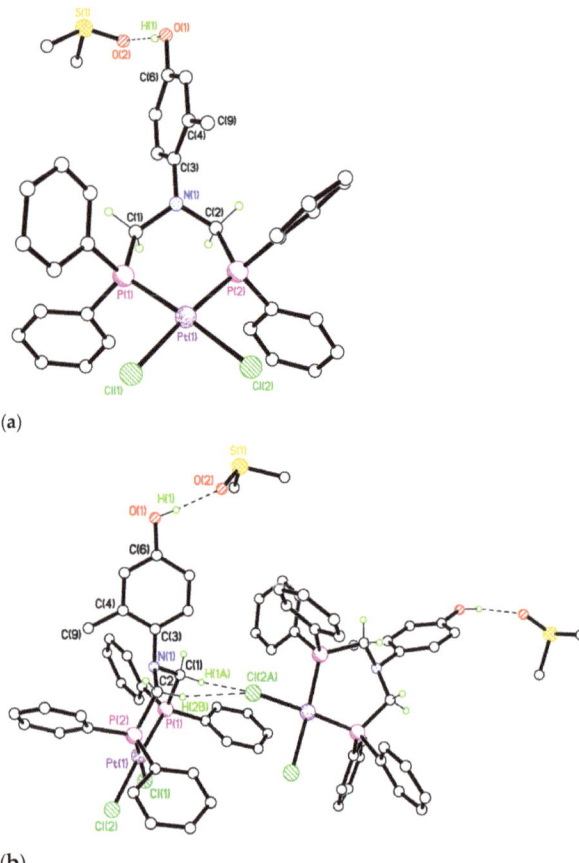

Figure 9. (a) Crystal structure of **4b**·(CH$_3$)$_2$SO showing the hydroxyl group H-bonding to the DMSO molecule of crystallisation. Most H-atoms removed for clarity. (b) Packing interactions in the crystal structure of **4b**·(CH$_3$)$_2$SO. Most H atoms omitted for clarity. Symmetry operator A = y − 1, 1 − x, $\frac{1}{4}$ + z.

For compound **4e**·$\frac{1}{2}$CHCl$_3$·$\frac{1}{2}$CH$_3$OH there are two independent Pt complexes, one CH$_3$OH, and one CHCl$_3$ in the asymmetric unit. Both Pt complexes form 1D chains aligned parallel to *b*, but these chains are different (Figure 10). The chain involving Pt(2) forms simple O–H···Cl H-bonds with the adjacent molecules via the *para* hydroxy group with d = 2.39(4) Å. For the chain involving the Pt(1)-containing molecules, the intermolecular H-bond has an inserted methanol molecule. The distances, *d*, are 2.32(5) and 1.82 Å for H(3)···Cl(2) and H(1A)···O(3), respectively. Atoms N(1)/N(2) and Pt(1)/Pt(2) lie 0.765(9)/0.798(9) and 0.424(5)/0.364(5) Å away from the P(1)/P(2)/C(1)/C(2) or P(3)/P(4)/C(34)/C(35) planes, respectively. So, as in **6**·(CH$_3$)$_2$SO, the core 6-membered Pt–P–C–N–C–P rings adopt more chair-shaped conformations. The hinge angles across the P(1)–P(2)/P(3)–P(4) vectors are 13.44(16)/12.47(16)°. The twist angles between planes P(1)/P(2)/C(1)/C(2) or P(3)/P(4)/C(34)/C(35) and rings C(3) > C(8) or C(36) > C(41) are 88.17(19)/54.62(15)°. So, while the other geometric parameters are similar between the two molecules, this twist angle is significantly different.

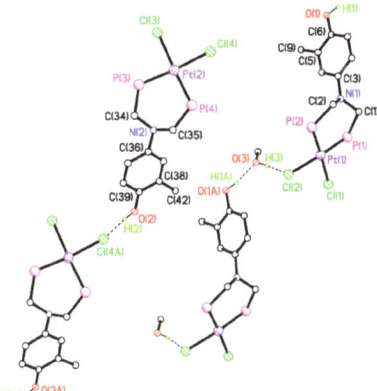

Figure 10. H-bonded packing motifs in the crystal structure of **4e**·$\frac{1}{2}$CHCl$_3$·$\frac{1}{2}$CH$_3$OH. Most H atoms, two Ph groups per P atom, and the disordered chloroform of crystallisation which is not involved in any significant intermolecular interactions, are omitted for clarity. Symmetry operators are x, y − 1, z and x, y + 1, z.

In **5c**, the amide and ring atoms from C(4) > C(11) are disordered over two sets of almost equally occupied positions. The disorder highlights two or more chain-forming possibilities for this structure, analogous to that observed in in **4e**·$\frac{1}{2}$CHCl$_3$·$\frac{1}{2}$CH$_3$OH, with one possibility being simple (hydroxyl)O–H···O(amide) links (Figure 11a), while the other, shown in Figure 11b, shows an alternative, water-inserted linkage. There is also likely to be some alternation of these motifs, given the random disorder and approx. 25% occupancy observed for water atom O(3). Unlike almost all of the other structures herein, the core 6-membered Pt–P–C–N–C–P ring adopts a conformation with atoms Pt(1)/P(1)/P(1)/C(2) being in a plane and atoms C(1) and N(2) being 1.021(6) and 1.237(6) Å, respectively, away from that plane. There is no C=O···HN intermolecular H-bonding observed between molecules. Instead, the amide N*H* forms a bifurcated H-bond with the two neighbouring acceptor atoms N(1) and the *ortho* hydroxyl O(2) with *d* = 2.37 and 2.28 Å, respectively, while *d* = 2.89 Å for H(2)···O(1A).

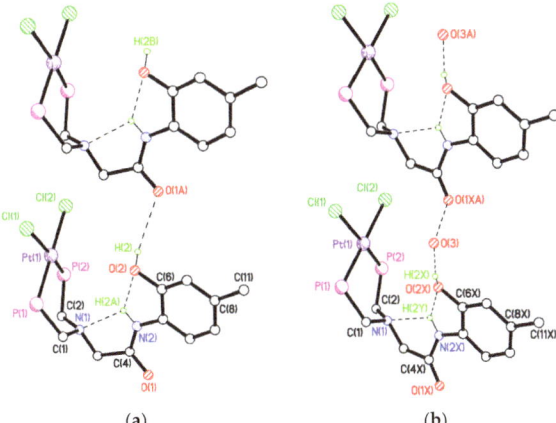

Figure 11. Most H atoms and 2 Ph groups per P atom have been omitted for clarity. (**a**) Packing motif 1 in the crystal structure of **5c**. Symmetry operator A = x + 1, y, z. (**b**) Packing motif 2 in the crystal structure of **5c**. The true structure is most likely an alternation of motifs 1 and 2. Symmetry operator A = x + 1, y, z.

In the second motif, adjacent molecules have an inserted water molecule in the H-bond pattern (Figure 11b). The amide N*H* again forms a bifurcated H-bond with the two neighbouring acceptor atoms N(1) and O(2X) with d = 2.14 and 2.25 Å, respectively, while d = 2.89 Å for H(2X)···O(3), which is a little long, and d for O(3)···O(1XA) = 2.21(3) Å, which is rather short. The distance d from water oxygen O(3) to O(1A), however, is entirely reasonable for an H-bond at 2.74 Å, suggesting a predominantly alternating pattern between the two disorder options is most likely.

Complex **5a**·½Et$_2$O was crystallised from a diethyl ether solution, including half a solvent molecule per complex molecule in the crystal lattice. There are two Pt complexes and two, half-occupied, Et$_2$O solvent molecules of crystallisation in the asymmetric unit. The packing adopted by this second complex with an *ortho* hydroxyl group is very different to **5c** (Figure 12). Here there is no intramolecular N–H···N H-bond, instead the *ortho* hydroxyl forms an intramolecular H-bond with the amide oxygen with d = 1.80 and 1.77(4) Å in the molecules containing Pt(1) and Pt(2), respectively. This does leave the two unique amide N*H* atoms free to form intermolecular interactions, which they do via highly asymmetric, bifurcated H-bonds with the coordinated chloride ligands on adjacent Pt complexes. From H(2) d = 2.60(11) and 2.95(13) Å to Cl(3) and Cl(4), respectively, while d = 2.52(7) and 3.12(15) Å from H(4) to Cl(1A) and Cl(2A), respectively. N(1)/N(3) and Pt(1)/Pt(2) lie 0.771(13)/0.781(14) and 0.349(8)/0.346(8) Å out of the planes P(1)/P(2)/C(1)/C(2) and P(3)/P(4)/C(37)/C(38), respectively. The twist angle between planes P(1)/P(2)/C(1)/C(2) and P(3)/P(4)/C(37)/C(38) relative to the rings C(5) > C(10) and C(41) > C(46) are 51.3(5) and 51.71(4)°, respectively. Hinge angles across P(1)–P(2) and P(3)–P(4) are 12.3(5) and 12.0(4)°, respectively. Differences between the two systems involving *ortho* hydroxyl groups are the position of the methyl ring substituent in the *meta* or *para* position, and the co-crystallised solvent being a small amount of water or Et$_2$O. Either, or both of these differences might account for the different intra- and intermolecular packing motifs observed. Selected hydrogen bonding parameters for **5a**·½Et$_2$O are shown in Table 9.

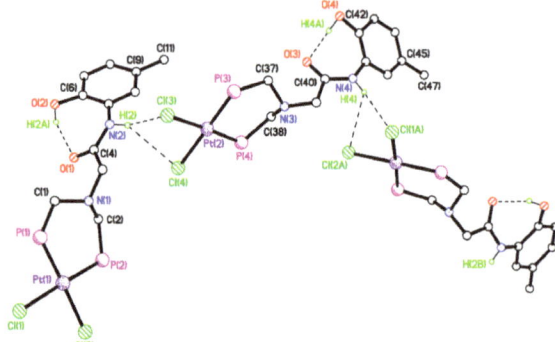

Figure 12. Packing motif in the crystal structure of **5a**·½Et$_2$O. Most H atoms, two Ph groups per P atom and the two, half-occupied, Et$_2$O molecules have been omitted for clarity.

Molecules of **5d**·Et$_2$O lie on a mirror plane, passing through Pt(1), between pairs of P and Cl atoms, and including the atoms from N(1) to the terminal hydroxy-substituted ring. Again, here the amide N*H* is involved in the 1D chain propagation (Figure 13), forming a symmetrical bifurcated H-bond with the two coordinated chloride ligands on the adjacent molecule with d = 2.66(15) Å. Supporting this is an additional (Ar)C–H(5)···Pt(1) interaction at 2.78 Å. The twist angle between the P(1)/P(1A)/C(1)/C(1A) plane and the ring C(4) > C(9) = 90° due to the imposed crystallographic symmetry. The hinge angle at P(1)–P(1A) = 29.5(5)°. Atoms N(1) and Pt(1) lie 0.79(2) and 0.782(14) Å away from the P(1)/P(2)/C(1)/C(2) plane, respectively. So, this is the most chair shaped core

Pt–P–C–N–C–P 6-membered ring. The *meta* hydroxyl group is not involved in the chain propagating intermolecular interactions and points into a cleft between a pair of Ph rings. It does not make an H-bond with the solvent of crystallisation.

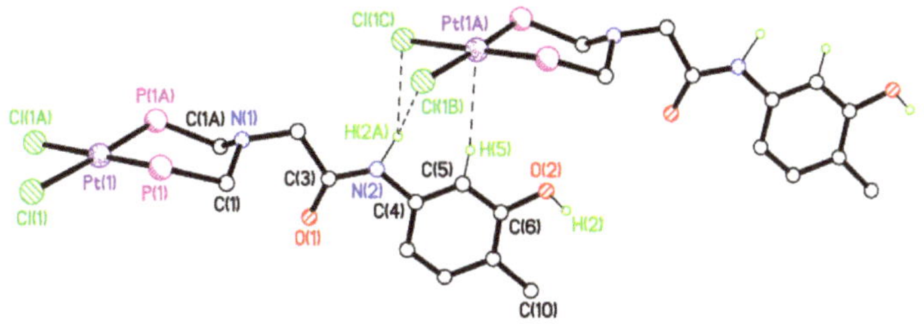

Figure 13. Packing plot of **5d**·Et$_2$O. Most H atoms, two Ph groups per P atom, and a diordered Et$_2$O molecule modelled by the Platon Squeeze procedure, are omitted for clarity. Symmetry operators: (i) for the mirror x, y, $-z + \frac{1}{2}$, (ii) for the chain direction x + 1, y, z.

For compound **5b**, the *para* position of the hydroxyl group facilitates 1D chain formation, forming an H-bond with one of the chloride ligands on an adjacent molecule with d = 2.09(6) Å (Figure 14). The amide NH here forms the familiar, but not universal, H-bond with the amine N(1) with d = 2.29(5) Å. The twist angle between the P(1)/P(2)/C(1)/C(2) plane and the ring C(5) > C(10) = 68.39(12)°. The hinge angle at P(1)–P(1A) = 4.95(10)°. Atoms N(1) and Pt(1) lie 0.810(4) and 0.164(3) Å away from the P(1)/P(2)/C(1)/C(2) plane, respectively.

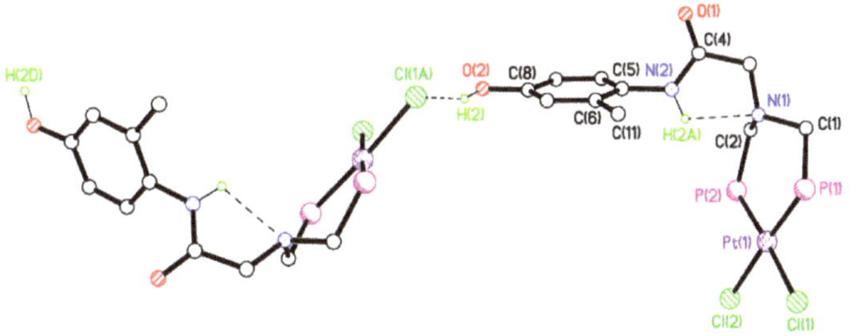

Figure 14. Packing plot in the crystal structure of **5b**. Phenyl groups and hydrogen atoms not involved in hydrogen bonding have been omitted for clarity.

3. Conclusions

In summary, we have shown that the position of the OH/CH$_3$ groups with respect to the N-arene, the inclusion of an amide spacer, and the solvent used in the crystallisation can dictate the solid-state packing behaviour of both non coordinated and *cis*-PtCl$_2$ bound diphosphine ligands. Unsurprisingly, the use of highly polar solvents (DMSO, CH$_3$OH) in this study has been shown to play an important role in disrupting packing behaviour. Our work reinforces the importance of substituent effects, not only those commonly associated with $-$PR$_2$ groups which may be alkyl or aryl based [37,38], but also those functional moieties positioned on the arene group of the central tertiary amine.

4. Materials and Methods

4.1. General Procedures

The synthesis of ligands **1a–e**, **2a–g**, and **3** were undertaken using standard Schlenk-line techniques and an inert nitrogen atmosphere. Ph$_2$PCH$_2$OH was prepared according to a known procedure [39]. All coordination reactions were carried out in air, using reagent grade quality solvents. The compound PtCl$_2$(η^4-cod) (cod = cycloocta-1,5-diene) was prepared according to a known procedure [40]. All other chemicals were obtained from commercial sources and used directly without further purification

4.2. Instrumentation

Infrared spectra were recorded as KBr pellets on a Perkin-Elmer Spectrum 100S (4000–250 cm^{-1} range) Fourier-Transform spectrometer. ^{1}H NMR spectra (400 MHz) were recorded on a Bruker DPX-400 spectrometer with chemical shifts (δ) in ppm to high frequency of Si(CH$_3$)$_4$ and coupling constants (J) in Hz. ^{31}P{^{1}H} NMR (162 MHz) spectra were recorded on a Bruker DPX-400 spectrometer with chemical shifts (δ) in ppm to high frequency of 85% H$_3$PO$_4$. NMR spectra were measured in CDCl$_3$ or (CD$_3$)$_2$SO at 298 K. Elemental analyses (Perkin-Elmer 2400 CHN Elemental Analyser) were performed by the Loughborough University Analytical Service within the Department of Chemistry.

4.3. Preparation of Ligands 1a–e, 2a–g, and 3

The following general procedure was used for the synthesis of 1a–e, 2a–g, and 3. A mixture of Ph$_2$PCH$_2$OH (2 equiv.) and the appropriate amine (1 equiv.) in CH$_3$OH (20 mL) was stirred under N$_2$ for 24 h. The volume of the solution was evaporated to ca. 2–3 mL, under reduced pressure, to afford the desired ligands which were collected by suction filtration (except 2a–c) and dried *in vacuo*. Isolated yields in range 38–97%. Characterising details are given in Table 1.

4.4. Preparation of cis-Dichloroplatinum(II) Phosphine Complexes 4a–e, 5a–g, and 6

The following general procedure was used for the synthesis of 4a–e, 5a–g, and 6. To a solution of PtCl$_2$(η^4-cod) (1 equiv.) in CH$_2$Cl$_2$ (5 mL) was added a solution of the appropriate ligand (1 equiv.) in CH$_2$Cl$_2$ (5 mL). The colourless (or pale yellow) solution was stirred for 30 min at r.t., evaporated to ca. 2–3 mL under reduced pressure, and diethyl ether (10 mL) added. The solids were collected by suction filtration and dried *in vacuo*. Isolated yields in range 73–99%. Characterising details are given in Table 4.

4.5. Single Crystal X-ray Crystallography

Suitable crystals of 1a, 1b·CH$_3$OH, 2f·CH$_3$OH, and 3 were obtained by slow evaporation of a CH$_3$OH solution whereas 2g was obtained by vapour diffusion of Et$_2$O into a CDCl$_3$/CH$_3$OH solution. Crystals of 4b·(CH$_3$)$_2$SO, 5a·½Et$_2$O, 5b, and 5c·¼H$_2$O were obtained by slow diffusion of Et$_2$O into a CDCl$_3$/(CH$_3$)$_2$SO/CH$_3$OH solution. Slow diffusion of hexanes [for 6·(CH$_3$)$_2$SO] into a CDCl$_3$/(CH$_3$)$_2$SO solution or vapour diffusion of Et$_2$O into a CHCl$_3$/(CH$_3$)$_2$SO/CH$_3$OH [for 4c·CHCl$_3$, 4e·½CHCl$_3$·½CH$_3$OH] or CH$_2$Cl$_2$/CH$_3$OH (for 5d·Et$_2$O)]. Slow evaporation of a CH$_2$Cl$_2$/Et$_2$O/hexanes solution gave suitable crystals of 4d·½Et$_2$O. Tables 2, 5 and 6 summarise the key data collection and structure refinement parameters. Diffraction data for compounds 1a, 1b·CH$_3$OH, 2f·CH$_3$OH 3, 4b·(CH$_3$)$_2$SO, 4c·CHCl$_3$, 4d 4e·½CHCl$_3$·½CH$_3$OH, 5d·Et$_2$O, and 6·(CH$_3$)$_2$SO, were collected using a Bruker or Bruker-Nonius APEX 2 CCD diffractometer using graphite-monochromated Mo-K$_\alpha$ radiation. Data for compounds 5b and 5c·¼H$_2$O, were collected using a Bruker APEX 2 CCD diffractometer using synchrotron radiation at Daresbury SRS Station 9.8 or 16.2 SMX for 5a·½Et$_2$O. Data for compound 2g was collected using a Bruker SMART 1000 CCD diffractometer using graphite-monochromated Mo-K$_\alpha$ radiation. All structures were solved by direct methods [except structures 4b·(CH$_3$)$_2$SO, 5a·½Et$_2$O, and 5b which were solved using Patterson synthesis] and refined by full-matrix least-squares methods on F^2. All CH atoms were placed in geometrically calculated positions and were refined using a riding model (aryl C–H 0.95 Å, methyl C–H 0.98 Å, methylene C–H 0.99 Å. Where data quality allowed, OH and NH atom coordinates and U_{iso} were freely refined, or refined with mild geometrical restraints; otherwise, they were placed geometrically with O/N–H = 0.84 Å. U_{iso}(H) values were set to be 1.2 times U_{eq} of the carrier atom for aryl CH and NH, and 1.5 times U_{eq} of the carrier atom for OH and CH$_3$. Throughout the text and tabulated data, where H atom geometry does not include a SU, the coordinates were constrained. Unless stated, all structural determinations proceeded without the need for restraints or disorder modelling. Where disorder was modelled it was supported with appropriate geometrical and U value restraints. In **1b**·CH_3OH, the methanol was modelled as disordered over two equally occupied sets of positions. In **2f**·CH_3OH the methanol was modelled using the Platon Squeeze procedure [41]. Compound **3** was found to contain a disordered methanol and was modelled over two sets of positions, each at half weight. In **4d**·$\frac{1}{2}Et_2O$, atoms C(1) > C(7) and N(1) were modelled with U value restraints. The Et_2O was modelled using Platon Squeeze due to significant disorder. In **4e**·$\frac{1}{2}CHCl_3$·$\frac{1}{2}CH_3OH$ the chloroform molecule was modelled over two sets of positions with major occupancy 57.1(7)% Restraints were applied to that molecule and also ring C(55) > C(60). In **5a**·$\frac{1}{2}Et_2O$ three Ph rings were modelled as disordered over two sets of positions with occupancies close to 50%. Restraints were applied to these rings and also the two half-occupancy Et_2O solvent molecules of crystallisation. In **5c**·$\frac{1}{4}H_2O$, atoms Cl(1) and C(3) > C(11), O(1), O(2) and N(1) were modelled as split over two sets of positions with major occupancy 56(4) and 50.9(6)%, respectively and restraints were applied. In **5d**·Et_2O the Et_2O was modelled as a diffuse area of electron density by the Platon Squeeze procedure and restraints were applied to atoms C(1) > C(10), C(11) > C(22) and N(2) O(2). In **6**·$(CH_3)_2SO$ the DMSO was modelled with restraints as disordered over two sets of positions with major component 71.0(5)% and with C(33) coincident for both components Programs used during data collection, refinement and production of graphics were Bruker SMART, Bruker APEX 2, SAINT, SHELXTL, COLLECT, DENZO and local programs [41–51]. CCDC 2101643-2101656 contain the supplementary crystallographic data for this paper. These data can be obtained free of charge from The Cambridge Crystallographic Data Centre via www.ccdc.cam.ac.uk/structures (accessed on 3 November 2021).

Author Contributions: Conceptualisation, M.B.S.; synthesis and characterisation of the compounds, N.M.S.-B., P.D.; single crystal X-ray crystallography, N.M.S.-B., M.R.J.E.; writing-original draft preparation, M.B.S.; writing-review and editing, M.R.J.E., M.B.S. All authors have read and agreed to the published version of the manuscript.

Funding: This research received no external funding.

Institutional Review Board Statement: Not applicable.

Informed Consent Statement: Not applicable.

Data Availability Statement: Not applicable.

Acknowledgments: We thank the EPSRC Centre for Doctoral Training in Embedded Intelligence under grant reference EP/L014998/1 for financial support (PD). Johnson Matthey are acknowledged for their kind donation of precious metals and the UK National Crystallography Service at the University of Southampton for three of the data collections. The STFC is thanked for the allocation of beam time at Daresbury Laboratory.

Conflicts of Interest: The authors declare no conflict of interest.

Sample Availability: Samples of the compounds in this article are not available from the authors.

References

1. Lehn, J.-M. Supramolecular Chemistry-Scope and Perspectives. Molecules, Supermolecules, and Molecular Devices (Nobel Lecture). *Angew. Chem. Int. Ed. Engl.* **1988**, *27*, 89–112. [CrossRef]
2. Jongkind, L.J.; Caumes, X.; Hartendorp, A.P.T.; Reek, J.N.H. Ligand Template Strategies for Catalyst Encapsulation. *Acc. Chem. Res.* **2018**, *51*, 2115–2128. [CrossRef] [PubMed]
3. James, S.L. Phosphines as building blocks in coordination-based self-assembly. *Chem. Soc. Rev.* **2009**, *38*, 1744–1758. [CrossRef]
4. Breit, B. Supramolecular Approaches to Generate Libraries of Chelating Bidentate Ligands for Homogeneous Catalysis. *Angew. Chem. Int. Ed.* **2005**, *44*, 6816–6825. [CrossRef] [PubMed]
5. Daubignard, J.; Detz, R.J.; de Bruin, B.; Reek, J.N.H. Phosphine Oxide Based Supramolecular Ligands in the Rhodium-Catalysed Asymmetric Hydrogenation. *Organometallics* **2019**, *38*, 3961–3969. [CrossRef]
6. Koshti, V.S.; Sen, A.; Shinde, D.; Chikkali, S.H. Self-assembly of P-chiral supramolecular phosphines on rhodium and direct evidence for Rh-catalyst-substrate interactions. *Dalton Trans.* **2017**, *46*, 13966–13973. [CrossRef]

7. Vasseur, A.; Membrat, R.; Palpacelli, D.; Giorgi, M.; Nuel, D.; Giordano, L.; Martinez, A. Synthesis of chiral supramolecular bisphosphinite palladcycles through hydrogen transfer-promoted self-assembly process. *Chem. Commun.* **2018**, *54*, 10132–10135. [CrossRef] [PubMed]
8. Romero-Nieto, C.; de Cózar, A.; Regulska, E.; Mullenix, J.B.; Rominger, F.; Hindenberg, P. Controlling the molecular arrangement of racemates through weak interactions: The synergy between p-interactions and halogen bonds. *Chem. Commun.* **2021**, *57*, 7366–7369. [CrossRef] [PubMed]
9. Carreras, L.; Serrano-Torné, M.; van Leeuwen, P.W.N.M.; Vidal-Ferran, A. XBphos-Rh: A halogen-bond assembled supramolecular catalyst. *Chem. Sci.* **2018**, *9*, 3644–3648. [CrossRef]
10. García-Márquez, A.; Frontera, A.; Roisnel, T.; Gramage-Doria, R. Ultrashort $H^{d+}\cdots H^{d-}$ intermolecular distance in a supramolecular system in the solid state. *Chem. Commun.* **2021**, *57*, 7112–7115. [CrossRef]
11. Blann, K.; Bollmann, A.; Brown, G.M.; Dixon, J.T.; Elsegood, M.R.J.; Raw, C.R.; Smith, M.B.; Tenza, K.; Willemse, A.; Zweni, P. Ethylene oligomerisation chromium catalysts with unsymmetrical PCNP ligands. *Dalton Trans.* **2021**, *50*, 4345–4354. [CrossRef] [PubMed]
12. De'Ath, P.; Elsegood, M.R.J.; Halliwell, C.A.G.; Smith, M.B. Mild intramolecular P-C(sp^3) bond cleavage in bridging diphosphine complexes of Ru^{II}, Rh^{III}, and Ir^{III}. *J. Organomet. Chem.* **2021**, *937*, 121704. [CrossRef]
13. Smith, M.B.; Dale, S.H.; Coles, S.J.; Gelbrich, T.; Hursthouse, M.B.; Light, M.E.; Horton, P.N. Hydrogen bonded supramolecular assemblies based on neutral square-planar palladium(II) complexes. *CrystEngCommun* **2007**, *9*, 165–175. [CrossRef]
14. Smith, M.B.; Dale, S.H.; Coles, S.J.; Gelbrich, T.; Hursthouse, M.B.; Light, M.E. Isomeric dinuclear gold(I) complexes with highly functionalised ditertiary phosphines: Self-assembly of dimers, rings and 1-D polymeric chains. *CrystEngCommun* **2006**, *8*, 140–149. [CrossRef]
15. Dann, S.E.; Durran, S.E.; Elsegood, M.R.J.; Smith, M.B.; Staniland, P.M.; Talib, S.; Dale, S.H. Supramolecular chemistry of half-sandwich organometallic building blocks based on $RuCl_2$(p-cymene)Ph_2PCH_2Y. *J. Organomet. Chem.* **2006**, *691*, 4829–4842. [CrossRef]
16. Durran, S.E.; Smith, M.B.; Slawin, A.M.Z.; Gelbrich, T.; Hursthouse, M.B.; Light, M.E. Synthesis and coordination studies of new aminoalcohol functionalised tertiary phosphines. *Can. J. Chem.* **2001**, *79*, 780–791. [CrossRef]
17. Jiang, M.-S.; Tao, Y.-H.; Wang, Y.-W.; Lu, C.; Young, D.J.; Lang, J.-P.; Ren, Z.-G. Reversible Solid-State Phase Transitions between Au-P Complexes Accompanied by Switchable Fluorescence. *Inorg. Chem.* **2020**, *59*, 3072–3078. [CrossRef]
18. Pandey, M.K.; Kunchur, H.S.; Mondal, D.; Radhakrishna, L.; Kote, B.S.; Balakrishna, M.S. Rare Au $\cdots$ H Interactions in Gold(I) Complexes of Bulky Phosphines Derived from 2,6-Dibenzhydryl-4-methylphenyl Core. *Inorg. Chem.* **2020**, *59*, 3642–3658. [CrossRef] [PubMed]
19. Bálint, E.; Tajti, Á.; Tripolszky, A.; Keglevich, G. Synthesis of platinum, palladium and rhodium complexes of α-aminophosphine ligands. *Dalton Trans.* **2018**, *47*, 4755–4778. [CrossRef] [PubMed]
20. Zhang, Y.-P.; Zhang, M.; Chen, X.-R.; Lu, C.; Young, D.J.; Ren, Z.-G.; Lang, J.-P. Cobalt(I) and Nickel(II) Complexes of a PNN Type Ligand as Photoenhanced Electrocatalysts for the Hydrogen Evolution Reaction. *Inorg. Chem.* **2020**, *59*, 1038–1045. [CrossRef]
21. Hou, R.; Huang, T.-H.; Wang, X.-J.; Jiang, X.-F.; Ni, Q.-L.; Gui, L.-C.; Fan, Y.-J.; Tan, Y.-L. Synthesis, structural characterisation and luminescent properties of a series of Cu(I) complexes based on polyphosphine ligands. *Dalton Trans.* **2011**, *40*, 7551–7558. [CrossRef]
22. Wang, X.-J.; Gui, L.-C.; Ni, Q.-L.; Liao, Y.-F.; Jiang, X.-F.; Tang, L.-H.; Zhang, Z.; Wu, Q. p-Stacking induced complexes with Z-shape motifs featuring a complimentary approach between electron-rich arene diamines and electron-deficient aromatic N-heterocycles. *CrystEngComm* **2008**, *10*, 1003–1010. [CrossRef]
23. Au, R.H.W.; Jennings, M.C.; Puddephatt, R.J. Supramolecular Organoplatinum(IV) Chemistry: Sequential Introduction of Amide Hydrogen Bonding Groups. *Organometallics* **2009**, *28*, 3754–3762. [CrossRef]
24. Coles, N.T.; Gasperini, D.; Provis-Evans, C.B.; Mahon, M.F.; Webster, R.L. Heterobimetallic Complexes of 1,1-Diphosphineamide Ligands. *Organometallics* **2021**, *40*, 148–155. [CrossRef]
25. Navrátil, M.; Císařová, I.; Alemayehu, A.; Škoch, K.; Štěpnička, P. Synthesis and Structural Characterisation of an N-Phosphanyl Ferrocene Carboxamide and its Ruthenium, Rhodium and Palladium Complexes. *ChemPlusChem* **2020**, *85*, 1325–1338. [CrossRef]
26. Navrátil, M.; Císařová, I.; Štěpnička, P. Intermolecular interactions in the crystal structures of chlorogold(I) complexes with N-phosphinoamide ligands. *Inorg. Chim. Acta* **2021**, *516*, 120138. [CrossRef]
27. Pachisia, S.; Kishan, R.; Yadav, S.; Gupta, R. Half-Sandwich Ruthenium Complexes of Amide-Phosphine Based Ligands: H-Bonding Cavity Assisted Binding and Reduction of Nitro-substrates. *Inorg. Chem.* **2021**, *60*, 2009–2022. [CrossRef] [PubMed]
28. Nasser, N.; Eisler, D.J.; Puddephatt, R.J. A chiral diphosphine as *trans*-chelate ligand and its relevance to catalysis. *Chem. Commun.* **2010**, *46*, 1953–1955. [CrossRef] [PubMed]
29. Elsegood, M.R.J.; Lake, A.J.; Smith, M.B.; Weaver, G.W. Ditertiary phosphines bearing a -N-C-C(O)-N(H)- linker and their corresponding dichloroplatinum(II) complexes. *Phosphorus Sulfur Silicon Relat. Elem.* **2019**, *194*, 540–544. [CrossRef]
30. Hoyos, O.L.; Bermejo, M.R.; Fondo, M.; García-Deibe, A.; González, A.M.; Maneiro, M.; Pedrido, R. Mn(III) complexes with asymmetrical N_2O_3 Schiff bases. The unusual crystal structure of [Mn(phenglydisal-3-Br,5-Cl)(dmso)] (H_3phenglydisal = 3-aza-*N*-{2-[1-aza-2-(2-hydroxyphenyl)vinyl]phenyl}-4-(2-hydroxyphenyl)but-3-enamide), a mononuclear single-stranded helical manganese(III) complex. *J. Chem. Soc. Dalton Trans.* **2000**, 3122–3127. [CrossRef]

31. Bermejo, M.R.; González, A.M.; Fondo, M.; García-Deibe, A.; Maneiro, M.; Sanmartín, J.; Hoyos, O.L.; Watkinson, M. A direct route to obtain manganese(III) complexes with a new class of asymmetrical Schiff base ligands. *New J. Chem.* **2000**, *24*, 235–241. [CrossRef]
32. Elsegood, M.R.J.; Smith, M.B.; Staniland, P.M. Neutral Molecular Pd$_6$ Hexagons Using k^3-P$_2$O Terdentate Ligands. *Inorg. Chem.* **2006**, *45*, 6761–6770. [CrossRef] [PubMed]
33. Etter, M.C.; MacDonald, J.C.; Bernstein, J. Graph-Set Analysis of Hydrogen-Bond Patterns in Organic Crystals. *Acta Crystallogr.* **1990**, *B46*, 256–262. [CrossRef]
34. Etter, M.C. Encoding and Decoding Hydrogen-Bond Patterns of Organic Compounds. *Acc. Chem. Res.* **1990**, *23*, 120–126. [CrossRef]
35. Bernstein, J.; Davis, R.E.; Shimoni, L.; Chang, N.-L. Patterns in Hydrogen Bonding: Functionality and Graph Set Analysis in Crystals. *Angew. Chem. Int. Ed. Engl.* **1995**, *34*, 1555–1573. [CrossRef]
36. Desiraju, G.; Steiner, T. *The Weak Hydrogen Bond*; Oxford University Press: Oxford, UK, 2001.
37. Klemps, C.; Payet, E.; Magna, L.; Saussine, L.; Le Goff, X.F.; Le Floch, P. PCNCP Ligands in the Chromium-Catalysed Oligomerisation of Ethylene: Tri-versus Tetramerization. *Chem. Eur. J.* **2009**, *15*, 8259–8268. [CrossRef]
38. Walsh, A.P.; Laureanti, J.A.; Katipamula, S.; Chambers, G.M.; Priyadarshani, N.; Lense, S.; Bays, J.T.; Linehan, J.C.; Shaw, W.J. Evaluating the impacts of amino acids in the second and outer coordination spheres of Rh-bis(diphosphine) complexes for CO$_2$ hydrogenation. *Faraday Discuss.* **2019**, *215*, 123–140. [CrossRef] [PubMed]
39. Hellman, H.; Bader, J.; Birkner, H.; Schumacher, O. Hydroxymethyl-phosphine, Hydroxymethyl-phosphoniumsalze und Chlormethyl-phosphoniumsalze. *Justus Liebigs Ann. Chem.* **1962**, *659*, 49–56. [CrossRef]
40. McDermott, J.X.; White, J.F.; Whitesides, G.M. Thermal Decomposition of Bis(phosphine)platinum(II) Metallocycles. *J. Am. Chem. Soc.* **1976**, *98*, 6521–6528. [CrossRef]
41. Spek, A.L. PLATON SQUEEZE: A tool for the calculation of the disordered solvent contribution to the calculated structure factors. *Acta Crystallogr. Sect. C-Struct. Chem.* **2015**, *71*, 9–18. [CrossRef]
42. Sluis, P.v.d.; Spek, A.L. BYPASS: An effective method for the refinement of crystal structures containing disordered solvent regions. *Acta Crystallogr.* **1990**, *A46*, 194–201. [CrossRef]
43. *Bruker SMART Version 5.611*; Bruker AXS Inc.: Fitchburg, WI, USA, 2001.
44. *Area-Detector Integration Software, APEX-II, Version V1*; Bruker-Nonius: Madison, WI, USA, 2004.
45. Denzo, Z.; Otwinowski, W. Processing of X-ray diffraction data in oscillation mode, Methods in Enzymology. In *Macromolecular Crystallography*; Carter, C.W., Jr., Sweet, R.M., Eds.; Academic Press: Cambridge, MA, USA, 1997; Volume 276, pp. 307–326.
46. Hooft, R.W.W. *COLLECT: Data Collection Software*; Nonius B.V.: Delft, The Netherlands, 1998.
47. *SAINT Software for CCD Diffractometers*; Bruker AXS Inc.: Madison, WI, USA, 2004.
48. Krause, L.; Herbst-Irmer, R.; Sheldrick, G.M.; Stalke, D.J. SADABS software. *J. Appl. Cryst.* **2015**, *48*, 3–10. [CrossRef] [PubMed]
49. Sheldrick, G.M. Crystal structure refinement with SHELXL. *Acta Crystallogr. Sect.* **2015**, *A71*, 3–8. [CrossRef]
50. Sheldrick, G.M. A short history of SHELX. *Acta Crystallogr. Sect.* **2008**, *A64*, 112–122. [CrossRef] [PubMed]
51. Sheldrick, G.M. *SHELXTL User Manual, Version 6.12*; Bruker AXS Inc.: Madison, WI, USA, 2001.

Article

Solvothermal Synthesis of a Novel Calcium Metal-Organic Framework: High Temperature and Electrochemical Behaviour

Russell M. Main [1], David B. Cordes [1], Aamod V. Desai [1,2], Alexandra M. Z. Slawin [1], Paul Wheatley [1], A. Robert Armstrong [1,2] and Russell E. Morris [1,2,*]

1 EaStCHEM School of Chemistry, Purdie Building, North Haugh, St Andrews KY16 9ST, UK; rmm29@st-andrews.ac.uk (R.M.M.); dbc21@st-andrews.ac.uk (D.B.C.); avd6@st-andrews.ac.uk (A.V.D.); amzs@st-andrews.ac.uk (A.M.Z.S.); psw@st-andrews.ac.uk (P.W.); ara@st-andrews.ac.uk (A.R.A.)
2 Harwell Science and Innovation Campus, The Faraday Institution, Quad One, Didcot OX11 0RA, UK
* Correspondence: rem1@st-andrews.ac.uk; Tel.: +44-133-446-3818

Abstract: The rapid growth in the field of metal-organic frameworks (MOFs) over recent years has highlighted their high potential in a variety of applications. For biological and environmental applications MOFs with low toxicity are vitally important to avoid any harmful effects. For this reason, Ca-based MOFs are highly desirable owing to their low cost and high biocompatibility. Useful Ca MOFs are still rare owing to the ionic character and large size of the Ca^{2+} ion tending to produce dense phases. Presented here is a novel Ca-based MOF containing 2,3-dihyrdoxyterephthalate (2,3-dhtp) linkers Ca(2,3-dhtp)(H$_2$O) (SIMOF-4). The material undergoes a phase transformation on heating, which can be followed by variable temperature powder X-ray diffraction. The structure of the high temperature form was obtained using single-crystal X-ray diffraction. The electrochemical properties of SIMOF-4 were also investigated for use in a Na ion battery.

Keywords: MOF; calcium MOF; electrochemistry; scXRD; VTXRD; bioMOF

1. Introduction

Metal-organic frameworks (MOFs) are a rapidly growing field in materials chemistry [1]. They are formed from metal ions or oxoclusters, called secondary building units (SBUs), connected by polydentate organic molecules creating interconnected 3D networks [1]. The vast array of metal ions and organic linkers available means that a huge variety of structures with a wide range of topologies have already been discovered, and the designability and tunability of MOFs have been much discussed [2]. By definition [3] MOFs must have potential void space within the framework and MOFs have been discovered with internal surface areas up to around 10,000 m^2/g [4]. The mix of organic and metallic components can provide a variety of novel functionalities [5–21], and while many applications for MOFs rely on their large porosities, recently more advanced MOFs have been developed that can respond to external stimuli (heat, pressure, light etc.,) [5] as well as MOFs that are stable in aggressive conditions [6]. Gas adsorption is one of the most well studied areas, with adsorption of fuels such as H$_2$ [7] and CH$_4$ [8], as well as carbon capture and storage [9]. Filtrations/separations are also possible [10], from improvements to industrial monomer refinements [11] to water and air purifications [12,13]. Other applications include, but are not limited to: catalysis [14], sensing [15] and anodes for batteries [16].

Of great interest to us is the use of MOFs in biological and environmental applications [17,18], particularly their use as drug carriers [19] and for the storage and release of biologically active NO gas [20,21]. To be used in biological systems it is important that the MOFs are non-toxic [17]. This is particularly true of the metal ions used: transition metals such as Ni and Cr are commonly used in MOF chemistry [1], but these metals are highly toxic [22]. Ca-based MOFs are highly desirable for biological applications because of their low cost and high biocompatibility [23]. Ca is important for bone health and many other biological processes, on average making up 1.4 wt.% of the human body [24]. Several Ca MOFs have been reported, however the high coordination number and ionic character of the Ca^{2+} ion tend to produce dense phases or Ca layers separated by the linkers [25,26]. We have previously reported a Ca MOF for NO delivery for use in biomedical applications [23].

The use of Ca MOFs is not just limited to biology, some Ca MOFs have been reported that show potential as anodic materials in Na ion batteries [27]. The potential of Na ion batteries has been widely discussed, with the aim of using them on a large scale for energy storage [28].

The stability of MOFs is also an important consideration, as they are rarely the thermodynamic product of a reaction and are instead kinetically stable intermediates [29]. The structures of MOFs are susceptible to collapse due to the application of heat, varying pH and the removal of internal solvent [6]. Some MOFs can undergo phase transitions on heating caused by rearrangement to a more thermodynamically stable product and/or the removal of internal solvent molecules [30,31]. Typically, these phase transitions cause structural damage owing to large changes in unit cell volume, the phase change going through an amorphous intermediate [32] and/or capillary forces caused by loss of solvent applying large stresses to the structure [1]. This can make structure determination of high temperature desolvated phases by scXRD challenging [32], as single crystals are almost inevitably damaged by this process. Often only a partial structure solution is possible.

Reported here is a new calcium MOF, SIMOF-4, based on 2,3-dihyrdoxyterephthalate (2,3-dhtp) linkers, with a 3D interconnected framework. Its crystal structure has been solved by scXRD, it has been characterised using PXRD and TGA. A high temperature transition to a layered material has been observed and its structure is determined using SCXRD on a damaged crystal. The structure was then confirmed using PXRD. Finally, its preliminary suitability for use as a battery anode have been assessed by measuring its electrochemical properties.

2. Results and Discussion

2.1. Synthesis Route

The linker 2,3-dihydroxyterephthalic acid (2,3-dhtpH$_2$) was synthesised from the carboxylation of catechol using the Kolbe-Schmidt method [33]. This involved heating dry catechol in a CO_2 atmosphere in the presence of excess potassium bicarbonate. This produced the potassium salt of 2,3-dihyrdoxyterephthalate from which 2,3-dhtpH$_2$ could be formed on acidification.

SIMOF-4 was synthesised using a solvothermal technique. The reactants were dissolved in solvent and heated in an autoclave to 120 °C for 3 days. Acetic acid was used as the modulator to aid crystal growth [1]. Adding p-xylene (pX) caused a reduction in yield but an increase in crystal size allowing for scXRD analysis to be completed. The nature of the effect of pX is still undetermined: it has been known to act as a templating agent [34], but in this case we suggest that it is only acting to hinder nucleation and therefore improve crystal size.

2.2. Description of Crystal Structure

SIMOF-4 crystallises in the triclinic space group $P\bar{1}$. Its structure consists of a three-dimensional network based on a [Ca(2,3-dhtp)(H$_2$O)]$_2$ asymmetric unit (Figure 1), which contains two unique Ca environments.

Ca2 is 7-coordinate with a distorted pentagonal bipyramidal geometry, and forms a dimer with a symmetry-related Ca via two bridging μ_2 water molecules (O23) (Figure 2). The intermetallic distance is 4.0123(7) Å and the O-Ca-O angle is 71.80(5)°. The Ca2 is further coordinated to both phenolic groups of a μ_4 dhtp, with a bite angle of 64.56(4)° and two μ_3 dhtp groups through a carboxylic acid oxygen, a water molecule makes up the remaining coordination site (Figure 3). All Ca-O bonds are between 2.2678(15) Å and 2.5167(14) Å, all O-Ca-O angles are between 64.56(4)° and 170.36(5)°.

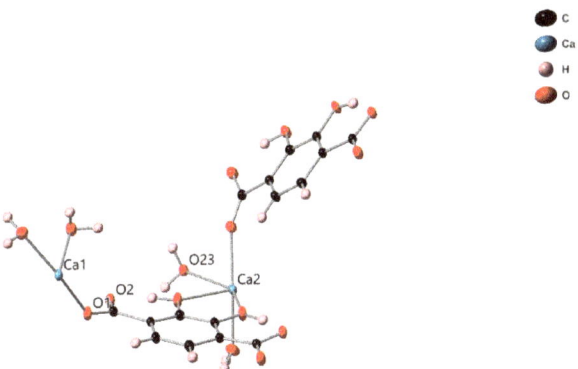

Figure 1. Thermal ellipsoid plot (50% probability ellipsoids) showing the asymmetric unit of SIMOF-4.

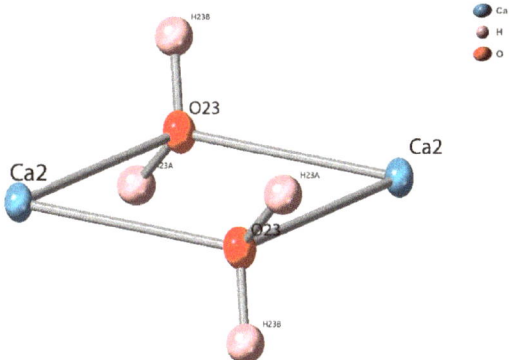

Figure 2. View of the Ca2 dimer of SIMOF-4 (50% probability ellipsoids).

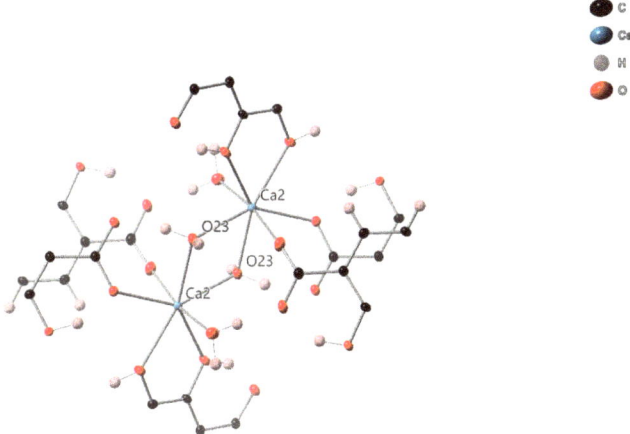

Figure 3. View of the bonding environment around the Ca2 dimer in SIMOF-4 (50% probability ellipsoids).

Ca1 is also 7 coordinates with a distorted pentagonal bipyramidal geometry, and forms a dimer with a symmetry-related Ca, in this case via two μ_4 dhtp carboxylic acid groups (O1, C1, O2), (Figure 4). The intermetallic distance is 5.035(11) Å. The Ca1 is further coordinated to both phenolic groups of a μ_3 dhtp, with a bite angle of 64.21(4)°, and two µ4 dhtp groups via carboxylic acid oxygens, a water molecule makes up the remaining coordination site (Figure 5). All Ca-O bonds are between 2.2917(15) Å and 2.4646(13) Å all O-Ca-O angles are between 64.21(4)° and 167.64(5)°.

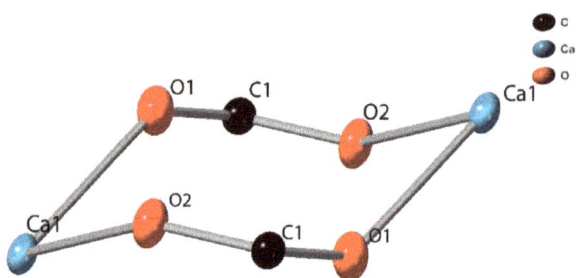

Figure 4. View of the Ca1 dimer unit of SIMOF-4 (50% probability ellipsoids).

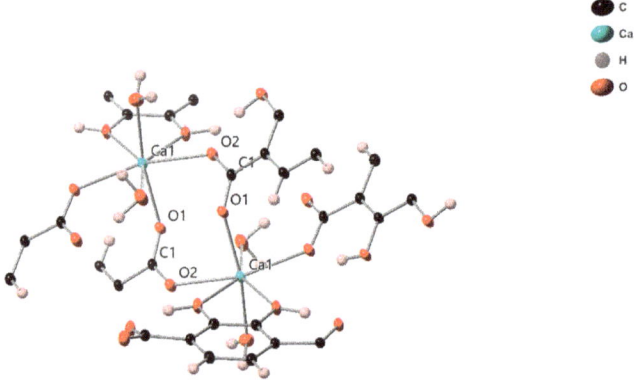

Figure 5. View of the bonding environment around the Ca1 dimer in SIMOF-4 (50% probability ellipsoids).

This high connectivity produces an interconnected 3D network (Figures 6, S1 and S2, Supplementary Materials). SIMOF-4 gains further stabilisation from hydrogen bonding, with each unbonded carboxylic oxygen hydrogen bonded to a water molecule with O···O separations between 2.6379(64) Å and 2.6947(67) Å. This structure appears non-porous. However, there are potential small voids occupied by bound water molecules. This indicates that removal of these water molecules could allow access of small molecules into the structure.

2.3. Characterisation

2.3.1. Powder X-ray Diffraction

The powder X-ray diffraction (PXRD) data pattern of SIMOF-4 is shown in Figure 7. The positions of the reflections are consistent with the theoretical pattern calculated from the single crystal X-ray structure.

2.3.2. Thermo-Gravimetric Analysis

Thermo-gravimetric analysis (TGA) shows multiple mass losses (Figure 8). The 12 wt.% mass loss between 135 and 215 °C, is endothermic and is consistent with loss of all water from the crystal structure. The two-step mass loss between 250 and 480 °C is exothermic and is consistent with degradation to $CaCO_3$. This is backed up by PXRD analysis after a 480 °C heat treatment, which matches theoretical peaks for $CaCO_3$ (Figure S3). The final mass loss at >600 °C is consistent with the decomposition to CaO as shown by

Karunadasa et al. [35]. The lack of mass loss before 135 °C indicates that there is no solvent within the structure, apart from the water bound to the metals.

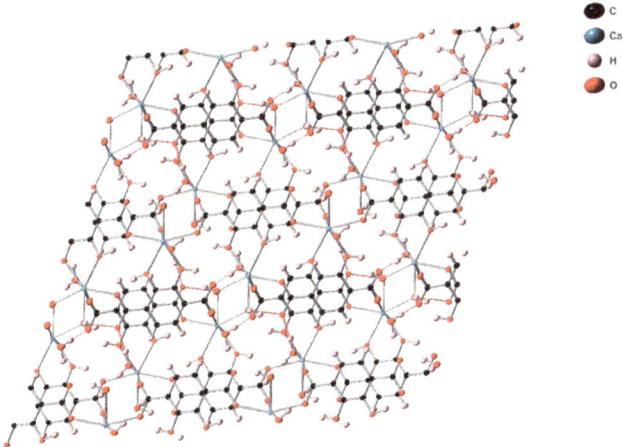

Figure 6. View of the 3D structure of SIMOF-4 (50% probability ellipsoids) as seen down the crystallographic *a*-axis, and showing hydrogen bonds (red).

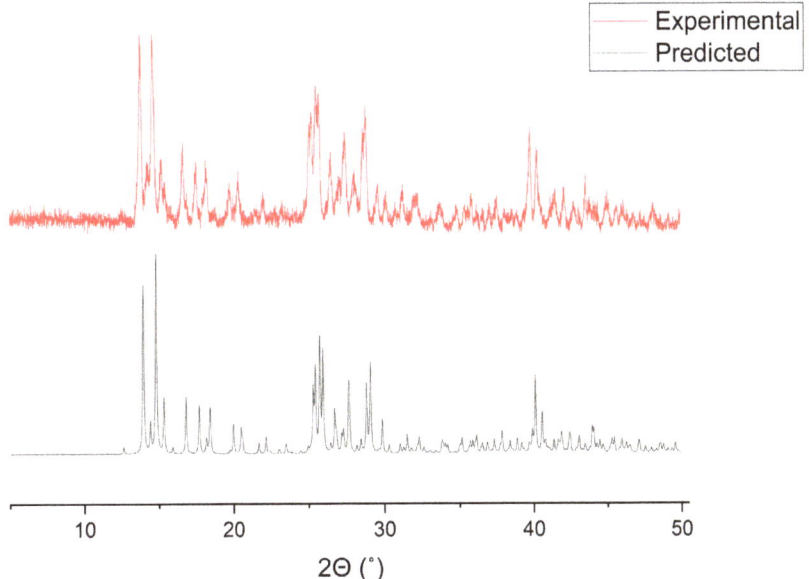

Figure 7. The simulated and experimental PXRD pattern of SIMOF-4 at room temperature.

2.3.3. Variable Temperature Behaviour

The water loss and first degradation step at 300 °C were explored further by use of variable temperature X-ray diffraction (VTXRD) (Figure 9). The VTXRD indicates the MOF is stable up to 100 °C, and that at 135 °C the pattern shows a subtle broadening indicating the start of a change in the structure, likely due to water loss. At 185 °C the pattern changes significantly: the major peak at 6.87° shifts to 6.95° and becomes much broader, the sharp peak at 13.62° is replaced by three broad peaks between 12.27 and 14.44°. This suggests that loss of water induces a change in the crystal structure, although the broadening of the peaks also suggests a reduction in crystallinity. Above 305 °C the only Bragg diffraction peaks visible come from the corundum sample holder indicating that the MOF has become amorphous, which shows that this peak in the DTA is destruction of the MOF. On cooling this amorphous structure remains. These results were confirmed by ex situ heat treatment and analysis (Figure S3).

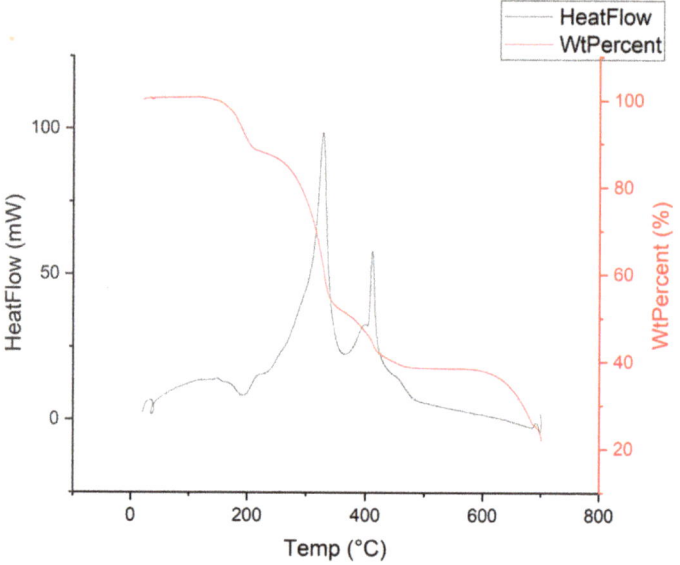

Figure 8. Thermo-gravimetric analysis of SIMOF-4, taken at a ramp rate of 5 °C/min.

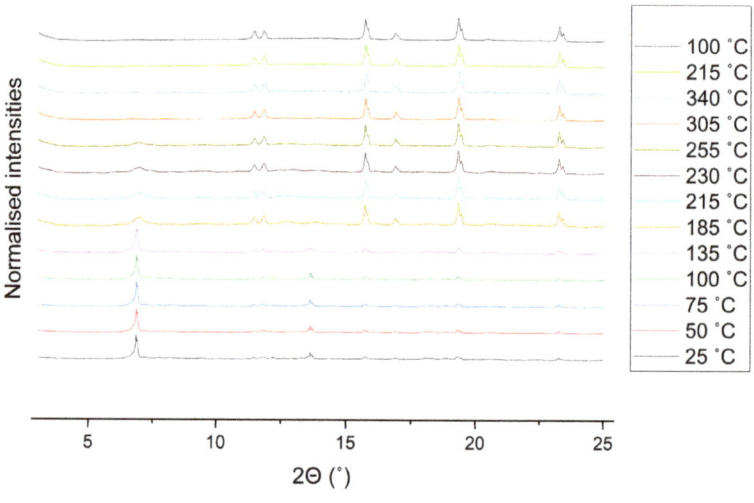

Figure 9. Main: normalised VTXRD patterns of SIMOF-4 taken on a corundum disc. All diffraction peaks in the patterns above 305 °C are from the corundum sample holder.

2.3.4. Single Crystal Structure of High Temperature Form

The phase transition at high temperature caused significant damage to the crystals; they underwent a change in morphology and became significantly less diffracting, this reduced their suitability for scXRD. This was true for all crystals we investigated. However, careful crystal selection and truncating the scXRD data did give diffraction that was good enough to solve, producing a plausible structure. This is not unusual in this type of science. As suggested by TGA, all water is lost, producing Ca(2,3-dhtp) (SIMOF-4-h), and this induces a significant change to the crystal structure. The unit cell remains triclinic but the unit cell parameters change, leading to a reduction in the unit cell volume by ≈18%. While we cannot infer any fine detail of the atomic arrangement and atomic displacement parameters, it is notable that this model does reproduce the experimental powder X-ray diffraction pattern well (Figure S4), and so some conclusions about the structure may be drawn from this model.

The phase change produces a dense interconnected structure that appears to show Ca and 2,3-dhtp layers in the *ab*-plane connected by further 2,3-dhtp linkers (Figure 10). The loss of water is detrimental to the stability of SIMOF-4 causing loss of crystallinity, as the structure changes to a layered phase. This collapse is most likely driven by loss of water from the Ca2 dimer units (Figure 2) causing structure rearrangements to maintain the 7 coordinate environment around the Ca ions.

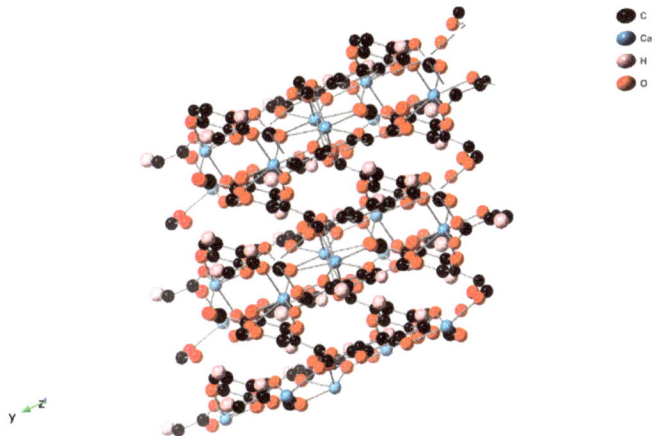

Figure 10. View of the crystal structure of SIMOF-4 (50% probability ellipsoids) after dehydration via heat treatment (down the crystallographic *c*-axis).

SIMOF-4-h tended to degrade in aqueous conditions and so rehydration was initially unsuccessful. However, prolonged exposure to air, approximately 3 months, caused a transition back to the hydrated form, SIMOF-4 (Figure 11).

2.3.5. Electrochemistry

Initial cyclic voltammetry (CV) was performed on SIMOF-4 between 0.01 and 2.5 V vs. Na$^+$/Na at a scan rate of 0.05 mV s^{-1}. For the first cycle a low-intensity reductive peak was observed at 1.95 V and another broad peak at ~0.15 V having a shoulder at ~0.56 V (Figure S6). No prominent oxidative peaks were observed, except a very broad profile between 1 and 1.7 V and another between 1.7 and 2.5 V. In the second cycle the reductive peak was shifted to ~1.48 V, while the remaining profile remained featureless. The shift continued in the 3rd cycle also along with a lowering of the peak height, which could be indicative of conversion to an amorphous phase [36] or structural changes, possibly to the dehydrated form above.

Subsequently, galvanostatic charge/discharge cycling was carried out at a current density of 50 mA g^{-1} (Figures 12 and 13). A high capacity of ~257 mAh g^{-1} was observed for the first discharge with a low coulombic efficiency (43%). This could be indicative of the contribution of SEI (solid-electrolyte interphase) layer formation [37] and changes to the structure or amorphization as anticipated from CV profiles. The load curves did not show long voltage plateaus and exhibited overlapping traces from the 3rd cycle onwards (Figure 12). The discharge capacity reduced to 144 mAh g^{-1} in the second cycle and had only a gradual fade and stabilised at ~110–120 mAh g^{-1} in the following cycles (Figure 13). This is a moderate discharge capacity [38] but one that does suggest some promise for this type of material.

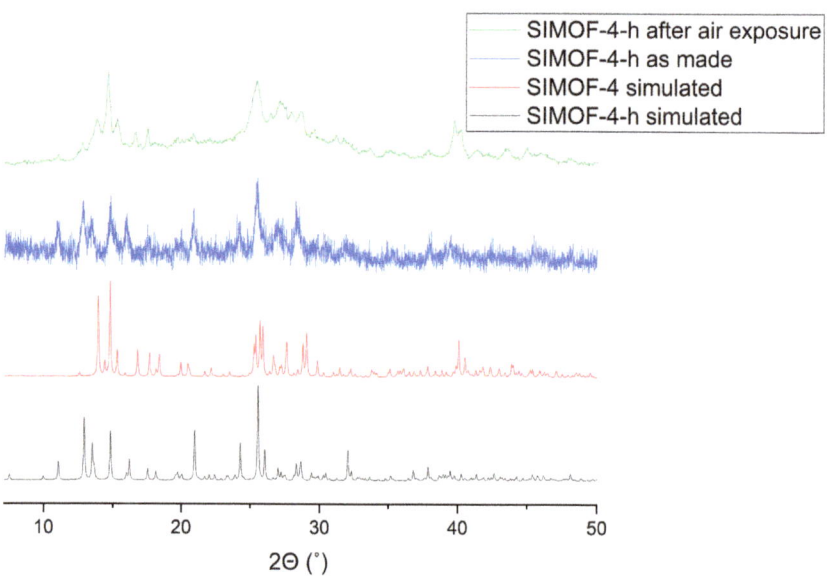

Figure 11. Simulated PXRD patterns of SIMOF-4 and SIMOF-4-h (high T form) and experimental patterns of SIMOF-4-h as synthesized and after prolonged exposure to air. All patterns obtained at room temperature.

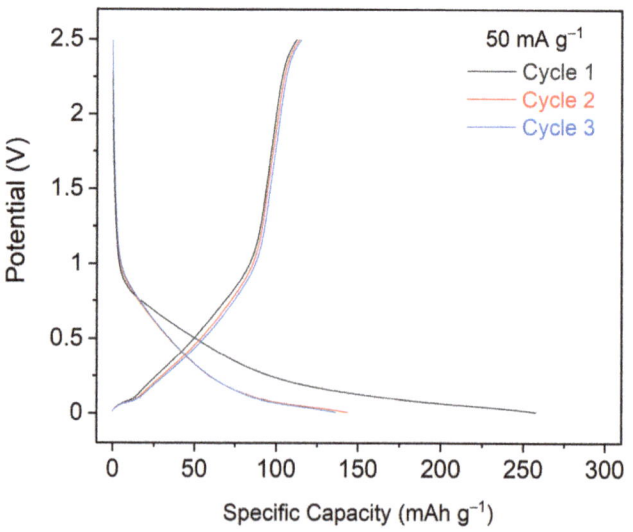

Figure 12. Galvanostatic charge/discharge curves for the first three cycles of SIMOF-4 recorded between 0.01 and 2.5 V at a current density of 50 mA g^{-1}.

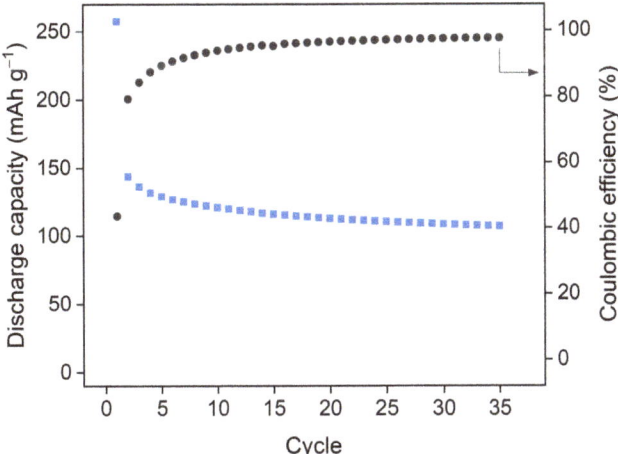

Figure 13. Discharge capacity and coulombic efficiency during galvanostatic cycling between 0.01 and 2.5 V at a current rate of 50 mA g^{-1}.

3. Materials and Methods

3.1. General Remarks

All chemicals were purchased commercially and applied directly. PXRD was performed on a STOE STADIP diffractometer (Germany) using Cu (Kα1) radiation monochromated with a curved Ge (111) crystal. PXRD of SIMOF-4-h was collected with Mo (Kα1) radiation monochromated with a primary beam monochromator. VTXRD was performed on a PANalytical Empyrean spectrometer (UK/The Netherlands) using Mo (Kα1,2) radiation, the heating profile is provided in the supplementary information (Figure S5). TGA was performed on a Stanton Redcroft STA-780 simultaneous TG-DTA (UK) with a ramping rate of 5 °C/min. N$_2$ adsorption isotherms were collected on a Micromeritics ASAP 2020 (USA), activation was performed at 25 °C, 100 °C, and 150 °C all under vacuum. NMR was performed on a Bruker AVII 400 instrument (Germany), with sample dissolved in DMSO-d_6.

3.2. Preparation of 2,3-Dhtp

Oven-dried catechol (7.5 g, 68.1 mmol) and potassium carbonate (20.46 g, 204 mmol) were placed in an autoclave. The autoclave was flushed with a vacuum/N$_2$ cycle three times and then charged with CO$_2$ to a pressure of 10 bar. The vessel was heated to 230 °C incrementally and left overnight. The product was cooled and the solid crushed and suspended in water (300 mL). The liquid was separated via centrifugation at 6000 rpm, and HCl (25 mL) was added. The resulting precipitate was filtered, washed with water and ethanol and dried in an oven overnight to produce a pink powder of 2,3-dhtpH$_2$. ^{1}H-NMR: 7.27 (2H, s). ^{13}C-NMR: 172.0 (1C, s), 151.2 (1C, s), 119.0 (1C, s), 117.2 (1C, s).

3.3. Preparation of SIMOF-4

Ca(NO$_3$)$_2$·4H$_2$O (2.36 mg, 1 mmol) and 2,3-dhtpH$_2$ (198 mg, 1 mmol) were dissolved via sonication in either 10 mL THF, 5 mL water and 5 mL ethanol for bulk product or 5 mL THF, 5 mL water, 5 mL ethanol and 5 mL p-xylene for single crystals. Acetic acid (5.25 mmol) was added, and the mixture was placed in an autogenous autoclave and heated to 120 °C for three days. The mixture was then cooled to room temperature and the product separated by filtration, washed with water and dried in an oven to produce grey crystals of the new Ca MOF (Yield: bulk 44%, single crystal 14%).

3.4. Preperation of SIMOF-4-h

Previously prepared SIMOF-4 was heated to 185 °C at 1 °C/min and the temperature was maintained for 3 h with exposure to air. The sample was then cooled to room temperature in an N_2 environment producing matte grey crystals of SIMOF-4-h.

3.5. X-ray Crystallography

Selected crystals of SIMOF-4 were analysed using a Rigaku FR-X Ultrahigh Brilliance Microfocus RA generator/confocal optics with XtaLAB P200 diffractometer [Mo Kα radiation (λ = 0.71073 Å)]. SIMOF-4-h crystals were analysed using a Rigaku MM-007HF High Brilliance RA generator/confocal optics with XtaLAB P100 diffractometer [Cu Kα radiation (λ = 1.54187 Å)]. Data collection was performed using CrystalClear [39], data reduction and cell refinement were performed using CrysAlisPro [40]. The heat treated crystals showed extremely weak diffraction and data were truncated at 1.06 Å. Structure solution was performed with SHELXT-2018/2 [41] version 2018/2, refinement was performed with SHELXL-2018/3 [42]. In SIMOF-4, non-hydrogen atoms were refined anisotropically, and CH hydrogen atoms were refined using a riding model. Hydroxyl hydrogen atoms were located from the difference Fourier map and refined isotropically subject to a distance restraint. In SIMOF-4-h, calcium atoms were refined anisotropically, while other non-hydrogen atoms were refined isotropically, and CH hydrogen atoms were refined using a riding model. Hydroxyl hydrogen atoms were placed in calculated positions, selected on the basis on neighbouring atoms, and refined with a riding model. Due to the poor data-quality in SIMOF-4-h, the benzene rings of the 2,3-dhtp ligands were constrained to ideality. All calculations were performed using the Olex2 [43] interface. Deposition numbers 2113160 and 2117049 contain the supplementary crystallographic data for this paper.

Crystal data (SIMOF-4). $C_8H_8CaO_8$, M = 272.22, triclinic, a = 7.1339(3), b = 11.5312(5), c = 13.1000(8) Å, α = 66.744(5), β = 80.811(4), γ = 80.834(4)°, U = 971.86(9) Å^3, T = 93 K, space group $P\bar{1}$ (no. 2), Z = 4, 11,622 reflections measured, 4197 unique (R_{int} = 0.0215), which were used in all calculations. The final R_1 [$I > 2\sigma(I)$] was 0.0403 and wR_2 (all data) was 0.1186.

Crystal data (SIMOF-4-h). $C_8H_4CaO_6$, M = 236.19, triclinic, a = 7.350(4), b = 9.540(6), c = 12.734(6) Å, α = 73.55(5), β = 73.92(5), γ = 71.84(5)°, U = 796.1(8) Å^3, T = 173 K, space group $P\bar{1}$ (no. 2), Z = 4, 2055 reflections measured, 1300 unique (R_{int} = 0.0549), which were used in all calculations. The final R_1 [$I > 2\sigma(I)$] was 0.1835 and wR_2 (all data) was 0.4867.

3.6. Electrochemical Testing

The working electrodes for SIMOF-4 were prepared in a mixture of active compound (55%), conducting carbon—Super C65 (35%) and binder—CMC (carboxymethyl cellulose, 10%). A slurry was prepared in water and cast on an aluminium foil (Advent Research Materials, UK) by a doctor blade. The electrodes were subsequently punched (~12 mm diameter) upon air drying and dried overnight in a vacuum oven at 110 °C and transferred to an argon-filled glovebox. The average mass loading of the active material per electrode was 1.54 mg cm^{-2}.

Electrochemical testing was done using coin cells (CR2325), that were assembled using sodium (Sigma-Aldrich) as the counter electrode and a glass fiber separator (Whatman GF/F). $NaPF_6$ in EC:DEC (1:1, ethylene carbonate—EC, diethyl carbonate—DEC) was used as the electrolyte and the process was carried out in the glovebox with oxygen and water content <1 ppm. Electrochemical measurements were recorded at 30 °C on a Biologic BCS-805 modular battery testing system in a potential window of 0.01–2.5 V (vs. Na$^+$/Na) and all data were analysed and processed using the BT-Lab software.

4. Conclusions

In this paper we have presented a new Ca MOF, SIMOF-4. The crystal structure of the as-made material has been determined using single-crystal X-ray diffraction, and its high temperature behaviour was studied using variable temperature PXRD and TGA. Despite significant damage to the single crystals on thermal treatment the structure of the high temperature phase was also determined. Initial electrochemistry experiments on the material's suitability for use as an anode in sodium-ion batteries showed that SIMOF-4 has moderate discharge capacity. These results present two possible avenues for future work. First, improving the discharge capacity to improve stability and creating a non-toxic anode for Na-ion batteries that may be viable in this type of material. Second, the replacement of water within the pores of the MOF with small biologically active molecules such as NO may yield non-toxic materials of potential utility.

Supplementary Materials: The following are available online. Figure S1: View of the 3D structure of SIMOF-4 (50% probability ellipsoids) as seen down the crystallographic *b*-axis, and showing hydrogen bonds. Figure S2: View of the 3D structure of SIMOF-4 (50% probability ellipsoids) as seen down the crystallographic *b*-axis, and showing hydrogen bonds. Figure S3: PXRD patterns of SIMOF-4 after different heat treatments, with simulated patterns of $CaCO_3$ and SIMOF-4 for comparison. All obtained using Cu(Kα1) radiation. Figure S4: PXRD pattern of heat treated SIMOF-4 simulated from scXRD structure compared with the experimental pattern. Obtained using Mo(Kα1) radiation. Figure S5: Cyclic voltammogram for SIMOF-4 recorded at a scan rate of 0.05 mV s^{-1} between 0.01–2.5 V. Figure S6: N_2 Adsorption isotherms of SIMOF-4 after three different activation protocols of 25 °C, 100 °C and 150 °C all under vacuum. Figure S7: Temperature time profile for the VTXRD experiment performed on SIMOF-4, showing programmed and measured temperature profiles.

Author Contributions: Conceptualization, R.M.M. and R.E.M.; methodology, R.M.M.; validation, R.M.M.; formal analysis, R.M.M., A.M.Z.S., D.B.C., A.V.D. and P.W.; investigation, R.M.M., A.M.Z.S., A.V.D., P.W. and D.B.C.; resources, R.E.M., A.R.A.; data curation, R.M.M. and R.E.M.; writing—original draft preparation, R.M.M. and A.V.D.; writing—review and editing, R.M.M., A.V.D., D.B.C., R.E.M. and A.R.A.; visualization, R.M.M. and A.V.D.; supervision, R.E.M. and A.R.A.; project administration, R.E.M. and A.R.A.; funding acquisition, R.E.M. and A.R.A. All authors have read and agreed to the published version of the manuscript.

Funding: This research was funded by the European Research Council AdG grant number ADOR-787073 and by the Faraday Institution (grant number FIRG018).

Institutional Review Board Statement: Not applicable.

Informed Consent Statement: Not applicable.

Data Availability Statement: CCDC-2113160 and 2117049 contains the supplementary crystallographic data for this paper. These data are provided free of charge by the joint Cambridge Crystallographic Data Centre and Fachinformationszentrum Karlsruhe Access Structures service www.ccdc.cam.ac.uk/structures. All other data can be obtained from the authors on request.

Acknowledgments: The authors would like to acknowledge Yuri Andreev for his technical assistance.

Conflicts of Interest: The authors declare no conflict of interest.

Sample Availability: Samples of the compounds are available from the authors on request.

References

1. Howarth, A.J.; Peters, A.W.; Vermeulen, N.A.; Wang, T.C.; Hupp, J.T.; Farha, O.K. Best practices for the synthesis, activation, and characterization of metal–organic frameworks. *Chem. Mater.* **2017**, *29*, 26–39. [CrossRef]
2. Stock, N.; Biswas, S. Synthesis of Metal-Organic Frameworks (MOFs): Routes to Various MOF Topologies, Morphologies, and Composites. *Chem. Rev.* **2012**, *112*, 933–969. [CrossRef]
3. Batten, S.R.; Champness, N.R.; Chen, X.M.; Garcia-Martinez, J.; Kitagawa, S.; Öhrström, L.; O'Keeffe, M.; Suh, M.P.; Reedijk, J. Terminology of metal-organic frameworks and coordination polymers (IUPAC recommendations 2013). *Pure Appl. Chem.* **2013**, *85*, 1715–1724. [CrossRef]
4. Guo, X.; Geng, S.; Zhuo, M.; Chen, Y.; Zaworotko, M.J.; Cheng, P.; Zhang, Z. The utility of the template effect in metal-organic frameworks. *Coord. Chem. Rev.* **2019**, *391*, 44–68. [CrossRef]
5. Schneemann, A.; Bon, V.; Schwedler, I.; Senkovska, I.; Kaskel, S.; Fischer, R.A. Flexible metal-organic frameworks. *Chem. Soc. Rev.* **2014**, *43*, 6062–6096. [CrossRef] [PubMed]

6. Yuan, S.; Feng, L.; Wang, K.; Pang, J.; Bosch, M.; Lollar, C.; Sun, Y.; Qin, J.; Yang, X.; Zhang, P.; et al. Stable Metal–Organic Frameworks: Design, Synthesis, and Applications. *Adv. Mater.* **2018**, *30*, 1704303. [CrossRef] [PubMed]
7. Song, P.; Li, Y.; He, B.; Yang, J.; Zheng, J.; Li, X. Hydrogen storage properties of two pillared-layer Ni(II) metal-organic frameworks. *Microporous Mesoporous Mater.* **2011**, *142*, 208–213. [CrossRef]
8. Tian, T.; Zeng, Z.; Vulpe, D.; Casco, M.E.; Divitini, G.; Midgley, P.A.; Silvestre-Albero, J.; Tan, J.C.; Moghadam, P.Z.; Fairen-Jimenez, D. A sol-gel monolithic metal-organic framework with enhanced methane uptake. *Nat. Mater.* **2018**, *17*, 174–179. [CrossRef]
9. Ding, M.; Flaig, R.W.; Jiang, H.L.; Yaghi, O.M. Carbon capture and conversion using metal-organic frameworks and MOF-based materials. *Chem. Soc. Rev.* **2019**, *48*, 2783–2828. [CrossRef]
10. Denny, M.S.; Moreton, J.C.; Benz, L.; Cohen, S.M. Metal-organic frameworks for membrane-based separations. *Nat. Rev. Mater.* **2016**, *1*, 16078. [CrossRef]
11. Dong, Q.; Zhang, X.; Liu, S.; Lin, R.; Guo, Y.; Ma, Y.; Yonezu, A.; Krishna, R.; Liu, G.; Duan, J.; et al. Tuning Gate-Opening of a Flexible Metal–Organic Framework for Ternary Gas Sieving Separation. *Angew. Chem.* **2020**, *59*, 22756–22762. [CrossRef]
12. DeCoste, J.B.; Peterson, G.W. Metal-organic frameworks for air purification of toxic chemicals. *Chem. Rev.* **2014**, *114*, 5695–5727. [CrossRef] [PubMed]
13. Kadhom, M.; Deng, B. Metal-organic frameworks (MOFs) in water filtration membranes for desalination and other applications. *Appl. Mater. Today* **2018**, *11*, 219–230. [CrossRef]
14. Llabrés i Xamena, F.X.; Abad, A.; Corma, A.; Garcia, H. MOFs as catalysts: Activity, reusability and shape-selectivity of a Pd-containing MOF. *J. Catal.* **2007**, *250*, 294–298. [CrossRef]
15. Kumar, P.; Deep, A.; Kim, K.H. Metal organic frameworks for sensing applications. *TrAC Trends Anal. Chem.* **2015**, *73*, 39–53. [CrossRef]
16. Dong, C.; Xu, L. Cobalt- and Cadmium-Based Metal–Organic Frameworks as High-Performance Anodes for Sodium Ion Batteries and Lithium Ion Batteries. *ACS Appl. Mater. Interfaces* **2017**, *9*, 7160–7168. [CrossRef] [PubMed]
17. McKinlay, A.C.; Morris, R.E.; Horcajada, P.; Férey, G.; Gref, R.; Couvreur, P.; Serre, C. BioMOFs: Metal-organic frameworks for biological and medical applications. *Angew. Chem. Int. Ed.* **2010**, *49*, 6260–6266. [CrossRef]
18. Desai, D.A.V.; Morris, P.R.E.; Armstrong, D.A.R. Advances in Organic Anode Materials for Na-/K-Ion Rechargeable Batteries. *ChemSusChem* **2020**, *13*, 4866. [CrossRef]
19. Abánades Lázaro, I.; Forgan, R.S. Application of zirconium MOFs in drug delivery and biomedicine. *Coord. Chem. Rev.* **2019**, *380*, 230–259. [CrossRef]
20. Hinks, N.J.; McKinlay, A.C.; Xiao, B.; Wheatley, P.S.; Morris, R.E. Metal organic frameworks as NO delivery materials for biological applications. *Microporous Mesoporous Mater.* **2010**, *129*, 330–334. [CrossRef]
21. Henkelis, S.E.; Vornholt, S.M.; Cordes, D.B.; Slawin, A.M.Z.; Wheatley, P.S.; Morris, R.E. A single crystal study of CPO-27 and UTSA-74 for nitric oxide storage and release. *CrystEngComm* **2019**, *21*, 1857–1861. [CrossRef]
22. Casalegno, C.; Schifanella, O.; Zennaro, E.; Marroncelli, S.; Briant, R. Collate literature data on toxicity of Chromium (Cr) and Nickel (Ni) in experimental animals and humans. *EFSA Support. Publ.* **2015**, *12*, 478E. [CrossRef]
23. Miller, S.R.; Alvarez, E.; Fradcourt, L.; Devic, T.; Wuttke, S.; Wheatley, P.S.; Steunou, N.; Bonhomme, C.; Gervais, C.; Laurencin, D.; et al. A rare example of a porous Ca-MOF for the controlled release of biologically active NO. *Chem. Commun.* **2013**, *49*, 7773–7775. [CrossRef] [PubMed]
24. Emsley, J. *The Elements*; Clarendon Press: Oxford, UK, 1998; ISBN 0198558198.
25. Volkringer, C.; Marrot, J.; Férey, G.; Loiseau, T. Hydrothermal Crystallization of Three Calcium-Based Hybrid Solids with 2,6-Naphthalene- or 4,4′-Biphenyl-Dicarboxylates. *Cryst. Growth Des.* **2007**, *8*, 685–689. [CrossRef]
26. Yang, Y.; Jiang, G.; Li, Y.Z.; Bai, J.; Pan, Y.; You, X.Z. Synthesis, structures and properties of alkaline earth metal benzene-1,4-dioxylacetates with three-dimensional hybrid networks. *Inorg. Chim. Acta* **2006**, *359*, 3257–3263. [CrossRef]
27. Xiao, F.; Gao, W.; Wang, H.; Wang, Q.; Bao, S.; Xu, M. A new calcium metal organic frameworks (Ca-MOF) for sodium ion batteries. *Mater. Lett.* **2021**, *286*, 129264. [CrossRef]
28. Slater, M.D.; Kim, D.; Lee, E.; Johnson, C.S. Sodium-Ion Batteries. *Adv. Funct. Mater.* **2013**, *23*, 947–958. [CrossRef]
29. Akimbekov, Z.; Katsenis, A.D.; Nagabhushana, G.P.; Ayoub, G.; Arhangelskis, M.; Morris, A.J.; Friščić, T.; Navrotsky, A. Experimental and Theoretical Evaluation of the Stability of True MOF Polymorphs Explains Their Mechanochemical Interconversions. *J. Am. Chem. Soc.* **2017**, *139*, 7952–7957. [CrossRef]
30. Martí-Rujas, J. Structural elucidation of microcrystalline MOFs from powder X-ray diffraction. *Dalton Trans.* **2020**, *49*, 13897–13916. [CrossRef]
31. Torresi, S.; Famulari, A.; Martí-Rujas, J.; Martí-Rujas, J. Kinetically Controlled Fast Crystallization of $M_{12}L_8$ Poly-[*n*]-catenanes Using the 2,4,6-Tris(4-pyridyl)benzene Ligand and $ZnCl_2$ in an Aromatic Environment. *J. Am. Chem. Soc.* **2020**, *142*, 9537–9543. [CrossRef]
32. Ohara, K.; Martí-Rujas, J.; Haneda, T.; Kawano, M.; Hashizume, D.; Izumi, F.; Fujita, M. Formation of a thermally stable, porous coordination network via a crystalline-to-amorphous-to-crystalline phase transition. *J. Am. Chem. Soc.* **2009**, *131*, 3860–3861. [CrossRef]
33. Lindsey, A.S.; Jeskey, H. *The Kolbe-Schmitt Reaction*; ACS Publications: Washington, DC, USA, 1957.
34. Zhou, D.D.; Chen, P.; Wang, C.; Wang, S.S.; Du, Y.; Yan, H.; Ye, Z.M.; He, C.T.; Huang, R.K.; Mo, Z.W.; et al. Intermediate-sized molecular sieving of styrene from larger and smaller analogues. *Nat. Mater.* **2019**, *18*, 994–998. [CrossRef] [PubMed]

35. Karunadasa, K.S.P.; Manoratne, C.H.; Pitawala, H.M.T.G.A.; Rajapakse, R.M.G. Thermal decomposition of calcium carbonate (calcite polymorph) as examined by in-situ high-temperature X-ray powder diffraction. *J. Phys. Chem. Solids* **2019**, *134*, 21–28. [CrossRef]
36. Zhao, R.; Liang, Z.; Zou, R.; Xu, Q. Metal-Organic Frameworks for Batteries. *Joule* **2018**, *2*, 2235–2259. [CrossRef]
37. Chen, H.; Ling, M.; Hencz, L.; Ling, H.Y.; Li, G.; Lin, Z.; Liu, G.; Zhang, S. Exploring Chemical, Mechanical, and Electrical Functionalities of Binders for Advanced Energy-Storage Devices. *Chem. Rev.* **2018**, *118*, 8936–8982. [CrossRef] [PubMed]
38. Xu, Y.; Zhou, M.; Lei, Y. Organic materials for rechargeable sodium-ion batteries. *Mater. Today* **2018**, *21*, 60–78. [CrossRef]
39. *CrystalClear-SM Expert v2.1 b45*; Rigaku Americas: The Woodlands, TX, USA; Rigaku Corporation: Tokyo, Japan, 2015.
40. *CrysAlisPro v1.171.40.14a and v1.171.41.82a*; Rigaku Oxford Diffraction; Rigaku Corporation: Oxford, UK, 2018–2020.
41. Sheldrick, G.M. SHELXT—Integrated space-group and crystal structure determination. *Acta Crystallogr. Sect. A* **2015**, *71*, 3–8. [CrossRef]
42. Sheldrick, G.M. Crystal structure refinement with SHELXL. *Acta Crystallogr. Sect. C* **2015**, *71*, 3–8. [CrossRef]
43. Dolomanov, O.V.; Bourhis, L.J.; Gildea, R.J.; Howard, J.A.K.; Puschmann, H. OLEX2: A complete structure solution, refinement and analysis program. *J. Appl. Cryst.* **2009**, *42*, 339–341. [CrossRef]

Article

Synthesis of a Hexameric Magnesium 4-pyridyl Complex with Cyclohexane-like Ring Structure via Reductive C-N Activation

Samuel R. Lawrence, Matthew de Vere-Tucker, Alexandra M. Z. Slawin and Andreas Stasch *

EaStCHEM School of Chemistry, University of St Andrews, North Haugh, St Andrews KY16 9ST, UK; sl264@st-andrews.ac.uk (S.R.L.); mdvt@st-andrews.ac.uk (M.d.V.-T.); amzs@st-andrews.ac.uk (A.M.Z.S.)
* Correspondence: as411@st-andrews.ac.uk; Tel.: +44-(0)-1334-463-382

Abstract: The reaction of [{(^{Ar}nacnac)Mg}$_2$] (^{Ar}nacnac = HC{MeC(NAr)}$_2$, Ar = 2,6-diisopropylphenyl, Dip, or 2,6-diethylphenyl, Dep) with 4-dimethylaminopyridine (DMAP) at elevated temperatures afforded the hexameric magnesium 4-pyridyl complex [{(^{Ar}nacnac)Mg(4-C$_5$H$_4$N)}$_6$] via reductive cleavage of the DMAP C-N bond. The title compound contains a large s-block organometallic cyclohexane-like ring structure comprising tetrahedral (^{Ar}nacnac)Mg nodes and linked by linear 4-pyridyl bridging ligands, and the structure is compared with other ring systems. [(^{Dip}nacnac)Mg(DMAP)(NMe$_2$)] was structurally characterised as a by-product.

Keywords: bond activation; low oxidation state complexes; magnesium; metallacycles; ring system; X-ray crystallography

1. Introduction

The study of self-assembly in metallacycles via coordination chemistry is dominated by transition metal complexes [1,2]. In these, the coordination geometry around transition metal centres is relatively strictly governed by their number of d-electrons and ligand field effects. Thus, the structure of metallacycles and coordination polymers can often be guided by the shape of connecting ligands and the position of donor functionalities together with suitable transition metal ions. In contrast, metal–ligand interactions in s-block metal complexes are predominantly electrostatic in nature [3]. This can lead to reversible metal–ligand coordination bonds with flexible coordination modes that are easy to distort from "ideal" geometry. Often, many complex species of similar energy are present in solution, that rapidly interconvert with low barriers, and are significantly influenced by ligand sterics, donor solvents and other factors. Thus, for a range of s-block metal complexes, equilibria between various oligomers, such as ring systems, can make it difficult to predict the range of possible structures. In addition to solution state spectroscopic studies, single crystal X-ray crystallography has been crucial in elucidating often surprising molecular structures, that were sometimes obtained serendipitously, and has allowed linking to solution state species. For coordination complexes with electropositive divalent metal centres, e.g., Mg^{2+} in organomagnesium or amidomagnesium compounds, ring systems have often been discovered when two types of anionic, often bridging ligands with unequal sizes are present, that induce a curvature and can lead to ring formation [4]. Relevant Mg complexes, **1**–**5**, with hexameric or dodecameric ring structures are shown in Figure 1. These examples show that smaller bridging ligands (hydride, ethyl, allyl) are located on the inside of the ring, whereas the more sterically demanding and solubilising ligands, most of them containing 2,6-diisopropylphenyl (Dip) substituents, are predominantly located towards the outside of the ring. The examples all contain sterically demanding monoanionic bridging (**1** [5], **2** [4], **3** [6], **4** [5]) or chelating (**5** [7]) ligands, plus small bridging ligands such as hydrides (**1** and **2**), ethyl groups (**3** and **4**) or allyl groups (**5**). Placing the larger ligands with solubilising groups on the outside of the ring will contribute to the solubility of the molecules, and thus may aid the formation of a preferred ring size through self-assembly. The general bond lability, and typical reversibility, of ionic metal–ligand interactions is a process for interconverting different oligomers to a thermodynamically preferred size and/or to those favoured by crystallisation from reaction mixtures under certain conditions. Very recently, and related to the work in here, a blue aluminium complex [{(MeCNDip)$_2$Al(4-C$_5$H$_4$N)}$_6$] **6** (see Figure 1) bearing a sterically demanding diazabutadiene-diide ligand, aluminium(III) centres and 4-pyridyl units, was prepared by pyridine CH activation with hydrogen elimination of a respective aluminium(II) complex with pyridine at elevated temperatures [8].

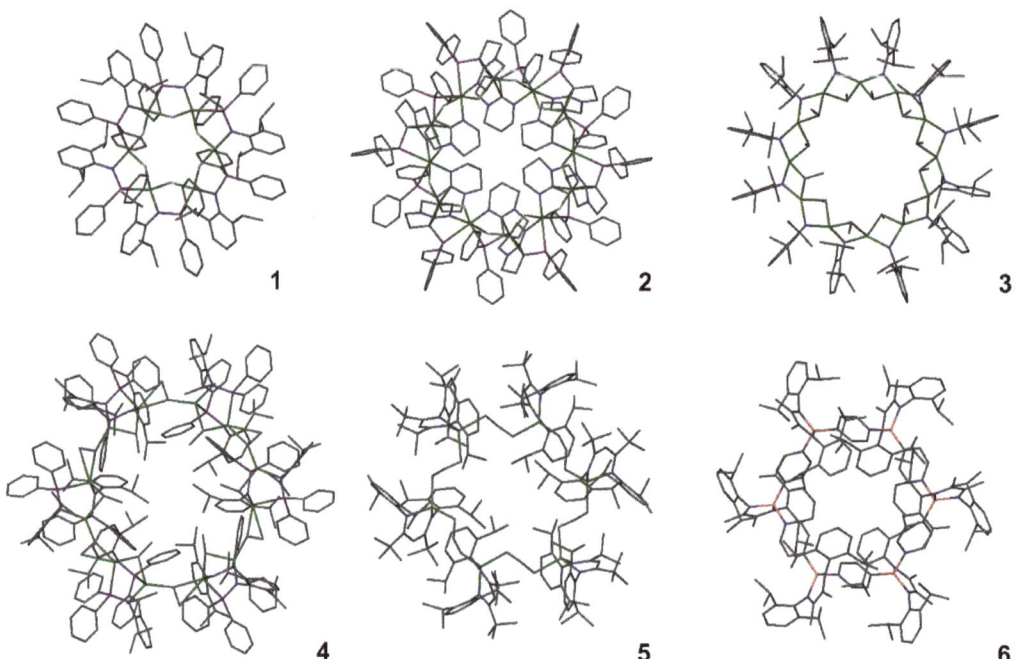

Figure 1. Overall molecular structures of complexes **1–6**. Hydrogen atoms, except MgH units, omitted for clarity. Colour code: P: purple, Al: pink, Mg: green, N: blue, C: grey, hydridic H: light grey.

In magnesium complex chemistry [9], dimagnesium(I) compounds of general type LMgMgL [10–12], where L is a monoanionic ligand such as in the β-diketiminate examples **7**, see Figure 2, are highly reducing and highly reactive reagents that can also serve as entry points to heteroleptic magnesium complexes with unusual anionic fragments. The magnesium atoms in LMgMgL can be coordinated by additional donor molecules, for example with substituted pyridines or cyclic ethers [13–16]. This can lead to complexes with significantly elongated Mg–Mg bonds, e.g., in the bis-DMAP adduct [{(^{Ar}nacnac)Mg(DMAP)}$_2$] **8**, where ^{Ar}nacnac = HC{MeC(NAr)}$_2$, with Ar = Dip (**8a**) [13] or Dep (2,6-diethylphenyl) (**8b**) [16], and DMAP = 4-dimethylaminopyridine, see Figure 2, with an elongation by ca. 12% in **8a** (Mg–Mg: 3.1962(14) Å) from ca. 2.85 Å in **7a** [13,17]. Coordination of ligands can be reversible, e.g., for cyclic ethers such as THF and dioxane, and the reactivity at the Mg centres of LMgMgL compounds is typically suppressed when fully coordinated. Mono-adducts of LMgMgL complexes, e.g., with DMAP or small N-heterocyclic carbene species, or partial donor addition, however, can induce reactivity not observed for the uncoordinated species. This has, for example, been demonstrated in the reductive coupling of CO [15,16].

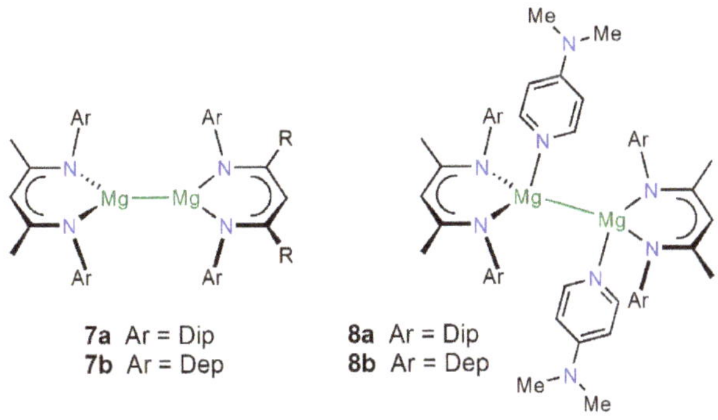

Figure 2. Dimagnesium(I) complexes 7 and 8. Dip = 2,6-diisopropylphenyl, Dep = 2,6-diethylphenyl.

2. Results and Discussion

2.1. Synthesis

The reaction of [{(^{Ar}nacnac)Mg}$_2$] 7, Ar = Dip (7a) [17] or Dep (7b) [18] with DMAP initially led to brown-red solutions of the adduct complexes 8. [13,16] Upon heating to 100 °C, the colour of the mixtures changed to purple, and after several hours at this temperature, the hexameric magnesium 4-pyridyl complexes [{(^{Ar}nacnac)Mg(4-C$_5$H$_4$N)}$_6$], Ar = Dip (9a), Dep (9b) were reproducibly afforded (see Scheme 1). The compounds were obtained as colourless crystals when performed in a non-stirred reaction mixture such as an NMR tube, or as a crystalline precipitate from a stirred reaction in low to moderate isolated yields. Complex 9 shows a large cyclohexane-like ring structure (Figure 3) that is described in the next section in more detail. Once crystallised, the complexes did not redissolve in hot deuterated benzene, hot deuterated THF or dichloromethane, and thus no meaningful NMR spectroscopic data could be obtained. For comparison, NMR spectra of aluminium complex 6 have been recorded in deuterated THF. [8] The bulk purity of compound 9 is supported by elemental analysis. In addition, hydrolysis of 9b with D$_2$O in aromatic solvents afforded only 4-deuteropyridine and the deuterated proligand ^{Dep}nacnacD in a 1:1 ratio, and no DMAP or other organic by-products according to ^{1}H and ^{2}H-NMR spectroscopy, plus a precipitate presumed to be Mg(OD)$_2$·D$_2$O. Similarly, the attempt to dissolve 9b in deuterated dimethyl sulfoxide afforded a 1:1 mixture of 4-deuteropyridine and one β-diketiminate complex, by implication [(^{Dep}nacnac)Mg{OS(CD$_3$)CD$_2$}], and no ^{Dep}nacnacD, as judged by ^{1}H and ^{2}H-NMR spectroscopy.

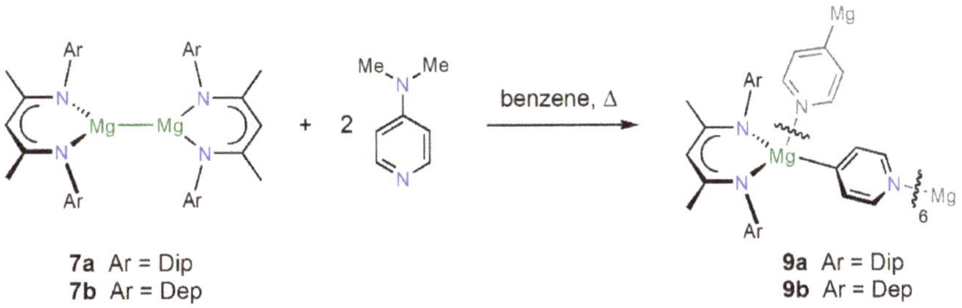

Scheme 1. Synthesis of complexes 9.

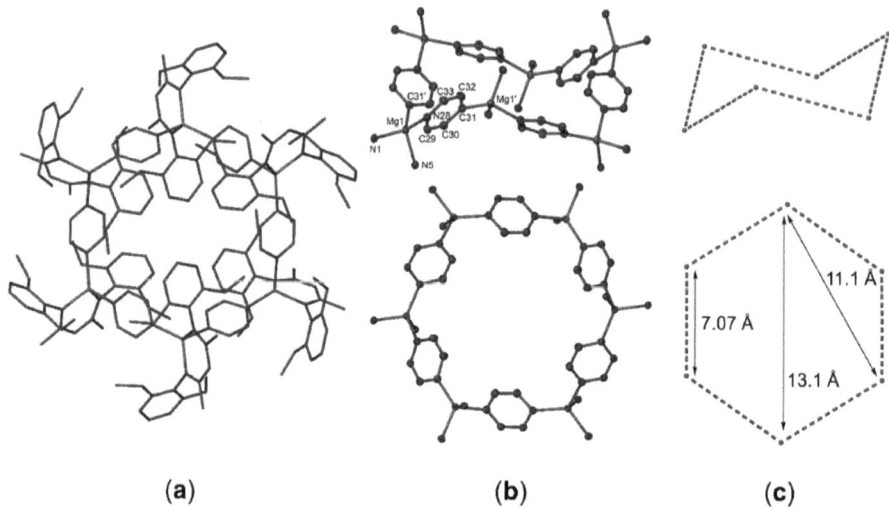

Figure 3. Molecular structure of [{(^(Dep)nacnac)Mg(4-C_5H_4N)}_6]·C_6H_6, **9b**·C_6H_6. Solvent molecule, hydrogen atoms and minor part of a disordered ethyl group omitted. Colour code: Mg: green, N: blue, C: grey. (**a**) Stick diagram of overall structure; (**b**) cut-back model as thermal ellipsoids (30%) in two views with atom labelling of asymmetric unit and nearest atoms; (**c**) only magnesium positions in two views and selected distances. Selected bond lengths (Å) and angles (°): Mg1–N1 2.0410(12), Mg1–N5 2.0505(12), Mg1–N28 2.1138(12), Mg1′–C31 2.1462(14), N28–C29 1.3437(18), C29–C30 1.379(2), C30–C31 1.4004(19), C31–C32 1.405(2), C32–C33 1.380(2), N28–C33 1.3411(18); N1–Mg1–N5 93.09(5), N28–Mg1–C31′ 106.84(5).

Both radical and diamagnetic reaction pathways can be considered for the observed reactivity and the reductive cleavage of the C-N bond by the Mg^I centres. The deeply coloured reaction mixtures, which decolourise rapidly when exposed to air, the high reaction temperature, and the extremely long and thus weakened Mg–Mg bond in the intermediate [{(^(Ar)nacnac)Mg(DMAP)}_2] **8** would support both possibilities, but point to a radical mechanism. ^{1}H-NMR spectroscopic studies of the supernatant solution, after product **9** precipitated, showed that a mixture of products remained, with overlapping NMR resonances from several β-diketiminate-containing complexes. Small quantities of complex **8** could be identified in these mixtures. Attempts were made to isolate by-products from these reactions and the new complex [(^(Dip)nacnac)Mg(DMAP)(NMe_2)] **10** was structurally characterised after crystallisation from *n*-pentane in a low yield, see Figure 4 (next section), and is an expected by-product from the DMAP C–N bond scission. For the formation of the related Al complex **6** derived from pyridine, a radical mechanism was proposed based on DFT studies that initiated with the homolytic cleavage of the Al–Al bond. [8] When we carried out analogous reactions of compounds **7** to **9** with pyridine instead of DMAP, we again observed highly coloured reaction mixtures at 100 °C (green to purple-blue) and no pure compounds could so far be isolated from these reactions. In situ NMR spectroscopic studies showed significant broadening and a decrease in intensities of resonances which could suggest the formation of radical species. No significant quantities of hydrogen were formed in these reactions (*c.f.* the synthesis of **6**). In transition metal chemistry, formation of the tetrameric β-diketiminate Fe 4-pyridyl species [{MeC{MeC(NXyl)}_2Fe(4-C_5H_4N)}_4], Xyl = 2,6-dimethylphenyl, from an iron(I) complex and DMAP, was shown to involve pyridyl radical anions and an analogous Fe complex to **10**, [MeC{MeC(NXyl)}_2Fe(DMAP)(NMe_2)], was also isolated. [19] Radical intermediates have previously been generated in β-diketiminate dimagnesium(I) chemistry, either as intermediates or from photochemically induced Mg–Mg bond cleavage [20,21]. Reductions in organic substrates with **7** can also form complexes with organic-based radicals, for example a structurally characterised purple-blue magnesium ketyl complex [22]. Supported by

DFT studies, the activation of CO via mono-adducts of **7** is, however, proposed to proceed via a diamagnetic pathway [15,16]. Overall, it is likely that the mechanism involves the Mg-induced reduction in the DMAP ligand to a radical anion and concomitant loss of the Mg–Mg bond, followed by reductive C–N cleavage and ultimately Mg–C bond formation.

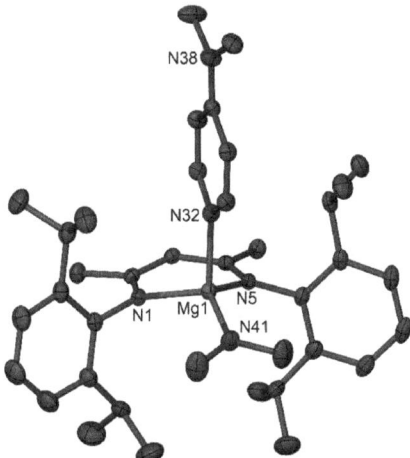

Figure 4. Molecular structure of [(^{Dip}nacnac)Mg(DMAP)(NMe$_2$)]·0.5 C$_5$H$_{12}$, **10**·0.5 C$_5$H$_{12}$, 30% thermal ellipsoids. Solvent molecule, hydrogen atoms and minor parts of disordered isopropyl groups omitted. Selected bond lengths (Å) and angles (°): Mg1–N1 2.0738(11), Mg1–N5 2.0756(11), Mg1–N32 2.1248(12), Mg1–N41 1.9468(12); N1–Mg1–N5 91.83(4), N41–Mg1–N32 108.39(5).

2.2. Molecular Structures

Complexes [{(^{Ar}nacnac)Mg(4-C$_5$H$_4$N)}$_6$] Ar = Dip (**9a**), Dep (**9b**), were characterised by single crystal X-ray diffraction, see Figure 3 for **9b**, and Figures S1 and S2 for **9a**. Both compounds show a similar overall arrangements of alternating (^{Ar}nacnac)Mg units and 4-pyridyl groups that form a large, cyclohexane-like ring system in a chair conformation where the (^{Ar}nacnac)Mg units serve as tetrahedral nodes and the 4-pyridyl groups act as linear linking units and coordinate via the pyridyl nitrogen to one Mg, and the 4-pyridyl carbon to another Mg centre (μ-κN:κC^4). Several datasets from different crystals were obtained for each compound and the data quality, refinement parameters and overall ordering in the crystal were repeatedly significantly better for **9b** (Ar = Dep; best R value ca. 4.5%) compared with **9a** (Ar = Dip; best R value ca. 15%). This is likely in part related to the differences in the substituents in the 2 and 6 position of the flanking aryl groups on the diketiminate ligand. The increased steric demand of the isopropyl groups in **9a** near the connecting 4-pyridyl groups appears to distort from the ideal geometry which likely translates to poor overall ordering and crystal quality. The ethyl groups in these positions in **9b** are more open, flexible, and accommodating. The other major difference is that solvates of **9b** included one benzene molecule per full formula unit, whereas those of **9a** contained an estimated 10 or 16 benzene molecules. Another contributing factor for the overall crystal quality in **9b** over **9a** could be that compounds with Dep substituents are typically more soluble than those with Dip substituents, and thus crystal formation and growth could be slower and thus provide higher quality crystals. The isolated yields were also higher for **9b** than for **9a** and that could relate to more facile ring formation.

The molecular structures of [{(^{Dep}nacnac)Mg(4-C$_5$H$_4$N)}$_6$]·C$_6$H$_6$, **9b**·C$_6$H$_6$, are broadly similar and only the highest quality example will be described in more detail. **9b**·C$_6$H$_6$ crystallised in a trigonal crystal system with one sixth of the molecule in the asymmetric unit and a disordered benzene molecule in the centre of the ring. Each Mg centre is four-coordinate with a 4-pyridyl Mg-C interaction (Mg1–C31' 2.1462(14) Å) to a pyridyl with approximate co-planar orientation relative to the diketiminate-Mg unit and a 4-pyridyl Mg-N interaction (Mg1–N28 2.1138(12)) to a pyridyl that is approximately perpendicular to the diketiminate-Mg unit. A similar overall arrangement of carbon versus nitrogen atom positions is also found in the molecular structure of Al complex **6**. [8] Exchanging the pyridyl carbon versus nitrogen atom positions in the structure of **9b** also leads to similar reasonable thermal ellipsoids and U_{iso} values for the atoms in question, but the final R value of the refinement increases by more than 1%. The bond lengths within the pyridyl unit further support the assignment of the nitrogen atom position with significantly shorter C-N bonds (1.342 Å mean) versus C-C bonds (1.403 Å mean). The pyridyl N-Mg-C angle in **9b** is 106.84(5)° and thus close to an ideal tetrahedral angle. The 4-pyridyl units do not provide a perfectly linear coordination motif between two Mg centres and show a Mg···4-pyridyl-midpoint···Mg angle of ca. 170°. In the related tetrameric iron complex [{MeC{MeC(NXyl)}$_2$Fe(4-C$_5$H$_4$N)}$_4$], the coordination geometry around the iron(II) centres is distorted towards trigonal pyramidal with smaller pyridyl-Fe-pyridyl angles of 102.0° (mean), and two alternating types of Fe···4-pyridyl-midpoint···Fe angles of 173.8° (mean) and 159.1° (mean) highlighting the distortion in this system [19]. In addition, transition metal-coordinated 4-pyridyl units have been employed in the formation of organometallic coordination polymers derived from bis(4-pyridyl)mercury [23,24].

The ring system with alternating tetrahedral (^{Ar}nacnac)Mg nodes and linear 4-pyridyl ligands is thus perfectly suited to form a large cyclohexane-like molecule in a chair configuration where the tetrahedrally coordinated Mg centres represent the carbon atoms and the linear 4-pyridyl ligands represent the C-C single bonds. In that sense, the overall structure of complex **9** is similar to that of the hexameric β-diketiminate magnesium allyl complex **5** [7], although the bridging allyl groups in **5** are expected to be coordinatively highly flexible, whereas the 4-pyridyl units in **9** provide a rigid bridging unit. Mg ring systems **1-4** appear to form due to the different sizes of bridging ligands and other oligomers are likely present in solution. For example, hexameric ring system **1** is a rare aggregate and a tetrameric structure was predominantly obtained [5].

Complex [(^{Dip}nacnac)Mg(DMAP)(NMe$_2$)]·0.5 C$_5$H$_{12}$, **10**·0.5 C$_5$H$_{12}$, crystallised with a full molecule in the asymmetric unit, see Figure 4. The overall arrangement of the neutral DMAP unit, approximately perpendicular to the diketiminate-magnesium plane, and the approximately co-planar arrangement of the anionic dimethylamide unit is typical for these types of complexes, for example in β-diketiminate magnesium butyl complexes with pyridine donor ligands [25–27], and similar to the arrangement in **9**. The dative Mg-N(DMAP) distance in **10** (2.1248(12) Å) is almost unchanged compared with that of the Mg-N(4-pyridyl) fragment in **9**, which is further support of the latter as a dative bond. The terminal Mg-amide bond in **10** (1.9468(12) Å) is significantly shorter.

3. Conclusions

In summary, the synthesis of the large hexameric magnesium 4-pyridyl ring system [{(^{Ar}nacnac)Mg(4-C$_5$H$_4$N)}$_6$] **9** via the reductive C-N cleavage of DMAP using dimagnesium(I) complexes at elevated temperatures is reported. [(^{Dip}nacnac)Mg(DMAP)(NMe$_2$)] **10** was structurally characterised as a by-product. Complex **9** shows a large cyclohexane-like ring structure that forms due to the alternating arrangement of tetrahedral (^{Ar}nacnac)Mg units connected via linear 4-pyridyl groups that act as a rigid bridging ligand. This forms a rare organometallic s-block metallacycle that was compared with examples from the p- and d-block. A minor modification of the 2,6-substituents (isopropyl for **9a**, ethyl for **9b**) on flanking aryl groups on the ligand repeatedly led to large differences in the crystal quality. This work highlights again the importance of single crystal X-ray diffraction for the elucidation of unusual molecular structures of s-block metal complexes.

4. Materials and Methods

4.1. Experimental Details

All manipulations were carried out using standard Schlenk line and glove box techniques under an atmosphere of high purity dinitrogen or argon. Benzene and *n*-pentane were dried and distilled from LiAlH$_4$ under inert gas. ^{1}H and ^{2}H-NMR spectra were recorded on a Bruker Avance 400 or AVIII 500 spectrometer in dried deuterated solvents and were referenced to the residual ^{1}H-NMR resonances, or in benzene and were locked and referenced to the residual D$_2$O peak (^{2}H-NMR spectra). Melting points were determined in sealed glass capillaries under argon and are uncorrected. Elemental analyses were performed by the Elemental Analysis Service at London Metropolitan University. The elemental analyses were affected by the highly air and moisture sensitive nature of the compounds. [{(^{Dip}nacnac)Mg}$_2$] **7a** [17] and [{(^{Dep}nacnac)Mg}$_2$] **7b** [18] were prepared according to literature procedures. DMAP was used as received from Fluorochem [Hadfield, UK]. DMSO-d_6 was dried over activated molecular sieves, degassed, and stored under inert gas.

4.2. Synthesis of [{(^{Dip}nacnac)Mg(4-C$_5$H$_4$N)}$_6$] 9a and Formation of [(^{Dip}nacnac)Mg(DMAP)(NMe$_2$)] 10

A J.Young flask was charged with [{(^{Dip}nacnac)Mg}$_2$] **7a** (200 mg, 226 µmol, 1 equiv.) and DMAP (55.3 mg, 453 µmol, 2 equiv.), and benzene (10 mL) was added. The resultant brown-red solution was heated to 100 °C for 16 h with stirring and resulted in a deep purple solution and an off-white precipitate. This solid was isolated by filtration and dried in vacuo, to give [{(^{Dip}nacnac)Mg(4-C$_5$H$_4$N)}$_6$] **9a** as a highly insoluble powder. Yield: 52.7 mg (22%, based on all Mg atoms). The filtrate was dried in vacuo and extracted into *n*-pentane (*ca.* 5 mL) to give a deep purple solution. Storage of this solution at room temperature for seven days yielded a crop of a mixture of compounds (according to ^{1}H-NMR spectroscopy) from which a yellow crystal of [(^{Dip}nacnac)Mg(DMAP)(NMe$_2$)]·0.5 C$_5$H$_{12}$, **10**·0.5 C$_5$H$_{12}$ was analysed by single crystal X-ray diffraction. Crystals of [{(^{Dip}nacnac)Mg(4-C$_5$H$_4$N)}$_6$]·x C$_6$H$_6$, **9a**·x C$_6$H$_6$, (x ≈ 10 or 16, see the X-ray section) were obtained directly from unstirred reaction mixtures. Data for **9a**: M.p.: 214–222 (decomposition to black solid); elemental analysis (C,H,N, combustion) for C$_{204}$H$_{270}$Mg$_6$N$_{18}$, found: C, 77.13; H, 8.45; N, 7.75; calc: C, 78.52; H, 8.72; N, 8.08.

4.3. Synthesis of [{(^{Dep}nacnac)Mg(4-C$_5$H$_4$N)}$_6$] 9b and Hydrolysis/Deuteronation

A J.Young flask was charged with [{(^{Dep}nacnac)Mg}$_2$] **7b** (200 mg, 259 µmol, 1 equiv.) and DMAP (63.3 mg, 518 µmol, 2 equiv.), and benzene (10 mL) was added. The resultant brown-red solution was heated to 100 °C for 16 h with stirring and resulted in a deep purple solution and an off-white precipitate. The solid was isolated by filtration and dried in vacuo, to give [{(^{Dep}nacnac)Mg(4-C$_5$H$_4$N)}$_6$] **9b**·as a highly insoluble powder. Yield: 85.0 mg (35%, based on all Mg atoms). Crystals of [{(^{Dep}nacnac)Mg(4-C$_5$H$_4$N)}$_6$]·C$_6$H$_6$, **9b**·C$_6$H$_6$, were obtained directly from unstirred reaction mixtures. Data for **9b**: M.p.: 199-207 (decomposition to black solid); elemental analysis (C,H,N, combustion) for C$_{186}$H$_{228}$Mg$_6$N$_{18}$, found: C, 77.30; H, 7.92; N, 8.37; calc: C, 78.06; H, 8.03; N, 8.81.

Hydrolysis/deuteronation experiments: D$_2$O (*ca.* 100 µL) was added to a sample of **9b** (5 mg) in an aromatic solvent. Subsequent sonication gave a colourless solution and a fine white precipitate, presumably Mg(OD)$_2$ D$_2$O. Analysis by ^{1}H (in deuterated benzene) and ^{2}H (in benzene) NMR spectroscopy only showed resonances for 4-deuteropyridine and ^{Dep}nacnacD (ND resonance in ^{2}H-NMR spectrum: δ 12.1 ppm) as organic products. NMR resonances for 4-deutero-pyridine: ^{1}H-NMR (400.1 MHz, C$_6$D$_6$) δ = 6.68–6.72 (m, vd, *J* = 4.5 Hz, 2H, 3-H); 8.49-8.53 (m, vd, *J* = 5.4 Hz, 2H, 2-H); ^{2}H-NMR (76.7 MHz, C$_6$H$_6$) δ = 7.00 (br, vtr, 4-D). In a J.Young NMR tube, **9b** (6 mg) was treated with DMSO-d_6 (0.5 mL). After shaking, the mixture dissolved, was analysed by ^{1}H and ^{2}HNMR spectroscopy, and showed the formation of 4-deuteropyridine and a highly symmetrical β-diketiminate product, likely [(^{Dep}nacnac)Mg(OS{CD$_3$}CD$_2$)]. NMR resonances for 4-deutero-pyridine:

^{1}H-NMR (499.9 MHz, DMSO-d_6) δ = 7.37-7.40 (m, vd, J = 5.0 Hz, 2H, 3-H), 8.56-8.59 (m, vd, J = 5.6 Hz, 2H, 2-H); ^{2}H-NMR (76.7 MHz, DMSO-d_6) δ = 7.77 (br, 4-D). NMR resonances for [(^{Dep}nacnac)Mg{OS(CD$_3$)CD$_2$}]: δ = 1.14 (t, J_{HH} = 7.5 Hz, 12H; CH$_2$CH$_3$), 1.45 (s, 6H; NCCH$_3$), 2.48 (dq, J_{HH} = 15.0, 7.5 Hz, 4H; CH$_2$CH$_3$), 2.64 (dq, J_{HH} = 15.0, 7.5 Hz, 4H; CH$_2$CH$_3$), 4.56 (s, 1H; γ-CH), 6.92-7.02 (m, 6H; Ar-H).

4.4. X-ray Crystallographic Details

Suitable crystals were mounted in paratone oil and were measured using either a Rigaku FR-X Ultrahigh brilliance Microfocus RA generator/confocal optics with XtaLAB P200 diffractometer (Mo Kα radiation) or a Rigaku MM-007HF High Brilliance RA generator/confocal optics with XtaLAB P200 diffractometer (Cu Kα radiation). Data for all compounds analysed were collected using CrystalClear. [28] Data were processed (including correction for Lorentz, polarization and absorption) using either CrystalClear [28] or CrysAlisPro.[29] Structures were solved by a dual-space method (SHELXT-2018/2) [30] and refined by full-matrix least-squares against F^2 using SHELXL-2018/3. [31] All non-hydrogen atoms were refined anisotropically except in selected cases as described below. Hydrogen atoms were placed in calculated positions (riding model). For solvates of **9a**, severely disordered solvent of crystallisation (benzene) was removed using the Platon/SQUEEZE routine [32] to estimate the solvent content and present images in the supporting information. All calculations were performed using CrystalStructure. [33] Details on individual crystal structure determinations and refinements are given below. Further experimental and refinement details are given in the CIF-files. CCDC 2119757-2119759 contains the supplementary crystallographic data for this paper. These data can be obtained free of charge via https://www.ccdc.cam.ac.uk/structures/.

Molecular structures of **9a**. Several datasets of benzene solvates of **9a** were collected of which information on two are briefly presented. Due to the poor overall diffraction and refinement data, only limited data are given here, and images are presented in the supporting information. In both cases, the Platon/SQUEEZE routine was used to estimate the disordered solvent content.

[{(^{Dip}nacnac)Mg(4-C$_5$H$_4$N)}$_6$]·10 C$_6$H$_6$, **9a**·10 C$_6$H$_6$. Crystallised with a sixth of the molecule in the asymmetric unit (Figure S1). C$_{264}$H$_{330}$Mg$_6$N$_{18}$, M = 3901.30, T = 123(2) K, Cu Kα, Cubic, Ia-3d, a = 46.1277(2) Å, b = 46.1277(2) Å, c = 46.1277(2) Å, V = 98148.9(13) Å^3, Z = 16, Reflections collected: 339735, Independent reflections: 4740 [R_{int} = 0.0934]; R_1 [$I > 2\sigma(I)$] ca. 15% (after use of SQUEEZE).

[{(^{Dip}nacnac)Mg(4-C$_5$H$_4$N)}$_6$]·16 C$_6$H$_6$, **9a**·16 C$_6$H$_6$. Crystallised with half of the molecule in the asymmetric unit (Figure S2). C$_{306}$H$_{366}$Mg$_6$N$_{18}$, M = 4442.01, T = 173(2) K, Mo Kα, Triclinic, P-1, a = 14.2254(3) Å, b = 21.0010(5) Å, c = 22.3351(8) Å, α = 79.966(3)°, β = 83.584(2)°, γ = 83.702(2)°, V = 6501.5(3) Å^3, Z = 1, Reflections collected: 204082, Independent reflections: 29318 [R_{int} = 0.0944]; R_1 [$I > 2\sigma(I)$] ca. 16.6% (after use of SQUEEZE).

Molecular structures of **9b**. Information on two similar datasets of **9b**·C$_6$H$_6$ are given, both contain one sixth of the molecule in the asymmetric unit. The first one was used in the discussion and is shown in Figure 3. In each case, one ethyl group is disordered and was modelled with two positions for the methyl group and their positions (73:27% or 71:29% parts) freely refined anisotropically using geometry restraints (DFIX).

[{(^{Dep}nacnac)Mg(4-C$_5$H$_4$N)}$_6$]·C$_6$H$_6$, **9b**·C$_6$H$_6$. CCDC 2119757, C$_{186}$H$_{228}$Mg$_6$N$_{18}$, M = 2861.80, T = 125(2) K, Cu Kα, Trigonal, R-3, a = 33.5457(4) Å, b = 33.5457(4) Å, c = 13.08330(16) Å, V = 12750.3(3) Å^3, Z = 3, ρ = 1.118 Mg/m^3, $F(000)$ = 4626, theta range: 2.634 to 75.779°, indices $-41 \leq h \leq 41$, $-41 \leq k \leq 41$, $-16 \leq l \leq 16$, Reflections collected: 50493, Independent reflections: 5822 [R_{int} = 0.0370], Completeness to theta (67.684°): 99.9%, Goof: 1.054, Final R indices [$I > 2\sigma(I)$]: R_1 = 0.0449, wR_2 = 0.1392, R indices (all data): R_1 = 0.0509, wR_2 = 0.1457, Largest diff. peak and hole: 0.24 and -0.24 e·Å^{-3}.

[{(^{Dep}nacnac)Mg(4-C$_5$H$_4$N)}$_6$]·C$_6$H$_6$, **9b**·C$_6$H$_6$. CCDC 2119758, C$_{186}$H$_{228}$Mg$_6$N$_{18}$, M = 2861.80, T = 173(2) K, Mo Kα, Trigonal, R-3, a = 33.599(3) Å, b = 33.599(3) Å, c = 13.0701(10) Å, V = 12778.0(19) Å^3, Z = 3, ρ = 1.116 Mg/m^3, $F(000)$ = 4626, theta range:

1.708 to 25.393°, indices $-40 \leq h \leq 40, -40 \leq k \leq 40, -15 \leq l \leq 15$, Reflections collected: 39441, Independent reflections: 5225 [R_{int} = 0.0481], Completeness to theta (25.242°): 99.9%, Goof: 1.378, Final R indices [$I > 2\sigma(I)$]: R_1 = 0.0527, wR_2 = 0.1582, R indices (all data): R_1 = 0.0758, wR_2 = 0.1820, Largest diff. peak and hole: 0.30 and -0.26 e·Å^{-3}.

Molecular structure of **10**·0.5 C_5H_{12}. A full molecule is present in the asymmetric unit. Two isopropyl groups are disordered and were modelled with two positions for each atom in the group (55:45% and 42:58% parts) and were freely refined anisotropically using geometry restraints (DFIX). The solvent molecule is disordered, was modelled using geometry restraints (DFIX, DANG), and was refined isotropically.

[(^{Dip}nacnac)Mg(DMAP)(NMe$_2$)]·0.5 C_5H_{12}, **10**·0.5 C_5H_{12}. CCDC 2119759, $C_{40.5}H_{63}$MgN$_5$, M = 644.28, T = 173(2) K, Mo Kα, Tetragonal, $I4_1/a$, a = 22.2072(6) Å, b = 22.2072(6) Å, c = 34.0161(9) Å, V = 16775.4(8) Å^3, Z = 16, ρ = 1.020 Mg/m^3, F(000) = 5648, theta range: 1.765 to 25.348°, indices $-24 \leq h \leq 26, -26 \leq k \leq 25, -40 \leq l \leq 40$, Reflections collected: 44417, Independent reflections: 7648 [R_{int} = 0.0138], Completeness to theta (25.242°): 99.6%, Goof: 1.019, Final R indices [$I > 2\sigma(I)$]: R_1 = 0.0442, wR_2 = 0.1303, R indices (all data): R_1 = 0.0485, wR_2 = 0.1361, Largest diff. peak and hole: 0.29 and -0.23 e·Å^{-3}.

Supplementary Materials: The following are available online, Figure S1: Molecular structure of [{(^{Dip}nacnac)Mg(4-C$_5$H$_4$N)}$_6$]·10 C$_6$H$_6$, **9a**·10 C$_6$H$_6$; Figure S2: Molecular structure of [{(^{Dip}nacnac)Mg(4-C$_5$H$_4$N)}$_6$]·16 C$_6$H$_6$, **9a**·16 C$_6$H$_6$. Solvent molecules and hydrogen atoms omitted.

Author Contributions: S.R.L. and M.d.V.-T. carried out the experiments and compound characterization and wrote the experimental section. A.M.Z.S. conducted the X-ray crystallographic analyses. A.S. conceived and supervised the project and wrote the main section of the manuscript with input from all authors. All authors have read and agreed to the published version of the manuscript.

Funding: The EPSRC is kindly acknowledged for research funding through a doctoral training grant (EP/N509759/1, S.R.L.) and the Centre for Doctoral Training in Critical Resource Catalysis (CRITICAT, EP/ L016419/1, M.d.V.-T.). We thank the University of St Andrews for support.

Institutional Review Board Statement: Not applicable.

Informed Consent Statement: Not applicable.

Data Availability Statement: X-ray crystallographic data are available via the CCDC; please see the X-ray Section 4.4.

Conflicts of Interest: The authors declare no conflict of interest.

References

1. Cook, T.R.; Stang, P.J. Recent Developments in the Preparation and Chemistry of Metallacycles and Metallacages via Coordination. *Chem. Rev.* **2015**, *115*, 7001–7045. [CrossRef] [PubMed]
2. Chakrabarty, R.; Mukherjee, P.S.; Stang, P.J. Supramolecular Coordination: Self-Assembly of Finite Two- and Three-Dimensional Ensembles. *Chem. Rev.* **2011**, *111*, 6810–6918. [CrossRef]
3. Harder, S. Chapter 1: Introduction to Early Main Group Organometallic Chemistry and Catalysis, in *Early Main Group Metal Catalysis*; Harder, S., Ed.; Wiley-VCH Verlag GmbH & Co. KgaA: Weinheim, Germany, 2020; pp. 1–29. [CrossRef]
4. Langer, J.; Maitland, B.; Grams, S.; Ciucka, A.; Pahl, J.; Elsen, H.; Harder, S. Self-Assembly of Magnesium Hydride Clusters Driven by Chameleon-Type Ligands. *Angew. Chem. Int. Ed.* **2017**, *56*, 5021–5025. [CrossRef] [PubMed]
5. Fohlmeister, L.; Stasch, A. Ring-Shaped Phosphinoamido-Magnesium-Hydride Complexes: Syntheses, Structures, Reactivity, and Catalysis. *Chem. Eur. J.* **2016**, *22*, 10235–10246. [CrossRef]
6. Olmstead, M.M.; Grigsby, W.J.; Chacon, D.R.; Hascall, T.; Power, P.P. Reactions between primary amines and magnesium or zinc dialkyls: Intermediates in metal imide formation. *Inorg. Chim. Acta* **1996**, *251*, 273–284. [CrossRef]
7. Bailey, P.J.; Liddle, S.T.; Morrison, C.A.; Parsons, S. The First Alkaline Earth Metal Complex Containing a μ-η^1:η^1 Allyl Ligand: Structure of [{HC[C(tBu)NC$_6$H$_3$(CHMe$_2$)$_2$]$_2$Mg(C$_3$H$_5$)}$_6$]. *Angew. Chem. Int. Ed.* **2001**, *40*, 4463–4466. [CrossRef]
8. Chen, W.; Liu, L.; Zhao, Y.; Xue, Y.; Xu, W.; Li, N.; Wu, B.; Yang, X.-J. Organometallo-macrocycle assembled through dialumane-mediated C–H activation of pyridines. *Chem. Commun.* **2021**, *57*, 6268–6271. [CrossRef] [PubMed]
9. Stasch, A. Chapter 3: Recent Advances in the Stoichiometric Chemistry of Magnesium Complexes, in *Catalysis with Earth-abundant Elements*; Schneider, U., Thomas, S., Eds.; The Royal Society of Chemistry: London, UK, 2021; pp. 55–80. [CrossRef]
10. Rösch, B.; Harder, S. New horizons in low oxidation state group 2 metal chemistry. *Chem. Commun.* **2021**, *57*, 9354–9365. [CrossRef]

11. Jones, C. Dimeric magnesium(I) β-diketiminates: A new class of quasi-universal reducing agent. *Nat. Rev. Chem.* **2017**, *1*, 0059. [CrossRef]
12. Stasch, A.; Jones, C. Stable dimeric magnesium(I) compounds: From chemical landmarks to versatile reagents. *Dalton Trans.* **2011**, *40*, 5659–5672. [CrossRef]
13. Green, S.P.; Jones, C.; Stasch, A. Stable Adducts of a Dimeric Magnesium(I) Compound. *Angew. Chem. Int. Ed.* **2008**, *47*, 9079–9083. [CrossRef]
14. Bonyhady, S.J.; Jones, C.; Nembenna, S.; Stasch, A.; Edwards, A.J.; McIntyre, G.J. β-Diketiminate-Stabilized Magnesium(I) Dimers and Magnesium(II) Hydride Complexes: Synthesis, Characterization, Adduct Formation, and Reactivity Studies. *Chem. Eur. J.* **2010**, *16*, 938–955. [CrossRef] [PubMed]
15. Yuvaraj, K.; Douair, I.; Paparo, A.; Maron, L.; Jones, C. Reductive Trimerization of CO to the Deltate Dianion Using Activated Magnesium(I) Compounds. *J. Am. Chem. Soc.* **2019**, *141*, 8764–8768. [CrossRef]
16. Yuvaraj, K.; Douair, I.; Jones, D.D.L.; Maron, L.; Jones, C. Sterically controlled reductive oligomerisations of CO by activated magnesium(I) compounds: Deltate vs. ethenediolate formation. *Chem. Sci.* **2020**, *11*, 3516–3522. [CrossRef]
17. Green, S.P.; Jones, C.; Stasch, A. Stable Magnesium(I) Compounds with Mg-Mg Bonds. *Science* **2007**, *318*, 1754–1758. [CrossRef]
18. Lalrempuia, R.; Kefalidis, C.E.; Bonyhady, S.J.; Schwarze, B.; Maron, L.; Stasch, A.; Jones, C. Activation of CO by Hydrogenated Magnesium(I) Dimers: Sterically, Controlled Formation of Ethenediolate and Cyclopropanetriolate Complexes. *J. Am. Chem. Soc.* **2015**, *137*, 8944–8947. [CrossRef] [PubMed]
19. MacLeod, K.C.; Lewis, R.A.; DeRosha, D.E.; Mercado, B.Q.; Holland, P.L. C-H and C-N Activation at Redox-Active Pyridine Complexes of Iron. *Angew. Chem. Int. Ed.* **2017**, *56*, 1069–1072. [CrossRef]
20. Gentner, T.X.; Rösch, B.; Ballmann, G.; Langer, J.; Elsen, H.; Harder, S. Low Valent Magnesium Chemistry with a Super Bulky β-Diketiminate Ligand, Low Valent Magnesium Chemistry with a Super Bulky β-Diketiminate Ligand. *Angew. Chem. Int. Ed.* **2019**, *58*, 607–611. [CrossRef]
21. Jones, D.D.L.; Douair, I.; Maron, L.; Jones, C. Photochemically Activated Dimagnesium(I) Compounds: Reagents for the Reduction and Selective C-H Bond Activation of Inert Arenes. *Angew. Chem. Int. Ed.* **2021**, *60*, 7087–7092. [CrossRef] [PubMed]
22. Jones, C.; McDyre, L.; Murphy, D.M.; Stasch, A. Magnesium(I) reduction of benzophenone and anthracene: First structural characterisation of a magnesium ketyl. *Chem. Commun.* **2010**, *46*, 1511–1513. [CrossRef] [PubMed]
23. Mocanu, T.; Rat, C.I.; Maxim, C.; Shova, S.; Tudor, V.; Silvestru, C.; Andruh, M. Bis(4-pyridyl)mercury–a new linear tecton in crystal engineering: Coordination polymers and co-crystallization processes. *Cryst. Eng. Comm.* **2015**, *17*, 5474–5487. [CrossRef]
24. Mocanu, T.; Kiss, L.; Sava, A.; Shova, S.; Silvestru, C.; Andruh, M. Coordination polymers and supramolecular solid-state architectures constructed from an organomettalic tecton, bis(4-pyridyl)mercury. *Polyhedron* **2019**, *166*, 7–16. [CrossRef]
25. Hill, M.S.; MacDougall, D.J.; Mahon, M.F. Magnesium hydride-promoted dearomatisation of pyridine. *Dalton Trans.* **2010**, *39*, 11129–11131. [CrossRef] [PubMed]
26. Hill, M.S.; Kociok-Köhn, G.; MacDougall, D.J.; Mahon, M.F.; Weetman, C. Magnesium hydrides and the dearomatisation of pyridine and quinoline derivatives. *Dalton Trans.* **2011**, *40*, 12500–12509. [CrossRef] [PubMed]
27. Balasanthiran, V.; Chisholm, M.H.; Choojun, K.; Durr, C.B.; Wambua, P.M. BDI*MgX(L) where X = ^{n}Bu and O^{t}Bu and L = THF, py and DMAP. The rates of kinetic exchange of L where BDI* = CH{C(^{t}Bu)N-2,6-^{i}Pr$_2$C$_6$H$_3$}$_2$. *Polyhedron* **2016**, *103*, 235–240. [CrossRef]
28. *CrystalClear-SM Expert*; v2.1; Rigaku Americas: The Woodlands, TX, USA,; Rigaku Corporation: Tokyo, Japan, 2015.
29. *CrysAlisPro*; v1.171.38.46; Rigaku Oxford Diffraction; Rigaku Corporation: Oxford, UK, 2015.
30. Sheldrick, G.M. SHELXT–Integrated space-group and crystal-structure determination. *Acta Cryst.* **2015**, *A71*, 3–8. [CrossRef]
31. Sheldrick, G.M. Crystal structure refinement with SHELXL. *Acta Cryst.* **2015**, *C71*, 3–8. [CrossRef]
32. Spek, A.L. PLATON SQUEEZE: A tool for the calculation of the disordered solvent contribution to the calculated structure factors. *Acta Cryst.* **2015**, *C71*, 9–18. [CrossRef]
33. *CrystalStructure*; v4.3.0; Rigaku Americas: The Woodlands, TX, USA; Rigaku Corporation: Tokyo, Japan, 2018.

Article

Synthetic and Structural Study of *peri*-Substituted Phosphine-Arsines

Brian A. Chalmers [1], D. M. Upulani K. Somisara [1], Brian A. Surgenor [2], Kasun S. Athukorala Arachchige [3], J. Derek Woollins [1,4], Michael Bühl [1], Alexandra M. Z. Slawin [1] and Petr Kilian [1,*]

1. EaStChem School of Chemistry, University of St Andrews, St Andrews KY16 9ST, UK; bac8@st-andrews.ac.uk (B.A.C.); dmuks@st-andrews.ac.uk (D.M.U.K.S.); jdw3@st-andrews.ac.uk (J.D.W.); mb105@st-andrews.ac.uk (M.B.); amzs@st-andrews.ac.uk (A.M.Z.S.)
2. Treatt Plc, Bury St Edmunds IP32 7FR, UK; brian.surgenor@treatt.com
3. Centre for Microscopy and Microanalysis, The University of Queensland, St Lucia, QLD 4072, Australia; kasun.athukorala@uq.edu.au
4. Department of Chemistry, Khalifa University, Abu Dhabi 127788, United Arab Emirates
* Correspondence: pk7@st-andrews.ac.uk

Abstract: A series of phosphorus-arsenic *peri*-substituted acenaphthene species have been isolated and fully characterised, including single crystal X-ray diffraction. Reactions of EBr_3 (E = P, As) with iPr_2PAcenapLi (Acenap = acenaphthene-5,6-diyl) afforded the thermally stable *peri*-substitution supported donor–acceptor complexes, iPr_2PAcenapEBr_2 **3** and **4**. Both complexes show a strong P→E dative interaction, as observed by X-ray crystallography and ^{31}P NMR spectroscopy. DFT calculations indicated the unusual As···As contact (3.50 Å) observed in the solid state structure of **4** results from dispersion forces rather than metallic interactions. Incorporation of the excess $AsBr_3$ in the crystal structure of **3** promotes the formation of the ion separated species $[iPr_2$PAcenapAsBr$]^+$Br$^-$ **5**. A decomposition product **6** containing the rare $[As_6Br_8]^{2-}$ heterocubane dianion was isolated and characterised crystallographically. The reaction between iPr_2PAcenapLi and $EtAsI_2$ afforded tertiary arsine $(BrAcenap)_2$AsEt **7**, which was subsequently lithiated and reacted with $PhPCl_2$ and Ph_2PCl to afford cyclic PhP(Acenap)$_2$AsEt **8** and acyclic EtAs(AcenapPPh$_2$)$_2$ **9**.

Keywords: *peri*-substitution; arsenic; organophosphorus; pnictine; single crystal X-ray structures

---

1. Dedication

This paper is dedicated to Prof Alex Slawin on the occasion of her 60th birthday, a world class crystallographer, a caring and compassionate colleague and mentor.

2. Introduction

Phosphines (R_3P) are known as prototypical neutral donors (Lewis bases) with the most common examples being their use as tunable ligands in coordination chemistry [1,2]. Bonding occurs through donation of the phosphorus lone pair into an acceptor orbital on the metal (σ-component), often combined with the acceptance of electron density from the metal orbitals to empty orbitals on the phosphine (π-component). Arsines are also used as ligands, although their use is somewhat limited by their greater toxicity.

Although possessing a lone pair of electrons, both phosphines and arsines can also act as electron pair acceptors (Lewis acids) when equipped with electron withdrawing substituents, such as halogens [3].

For this paper, dative species in which a phosphine acts as donor, and another phosphine or arsine acts as acceptor, are of interest. Holmes experimented with various pnictogen compounds to form pnictogen–pnictogen donor–acceptor complexes, and in all cases, decomposition occurred at ambient temperatures [4,5]. For example, the reaction between Me_3P and PCl_3 (both colourless liquids) afforded a white solid of the composition

(Me$_3$P)$_2$·PCl$_3$ at temperatures below freezing, the solid however decomposed upon warming [5]. Sisler then investigated the decomposition pathways and discovered redox degradation occurred readily at ambient conditions (e.g., Et$_3$P + MePCl$_2$ → Et$_3$PCl$_2$ + (MeP)$_5$) [6].

It was not until 2001 that the first examples of phosphine–phosphine donor–acceptor species, Me$_3$P→PBr$_3$ and Me$_3$P→Bz′PBr$_2$ (Bz′ = 3,5-dimethylbenzyl), were structurally characterised by Müller and Winkler (Figure 1) [7]. Even so, these two adducts still decomposed at ambient conditions.

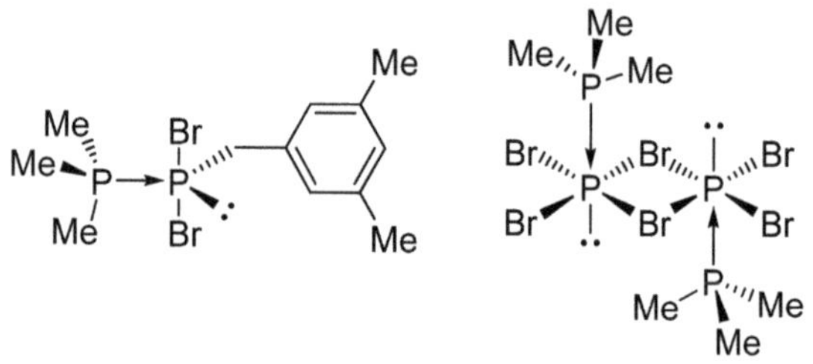

Figure 1. The first structurally characterised phosphine–phosphine donor–acceptor complexes.

The first room temperature stable phosphine–phosphine donor–acceptor complex, (iPr$_2$PAcenapPCl$_2$ (Acenap = acenaphthene-5,6-diyl), was isolated almost a decade later by us [8]. With *peri*-substitution enforcing the interaction between the two phosphorus atoms, the resulting strong dative interaction and reduced flexibility makes the redox decomposition pathways less accessible, allowing isolation and manipulation of these dative species at ambient temperatures.

Donor–acceptor species with heavier acceptor pnictogens were reported to be more stable owing to their reduced propensity to the redox decomposition pathways. Thus, while Me$_3$P→PCl$_3$ decomposes above −20 °C, [9] (Me$_3$P)$_2$→SbCl$_3$ and Me$_3$As→SbCl$_3$ are stable at ambient conditions and melt with decomposition at 110 °C [10] and 145 °C [4,11], respectively. In his early study, Sisler reported on the prototypical R$_3$Pn′→PnX$_3$ pnictogen–pnictogen donor–acceptor complexes. Their stability was increased for heavier halogens in the acceptor species (I > Br > Cl > F) and followed the trend Sb > As > P for the Pn in the acceptor species (PnX$_3$), while the stability trend As > P > Sb was observed for Pn′ in the donor species (R$_3$Pn′) [9]. The stability of these and related dative species is likely to be driven by the difference in the Lewis acidity and basicity of the donor and acceptor (i.e., the strength of the dative bond), combined with the redox properties of the two components (i.e., the reducing and oxidising power of these). Comprehensive accounts on neutral [12] and related cationic species [13,14] have been published recently, encompassing the variety of the structural modes adopted by main group pnictine complexes.

In the last two decades, we and others focused our attention on the *peri*-substituted pnictogen–pnictogen donor–acceptor complexes as shown in Figure 2.

Among complexes with dihalopnictines as acceptors, four structural types have been observed, including two molecular modes (**A**, **B**), µ-dichloro-bridged dimer **C** and fully ionic species **D**. Monohalopnictogen adducts attain two structural types, **D** (ionic) and **B** (molecular).

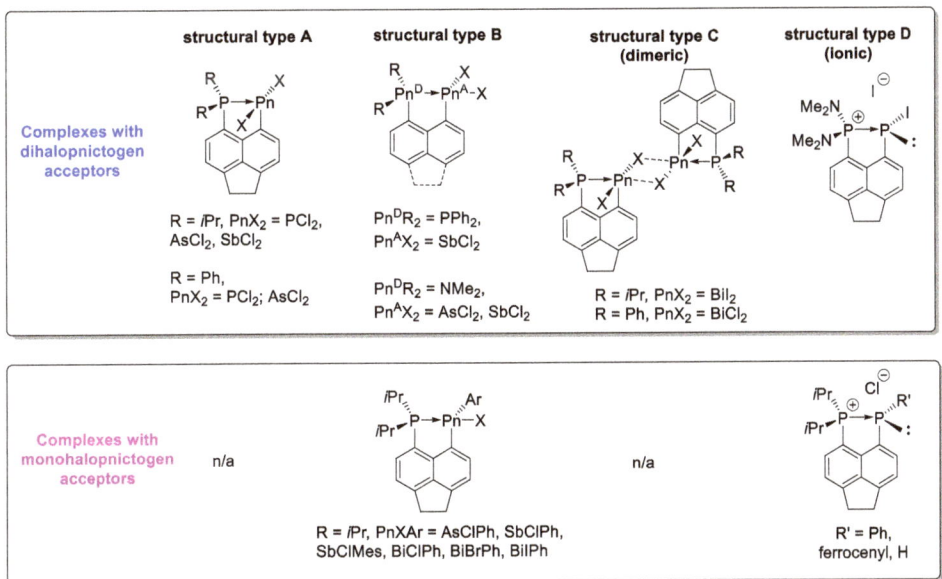

Figure 2. Selected structurally characterised *peri*-substituted pnictine-pnictine donor–acceptor complexes, indicating their structural diversity [8,15–23].

Motivated by the utility of dihalopnictines as precursor molecules in the development of new radical C–H coupling reactions [21], our attention turned to related species with dibromopnictine acceptor groups. Replacement of chlorine atoms with bromine was expected to alter thermal stability (through altering redox properties) and solubility in organic solvents (through altering the aggregation properties); both thermal stability and solubility are important for the synthetic utility of these dihalides. We also report on the chemistry of a geminal bis(acenaphthene)arsine, specifically its ability to tolerate treatment with *n*BuLi without cleavage of the As–C bonds. This opens up new synthetic pathways to further P and As *peri*-substituted species, some of which may serve as the C–H coupling precursors as mentioned above.

3. Results & Discussion

3.1. Reactions Involving PBr$_3$ and AsBr$_3$

In the syntheses described herein, 5,6-dibromoacenaphthene, **1**, was used as the principal precursor. Two subsequent lithium–halogen exchanges, followed by reactions with pnictogen halide electrophiles, were used to access the variety of *peri*-substitution patterns shown in Scheme 1.

5-Bromo-6-(diisopropylphosphino)acenaphthene **2** was prepared using our literature procedure [8]. Lithiation of **2** using *n*BuLi, followed by a slow addition of the formed *i*Pr$_2$PAcenapLi to a fivefold excess PBr$_3$, gave the phosphonium-phosphoranide **3** (*i*Pr$_2$PAcenapPBr$_2$) in a good yield (65%). The related phosphonium-arsoranide **4** (*i*Pr$_2$PAcenapAsBr$_2$) was made using the same procedure in ~75% yield; however, in this case, only one molar equivalent of AsBr$_3$ was used in the reaction. The selectivity of the reactions of aryllithiums towards pnictogen halides PnX$_3$ (Pn = pnictogen) is affected greatly by the order and rate of the reactant addition and the stoichiometric ratio of the reagents. A large excess of the pnictogen halide is often necessary in order to promote monosubstitution giving R$_2$PAcenapPnX$_2$, i.e., to avoid geminal di- or trisubstitution, leading to the formation of (R$_2$PAcenap)$_2$PnX or (R$_2$PAcenap)$_3$Pn [8,15,17,22,24,25]. In the preparation of **3**, the excess PBr$_3$ was necessary to achieve good yields; the excess PBr$_3$ was removed by careful washing with THF and diethyl ether.

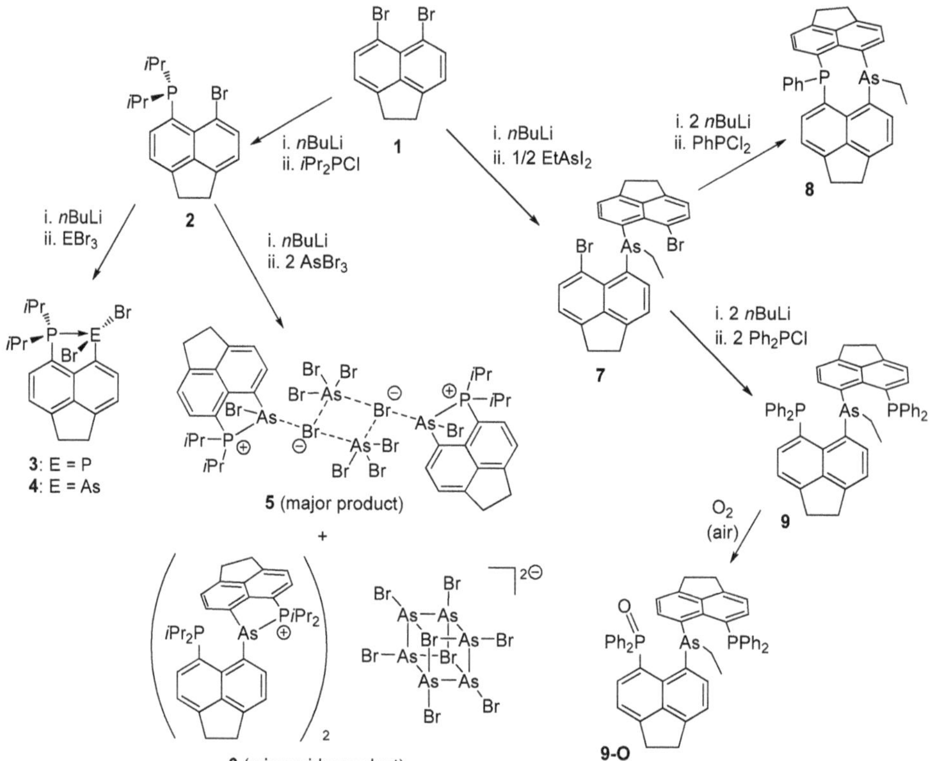

Scheme 1. Syntheses reported in this paper.

Compounds **3** and **4** represent an expansion of the small number of phosphine–phosphine and phosphine–arsine donor–acceptor complexes, that are stable (isolable) at ambient temperatures. The unsupported phosphine–phosphine complexes undergo redox degradation at temperatures above −40 °C or at even lower temperatures [7]. Both **3** and **4** can be stored indefinitely under inert atmosphere at ambient temperature in the solid form, but **3** undergoes decomposition slowly (over the course of a week) in common organic solvents (thf, dcm, chloroform). The arsenic congener **4** immediately decomposes in chlorinated solvents, giving a mixture of insoluble products. Both **3** and **4** are air sensitive. It should be noted that the thermal stability and air sensitivity of **3** and **4** is rather similar to that of the chlorine congeners iPr$_2$PAcenapECl$_2$ (E = P, As) [8,15].

The ^{31}P{^{1}H} NMR spectrum of **3** consists of two doublets at δ_P 65.3 (iPr$_2$P) and 32.9 ppm (PBr$_2$) with a large $^1J_{PP}$ magnitude of 353.7 Hz (AX spin system). This is in good agreement with the relevant data of iPr$_2$PAcenapPCl$_2$ (δ_P 68.8 and 40.4 ppm, $^1J_{PP}$ 363 Hz) [8]. The ^{31}P{^{1}H} NMR spectrum of **4** (singlet at δ_P 56.6 ppm) is in agreement with the previously published data for iPr$_2$PAcenapAsCl$_2$ (δ_P 65.3 ppm) [15]. This strongly suggests the structures of **3** and **4** in the solution are similar to those of their chlorine congeners. Significant deshielding of the donor (iPr$_2$P) phosphine group in **3** and **4** compared to that in **2** (δ_P −2.2 ppm) strongly suggests major sequestration of the lone pair electron density takes place by the pnictogen halide (PnX$_2$) acceptor group. Compounds **3** and **4** were fully characterised including ^{1}H, ^{13}C{^{1}H} and ^{31}P{^{1}H} NMR, IR and MS.

The crystal structures of **3** and **4** are shown in Figure 3 and Table 1. The diffraction data indicate a strong dative interaction is formed between the two phosphorus atoms in **3**, with a P–P bond length of 2.2701(16) Å, which is just within a range of standard P–P single bond (2.22 ± 0.05 Å). Two molecules of **4** are present in the asymmetric unit. The P–As distances in these (2.405(3) and 2.407(3) Å) are crystallographically identical, consistent with a strong dative bond, and comparable to a standard P–As single bond length. Based on the electronegativity considerations, the ECl$_2$ group is expected to be better acceptor than the respective EBr$_2$ group (E = P, As). However, no dramatic lengthening of the P→EBr$_2$ vs. the P→ECl$_2$ dative bond is observed; the P–As bond length in the chlorine congener iPr$_2$P-Acenap-AsCl$_2$ (2.4029(7) Å) [15] is crystallographically identical to that in **4**, while the P–P bond length in iPr$_2$PAcenapPCl$_2$ (2.2570(14) Å) [8] is only marginally shorter than that in **3**.

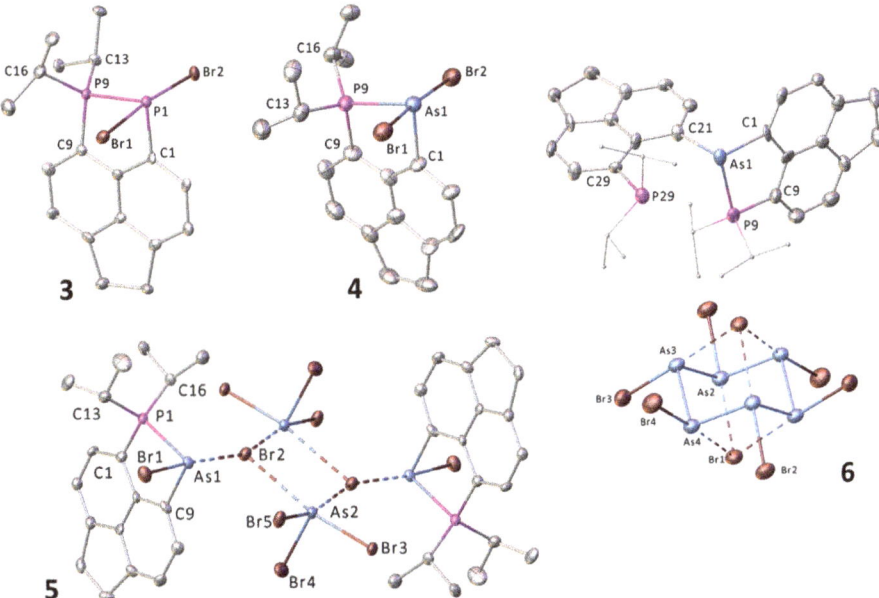

Figure 3. Crystal structures of **3–6**. Hydrogen atoms and solvating molecules (THF in **3** and benzene in **4**) are omitted. The second molecule in the asymmetric unit of **4** is omitted. **5** forms centrosymmetric dimer, two asymmetric units are shown in the figure. Monocation of **6** is shown at the top, the [As$_6$Br$_8$]$^{2-}$ dianion is at the bottom. Thermal ellipsoid ds are plotted at the 40% level.

Although neither **3** nor **4** form a halogen bridged dimeric assemblies in the crystal, the P–Br bond lengths are nonidentical within **3** (P1–Br1 2.6864(13) Å; P1–Br2 2.4376(13) Å). In contrast, the As–Br bond lengths are almost identical in **4**: As1–Br1 2.6337(19) [2.6339(19)] Å, As1–Br2: 2.6400(18) [2.664(2)] Å (values in square brackets are for the second molecule in the asymmetric unit). The P–Br and As–Br bond lengths in **3** and **4** indicate an intermediate character between ionic and covalent bonds; their elongation is rather significant when compared to the bond lengths in the covalent PBr$_5$ (P–Br: 2.221 and 2.158 Å) [26] and AsBr$_3$ (in adduct with hexaethylbenzene, P–As 2.322(1) Å) [27]. The acceptor pnictogen atoms in **3** (P1) and **4** (As1) adopt a pseudo-trigonal bipyramidal geometry, with the two bromine atoms occupying the axial positions (Br–E–Br 172.22(5)° in **3** and 173.95(6)° [172.27(6)°] in **4**). The P9 and C1 atoms occupy the equatorial positions, with the 3rd position taken up by the lone pair. The inequality and elongation of the phosphorus–halogen bonds are phenomena observed in all relevant structurally characterised phosphine–phosphine and phosphine–arsine donor–acceptor complexes, and presumably stem mainly from packing effects [7,8,17]. The equal length of As–Br bonds, observed within **4**, is likely a result of the symmetrical nature of the local polarity effects around the bromine atoms within the crystal structure of **4**. The acenaphthene framework in both **3** and **4** encounters only minimal

161

in-plane and out-of-plane distortions from the ideal (planar) geometry, consistent with a relaxed *peri*-region (P–Pn bonded) structures.

Table 1. Selected bond lengths [Ångströms (Å)] and angles (degrees, °) for 3 to 9-O.

Compound	3·THF (E = P)	4·C$_6$H$_6$ [b] (E = As)	5 (E = As)	6 [c] (E = As)
		peri-region bond distances		
P9–E1	2.2701(16)	2.405(3) [2.407(3)]	2.3978(15)	2.365(4) [3.145(4)]
E1–Br1	2.6864(13)	2.6337(19) [2.6639(19)]	2.4375(9)	–
E1–Br2	2.4376(13)	2.6400(18) [2.634(2)]	3.000(9)	–
		peri-region angles		
C9–P9–E1	97.92(14)	97.9(3) [98.4(3)]	97.78(18)	98.7(4)
C1–E1–P9	88.82(13)	84.9(3) [85.3(3)]	86.25(17)	85.7(3)
Br1–E1–Br2	172.22(5)	173.95(6) [172.27(6)]	175.3(2)	–
P9–E1–Br1	91.13(5)	96.01(9) [91.02(9)]	95.57(4)	–
P9–E1–Br2	96.38(5)	90.00(9) [95.95(9)]	88.77(4)	–
splay angle [a]	−7.8(5)	−5(1) [−5(1)]	−4(1)	−5(1) [15(1)]
		out-of-plane displacements		
P9	0.23	0.07 [0.10]	0.09	0.21 [0.12]
E1	−0.03	−0.36 [−0.26]	−0.01	−0.11 [−0.05]
		dihedral angles		
P9–C9···C1–E	7.7(2)	10.9(4) [9.2(4)]	2.4(2)	2.8(5) [1.2(5)]

Compound	7 1/2CH$_2$Cl$_2$ (E = Br)	8·CH$_2$Cl$_2$ (E = P)	9-O (E = P) [f]
	peri-region bond distances		
As1–E9	3.23(1)	3.004(6)	3.176(5)
As1–E29	3.28(1)	–	3.273(5)
	peri-region angles		
splay angle [a,d]	17(1)	11.7(4)	16.1(8)
splay angle [a,e]	19(1)	11.4(4)	15.2(8)
	out-of-plane displacements		
As1	0.35, 0.02	0.19, 0.02	0.04, 0.71
E9	−0.08	−0.03	−0.40
E29	−0.01	−0.23	0.41
	dihedral angles		
As1–C1···C9–E9	12(1)	3.3(6)	8(1)
As1–C21···C29–E29	1(1)	5.0(6)	27(1)

[a] splay angle = sum of the bay region angles—360. [b] values in square parentheses are for the 2nd molecule in the asymmetric unit. [c] values in square parentheses refer to the P29 acenaphthene unit. [d] C10 acenaphthene unit. [e] C30 acenaphthene unit. [f] E29 in this structure represents P39.

An interesting feature was observed in the crystal packing of **4**. Rather than forming a Br···As intermolecular close contacts, commonly seen in halogen-bridged dimers [7,16], pairs of molecules of **4** are oriented so that a short (intermolecular) As···As contact (3.50 Å) is formed. To gain some insight into significance of this, we performed density functional theory (DFT) computations on a 'dimer' motif, carved out of the crystal. Because the interaction between the two molecules was expected to be weak, a level was chosen that contains corrections for attractive van der Waals forces (dispersion) and basis-set superposition error (BSSE; an intrinsic error for intermolecular interaction energies with finite basis sets), denoted B3LYP-D3/6-31(+)G* (see SI for details; B3LYP functional and basis sets are the same as in our previous studies on *peri*-acenaphthene derivatives) [16,21]. Interestingly, full geometry optimisation of the dimer at that level affords an even shorter As···As contact (3.28 Å) than that observed in the solid, and a substantial binding energy ($\Delta E = -15.7$ kcal mol^{-1}, estimated ΔH^{298} −14.4 kcal mol^{-1} and ΔG^{298} −3.43 kcal mol^{-1}). Despite this sizeable interaction energy, only small covalent contributions to the binding are apparent from the computed intermolecular Wiberg bond indices (WBIs) [28], which do not exceed 0.05 for both As···As and As···Br contacts. Closer inspection of the contributions to the binding energy ΔE reveals that it is entirely dominated by dispersion interactions: the

-D3 contribution in the optimised structure amounts to −16.5 kcal mol^{-1}, i.e., the interaction would be expected to be repulsive at the B3LYP level without explicit dispersion correction. Indeed, if the structure is relaxed at the B3LYP/6-31(+)G* level (still BSSE-corrected), the dimer essentially falls apart. In this scenario, a much longer As⋯As distance ensues (3.87 Å) with a very weak binding energy ΔE = −2.1 kcal mol^{-1}, presumably due to weak electrostatic interactions. It is remarkable that the pairs of molecules are held together by such strong dispersion forces, as it brings the arsenic atoms (with their lone pairs pointing toward each other) to an essentially repulsive distance, as shown in Figure 4.

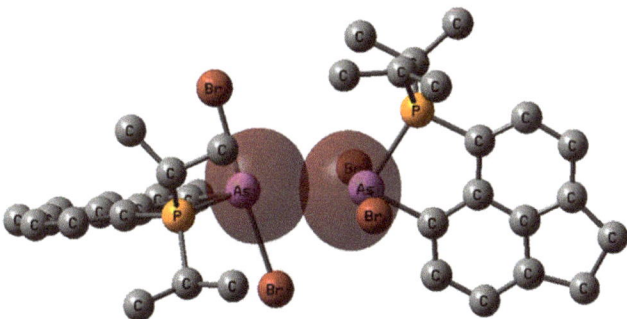

Figure 4. Plot demonstrating overlap of the lone pairs on arsenic atoms in a pair of molecules **4**. Natural bond orbitals, B3LYP/6-31(+)G* level on B3LYP-D3/6-31(+)G* optimised geometry, isovalue for plot 0.04 a.u.; hydrogen atoms omitted for clarity.

Conducting the reaction of *i*Pr$_2$PAcenapLi with 2 molar equivalents of AsBr$_3$ gave **5** as a pale-yellow solid, which was contaminated with residual LiBr. Crystals suitable for diffraction work were grown from acetonitrile. The crystal structure of **5** is shown in Figure 3 and Table 1. Compound **5** is formed by an ion pair [*i*Pr$_2$PAcenapAsBr]$^+$Br$^-$, co-crystallised with an AsBr$_3$ molecule (see Scheme 1), with two acenaphthene units and PBr$_3$ molecules forming a centrosymmetric assembly. All Br⋯As distances around the μ^3-bridging Br2 atom (3.00–3.04 Å) are significantly elongated compared to all other As–Br bonds (2.3370(9)–2.4375(9) Å), supporting the interpretation of the bonding as ionic. Comparing the structures of **4** and **5** shows no significant change in the P–As bond length (2.3978(15) Å in **5**). The most substantial changes are the shortening of the As1–Br1 bond to 2.4375(9) Å and the elongation of the other distance, As1⋯Br2, to 3.000(9) Å (c.f. As–Br bond lengths 2.634(2)–2.6639(19) Å in **4**).

In one incidence, while attempting to grow crystals of **5** from boiling acetonitrile, a small crop of colourless crystals was obtained. These were subjected to single-crystal X-ray diffraction and were found to be compound **6**, consisting of a complex phosphonium cation and an unusual octabromohexaarsenate ([As$_6$Br$_8$]$^{2-}$) dianion (Scheme 1, Figure 3 and Table 1). The cation of **6** consists of two acenaphthene groups geminally attached to an arsenic atom. The two P–As distances are very disparate, indicating the presence of a standard P9–As1 bond (2.365(4) Å), and an onset of 3-center 4-electron P⋯As–C interaction (P29⋯As1 3.145(4) Å; P29⋯As1–C1 angle is 176.9(3)°). A tentative mechanism of the cation formation involves an attack of the second molecule of *i*Pr$_2$PAcenapLi on **4** or **5**. Only one previous incidence of crystallographic characterisation of octabromohexaarsenate dianion was found in the literature; this was in the form of its tetraphenylphosphonium salt [29]. The dianion of **6** is essentially isostructural to the previously reported example, both can be seen as an (AsBr)$_6$ oligoarsine ring in a chair conformation, capped by two bromide anions to form a heterocubane cluster. A tentative mechanism of formation of **6** involves thermally induced reduction of AsBr$_3$ (present in the structure of **5**), leading to the formation of cyclohexaarsine (AsBr)$_6$. This is followed by coordination by bromide anions to give the [As$_6$Br$_8$]$^{2-}$ dianion. The reduction step may be analogous to that seen throughout the chemistry of pnictine–pnictine complexes, first described by Sisler [6], in which the trihalide halogenates the tertiary phosphine moiety (to halophosphonium [R_3P^VCl]$^+$Cl$^-$), reducing itself to ($As^IX)_n$. This proposed mechanism is distinct (but related) to that proposed by Macdonald for formation of octaiodocyclohexaarsenates, where the diphosphine ligand (L) cleavage from pre-assembled arsenic(I) species [LAs][I] was suggested [30]. Further characterisation of **6** was not possible due to the small amount of crystals available. Attempts to repeat synthesis of **6** on a larger scale were not successful.

Our attempts to prepare the iodine congener iPr$_2$PAcenapPI$_2$ via reaction of iPr$_2$PAcenapLi with PI$_3$ or P$_2$I$_4$ gave complex mixtures as judged by ^{31}P NMR. On a similar note, reactions of iPr$_2$PAcenapLi with SbBr$_3$ gave inseparable mixtures, presumably via undesirable redistribution and redox processes [9]. This is in stark contrast to the analogous reaction of iPr$_2$PAcenapLi with SbCl$_3$, which afforded the desired donor–acceptor complex iPr$_2$PAcenapSbCl$_2$ in an excellent yield [16].

3.2. Reactions Involving EtAsI$_2$

To expand the range of arsenic *peri*-substituted species, reactions with ethyldiiodoarsine have been studied. Monolithiation of 5,6-dibromoacenaphthene, followed by an addition of the formed iPr$_2$PAcenapLi to one molar equivalent of EtAsI$_2$, was expected to yield iodoarsine Et(I)AsAcenapBr. However, the ^{1}H NMR spectrum of the white powder obtained after workup indicated a product of a double (geminal) substitution, a tertiary arsine **7**, has been formed solely (Scheme 1). The reaction was repeated with highly diluted EtAsI$_2$ and very slow addition rate to limit the double substitution, however even then, **7** was the sole product of the reaction.

The crystal structure of **7** is shown in Figure 5 and Table 1. The arsenic atom adopts trigonal pyramidal geometry. The As···Br distances (3.23(1) and 3.28(1) Å, 82–83% of the sum of the respective Van der Waals radii [31]) as well as the large splay angles (17(1) and 19(1)°) indicate repulsive interactions in the *peri*-region, with As···Br interactions forced through *peri*-geometry.

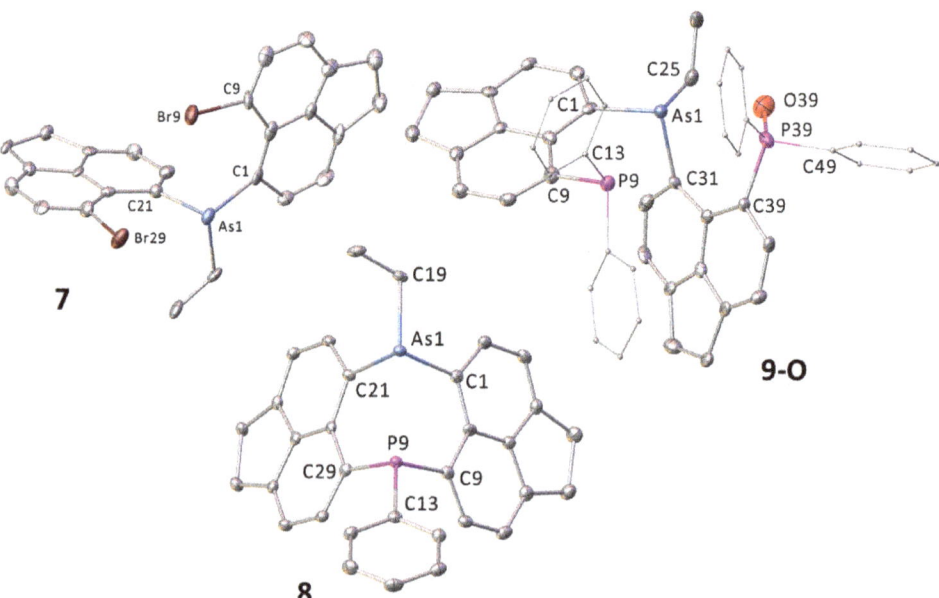

Figure 5. Crystal structures of **7**, **8** and **9-O**. Hydrogen atoms and solvating molecules (CH$_2$Cl$_2$ in **7** and **8**) are omitted. Thermal ellipsoids are plotted at the 40% level.

Following the clean synthesis of **7**, we were interested in its use as a synthon, particularly its ability to undergo lithium–halogen exchange, which would open many options for its further functionalisation. Initially, we focussed on introducing phosphorus moieties to the *peri*-positions. Synthesis of two examples, cyclic (**8**) and acyclic (**9**), resulted from these efforts (Scheme 1). Using standard low-temperature lithium–halogen exchange conditions, double lithiation of **7** was incomplete after several hours at −78 °C, despite using a slight excess over two equivalents of *n*-butyllithium. To promote the double lithiation we have adopted conditions similar to those reported by Kasai for double lithiation of 5,6-dibromoacenaphthene [32], which included the addition of chelating base tetramethylethylenediamine, TMEDA. In addition, we have switched to a more polar reaction solvent (thf). These changes sufficiently stabilised the dilithiated species EtAs(AcenapLi)$_2$, as the subsequent reaction with PhPCl$_2$ gave cyclic species **8** in a good yield (ca. 70%). Unfortunately, the workup did not allow for complete removal of the salt impurities as indicated by the results of the elemental analysis, meaning the yield mentioned above is only approximate. **8** showed a sharp singlet at δ_P −26.5 ppm and was only barely soluble in common organic solvents, which precluded acquisition of good quality ^{13}C{^{1}H} NMR spectra.

Reacting the in situ formed dilithiated species EtAs(AcenapLi)$_2$ with two molar equivalents of chlorodiphenylphosphine in thf afforded **9** as the sole phosphorus-containing product as confirmed by ^{31}P{^{1}H} NMR spectroscopy (singlet at δ_P −19.9 ppm). **9** was found to be insoluble in many common organic solvents and only sparingly soluble in dichloromethane. In an attempt to grow crystals of **9** suitable for X-ray work, colourless needle shaped crystals were obtained by diffusion of diethyl ether into a saturated solution of **9** in dichloromethane. The X-ray diffraction showed that partial oxidation (presumably by air) has taken place in this sample, giving the phosphine oxide species **9-O**. Notably, the presence of the oxidised species **9-O** was not evident in the bulk of the reaction product **9** by any other characterisation results such as elemental analysis and mass spectroscopy. Compounds **7**–**9** were fully characterised, including ^{1}H, ^{13}C{^{1}H} NMR (where solubility allowed this), IR, Raman and MS.

The crystal structures of **8** and **9-O** are shown in Figure 5 and Table 1. The central part of the molecule of **9** adopts the shape of a puckered (8-membered) ring, with the angle between the two acenaphthene planes being 89°. The molecule of **9-O** is rather crowded with large groups attached to the two *peri*-regions. As mentioned, one of the phosphorus environments in **9-O** is partially oxidised (50% occupancy) to the phosphine oxide (from accidental exposure to air).

The P···As distance comparison in **8** and **9-O** is rather interesting. The As1···P9 distance in **9-O** is 3.176(5) Å, while the relevant distance of 3.004(6) was observed in **8**. This indicates the buttressing effect of the double *peri*-strain in **8** is rather significant, shortening the P···As distance by 0.17 Å, i.e., by 4.2% of the sum of the respective Van der Waals radii [31]. The large positive splay angles indicate repulsive interactions in the *peri*-regions of **7**, **8** and **9-O**, with the in-plane distortions being the major mechanism of strain relaxation, although in **9-O** the arsine group (As1) shows also large out-of-plane displacement (0.71 Å) from one of the acenaphthene mean planes.

4. Experimental

4.1. General Considerations

All reactions and manipulations were carried out under an atmosphere of nitrogen using standard Schlenk techniques or under an argon atmosphere in a Saffron glove box. Dry solvents were either collected from an MBraun Solvent Purification System, or dried and stored according to common procedures [33]. Compounds **1** and **2** were prepared according to literature procedures [8,34]. Arsenic tribromide was prepared using a modified version of the published procedure (see Supplementary Materials) [35]. EtAsI$_2$ was prepared as described in the literature [36]. Arsenic oxide (>99.9%) was purchased from Alfa Aesar and used as received. Other chemicals were purchased from commercial sources and used as received. Further experimental details are provided in SI. NMR numbering scheme is shown in Figure 6.

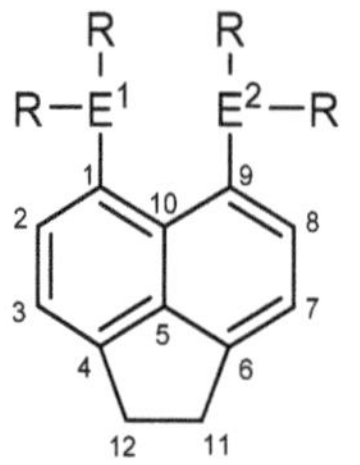

3: E^1R_2 = PiPr$_2$, E^2R_2 = PBr$_2$
4: E^1R_2 = PiPr$_2$, E^2R_2 = AsBr$_2$
7: E^1R_2 = AsEtAce, E^2R_2 = Br
8: E^1R_2 = AsEtAce, E^2R_2 = PPhAce
9: E^1R_2 = AsEtAce, E^2R_2 = PPh$_2$

Figure 6. General NMR numbering scheme.

Caution! Arsenic halides are highly toxic powerful vesicants, which cause severe irritation and blistering if allowed to come in contact with skin. Suitable precautions, including the use of neoprene or rubber gloves, should be taken when handling these.

4.2. Synthetic Methods

4.2.1. iPr$_2$PAcenapPBr$_2$ (**3**)

To a cooled (−78 °C) rapidly stirring solution of **2** (2.60 g, 7.4 mmol) in diethyl ether (60 mL), *n*-butyllithium (3.0 mL, 2.5 M in hexane, 7.5 mmol) was added dropwise over 1 h and the mixture was left to stir for 2 h at the same temperature. The resulting suspension of *i*Pr$_2$PAcenapLi was added via cannula (in small batches) over 1 h to a rapidly stirring solution of phosphorus tribromide (9.90 g, 3.5 mL, 36.7 mmol) in diethyl ether (30 mL), cooled to −78 °C. The reaction mixture was left to stir and warm to ambient temperature overnight. The orange suspension was filtered and the solid collected was washed with THF (30 mL) followed by diethyl ether (30 mL) to remove excess PBr$_3$ and partially also LiBr. After drying in vacuo, **3** (contaminated with LiBr) was obtained as a yellow powder (2.20 g, ~65%). Note that it is important when filtering the compound not to let the solid settle and compact as this will prevent removal of all PBr$_3$. Crystals of **3** suitable for X-ray diffraction were grown from thf. Small scale recrystallisation from chloroform afforded analytically pure material. M. p. 168–172 °C. **Elemental Analysis** Calcd. (%) for C$_{18}$H$_{22}$P$_2$Br$_2$: C 46.99, H 4.82; Found: C 47.07, H 4.87. **^{1}H NMR**: δ_H (400.1 MHz, CDCl$_3$) 8.82 (1H, dd (~t), $^3J_{HH}$ = 6.8, $^3J_{HP}$ = 1.2 Hz, H-8), 8.17 (1H, dd, $^3J_{HH}$ = 7.2, $^3J_{HP}$ = 5.1 Hz, H-2), 7.73 (1H, dd, $^3J_{HH}$ = 6.8, $^4J_{HP}$ = 2.4 Hz, H-7), 7.62 (1H, dd, $^3J_{HH}$ = 7.2, $^4J_{HP}$ = 2.8 Hz, H-3), 4.29–4.14 (2H, m (~septet), $^3J_{HH}$ = 7.0 Hz, C$\underline{H}$(CH$_3$)$_2$), 3.58 (4H, br s, H-11 and H-12), 1.57 (6H, $^2J_{HP}$ = 19.2, $^3J_{HH}$ = 7.0 Hz, 2 × CH$_3$), 1.50 (6H, dd, $^2J_{HP}$ = 19.5, $^3J_{HH}$ = 7.0 Hz, 2 × CH$_3$). **^{13}C{^{1}H} NMR**: δ_C (100.6 MHz, CDCl$_3$) 154.4 (s, qC-6), 151.8 (s, qC-4), 140.3 (dd, $^2J_{CP}$ = 20.5, $^2J_{CP}$ = 3.4 Hz, qC-10), 139.0 (dd, $^3J_{CP}$ = 11.0, $^3J_{CP}$ = 1.5 Hz, qC-5), 137.5 (br s, C-8), 134.8 (dd, $^2J_{CP}$ = 31.5, $^3J_{CP}$ = 8.5 Hz, C-2), 129.8 (dd, $^1J_{CP}$ = 46.4, $^2J_{CP}$ = 4.7 Hz, qC-1), 122.6 (s, C-3), 122.5 (s, C-7), 112.8 (dd, $^1J_{CP}$ = 55.7, $^2J_{CP}$ = 8.0 Hz, qC-9), 31.6 (s, C-11 or C-12), 31.2 (s, C-11 or C-12), 27.5 (dd, $^1J_{CP}$ = 28.5, $^2J_{CP}$ = 4.5 Hz, C$\underline{H}$(CH$_3$)$_2$), 18.8 (br s, 2 × CH$_3$), 18.3 (br s, 2 × CH$_3$). **^{31}P{^{1}H} NMR**: δ_P (162.0 MHz, CDCl$_3$) 65.3 (d, P$_{PBr2}$), 32.9 (d, P$_{PiPr2}$), $^1J_{PP}$ = 353.7 Hz. **IR** (KBr disc, cm^{-1}) v = 2963 vs, 2934 vs, 2862 s, 1457 s, 1443 s, 839 m, 640 s. **HRMS (APCI+)**: *m/z* (%); Calcd. for C$_{18}$H$_{22}$P$_2$Br (M−Br): 379.0375, found 379.0376 (100).

4.2.2. iPr$_2$PAcenapAsBr$_2$ (4)

Compound **4** was prepared using the same procedure as per compound **3** except using the following: **2** (2.00 g, 5.7 mmol) in diethyl ether (40 mL), n-butyllithium (2.3 mL, 2.5 M in hexanes, 5.7 mmol) and arsenic tribromide (1.80 g, 5.7 mmol) in diethyl ether (50 mL). Addition time of iPr$_2$PAcenapLi to AsBr$_3$ was 3 h. **4** was isolated as a yellow powder (2.17 g, ~75%). M. p. 159–161 °C. Crystals suitable for X-ray diffraction work were grown from benzene at room temperature. Contamination with residual LiBr prevented satisfactory microanalysis and accurate determination of the yield. **^{1}H NMR**: δ_H (500.1 MHz, C$_6$D$_6$) 7.92 (1H, d, $^3J_{HH}$ = 7.1 Hz, H-8), 7.37 (1H, dd, $^3J_{HP}$ = 8.6, $^3J_{HH}$ = 7.2 Hz, H-2), 7.01 (1H, dt, $^3J_{HH}$ = 7.2, $^4J_{HH}$ = 1.5 Hz, H-7), 6.99 (1H, dt, $^3J_{HH}$ = 7.2, $^4J_{HH}$ = 1.4 Hz, H-3), 3.61 (2H, septet, $^3J_{HH}$ = 7.2 Hz, C$\underline{H}$(CH$_3$)$_2$), 2.89–2.73 (4H, m, H-11 and H-12), 1.36 (6H, dd, $^3J_{HP}$ = 18.9, $^3J_{HH}$ = 7.2 Hz, 2 × CH$_3$), 1.14 (6H, dd, $^3J_{HH}$ = 16.6, $^3J_{HH}$ = 7.3 Hz, 2 × CH$_3$). **^{13}C{^{1}H} NMR**: δ_C (125.8 MHz, C$_6$D$_6$) 153.2 (s, qC-4), 146.5 (s, qC-9), 145.8 (s, qC-5), 140.7 (d, $^1J_{CP}$ = 22.3 Hz, qC-1), 134.1 (s, C-2), 131.2 (d, $^3J_{CP}$ = 7.5 Hz, C-8), 122.2 (s, C-7), 119.5 (d, $^3J_{CP}$ = 9.5 Hz, C-3), 30.8 (s, C-11 or C-12), 30.2 (s, C-11 or C-12), 27.6 (d, $^1J_{CP}$ = 26.3 Hz, C$\underline{H}$(CH$_3$)$_2$), 20.0 (s, 2 × CH$_3$), 17.7 (d, $^2J_{CP}$ = 5.7 Hz, 2 × CH$_3$). **^{31}P{^{1}H} NMR**: δ_P (202.5 MHz, C$_6$D$_6$) 56.6 (s). **MS** (EI): m/z (%) 422.97 (20) [M−Br], 344.04 (95) [M−2Br], 300.99 (90) [iPrPAcenapAs], 257.95 (95) [AcenapPAs], 234.74 (100) [AsBr$_2$].

4.2.3. [iPr$_2$PAcenapAsBr]$^+$Br$^-$·AsBr$_3$ (5) and [(iPr$_2$PAcenap)$_2$As]$_2$[As$_6$Br$_8$] (6)

To a cooled (−78 °C) rapidly stirring solution of **2** (2.00 g, 5.7 mmol) in diethyl ether (40 mL), n-butyllithium (2.3 mL, 2.5 M in hexane, 5.7 mmol) was added dropwise over 1 h and the mixture was left to stir for 2 h at the same temperature. The resulting suspension of iPr$_2$PAcenapLi was added via cannula (in small batches) over 1 h to a rapidly stirring solution of arsenic tribromide (3.60 g, 11.5 mmol) in diethyl ether (80 mL), cooled to −78 °C. The reaction mixture was left to stir and warm to ambient temperature overnight. The suspension was filtered and the solid collected was washed with diethyl ether (30 mL). After drying in vacuo, **5** (contaminated with LiBr) was obtained as a pale-yellow powder (3.84 g, ca. 81%). Crystals of **5** suitable for X-ray diffraction work were grown from THF at room temperature. Contamination with LiBr prevented satisfactory microanalysis and accurate determination of the yield. Solution NMR spectra of **5** are identical to those of **4**.

Few crystals of **6** suitable for diffraction work were obtained from recrystallisation of the crude product **5** from hot acetonitrile. No further characterisation was possible due to small amount available.

4.2.4. (BrAcenap)$_2$AsEt (7)

A solution of n-butyllithium (12.8 mL of 2.5 M solution in hexanes, 32 mmol) was added dropwise to a rapidly stirred suspension of 5,6-dibromoacenaphthene **1** (10.0 g, 32 mmol) in THF (120 mL) at −78 °C. The mixture was maintained for 2 h at this temperature. To this, a solution of ethyldiiodoarsine (1.95 mL, 16 mmol) in THF (20 mL) was added dropwise over one hour at −78 °C. The resulting suspension was left to warm to room temperature overnight. The volatiles were removed in vacuo. Dichloromethane (50 mL) was added, and the resulting suspension was filtered. The product was obtained as a white powder after removal of the volatiles from the filtrate in vacuo. Recrystallisation of the crude material from dichloromethane gave **7** as colourless needle crystals (5.37 g, 82%), some of these were suitable for X-ray diffraction work. M. p. 199–200 °C. **Elemental analysis**: Calcd. (%) for C$_{26}$H$_{21}$AsBr$_2$: C 54.96, H 3.72; found: C 54.86, H 3.65. **^{1}H NMR**: δ_H (400.1 MHz, CD$_2$Cl$_2$) 7.60 (2H, d, $^3J_{HH}$ = 7.4 Hz, H-8), 7.48 (2H, d, $^3J_{HH}$ = 7.2 Hz, H-2), 7.08 (2H, d, $^3J_{HH}$ = 7.2 Hz, H-3), 7.00 (2H, d, $^3J_{HH}$ = 7.4 Hz, H-7), 3.29–3.19 (8H, m, H-11 and H-12), 2.04 (2H, q, $^3J_{HH}$ = 7.7 Hz, As-CH$_2$), 1.19 (3H, t, $^3J_{HH}$ = 7.6 Hz, CH$_3$). **^{13}C{^{1}H} NMR**: δ_C (100.6 MHz, CD$_2$Cl$_2$) 146.2 (s, qC), 145.7 (s, qC), 140.7 (s, qC), 135.6 (s, qC), 134.6 (s, C-2), 133.0 (s, C-8), 132.3 (s, qC), 119.4 (s, C-3), 119.2 (s, C-7), 115.5 (s, qC), 29.1 (s, C-11 or C-12), 28.8 (s, C-11 or C-12), 23.0 (s, As-CH$_2$), 10.4 (s, CH$_3$). **IR Data** (KBr disc, cm^{-1}): ν = 3025 w (ν_{ArH}), 2920 s, 2863 w (ν_{CH}), 1603 s, 1411 s, 1320 vs, 1251 m, 1200 m, 1100 m,

1044 m, 836 vs, 738 m, 602 m, 543 m. **Raman** (glass capillary, cm^{-1}) ν = 3063 m (ν$_{ArH}$), 2952 m, 2932 s, (ν$_{CH}$), 1608 m, 1561 vs, 1442 vs, 1412 s, 1345 m, 1320 vs, 815 m, 702 s, 577 vs, 557 s, 290 vs. **MS (ES+)** m/z (%) 591 (100) [M + Na]. **HRMS:** m/z Calcd. for C$_{26}$H$_{21}$Br$_2$AsNa (M + Na): 590.9103, found 590.9094.

4.2.5. Cyclo-PhP(Acenap)$_2$AsEt (8)

N,N,N′,N′–tetramethylethylene-1,2-diamine (TMEDA, 0.51 mL, 3.4 mmol) was added to a suspension of **7** (0.75 g, 1.32 mmol) in THF (35 mL) and the mixture was cooled to −78 °C. A solution of n-butyllithium (1.26 mL, 2.5 M in hexanes, 3.2 mmol) was added dropwise with stirring over 1 h. The solution was maintained at −78 °C for 30 min before being warmed up and maintained at 0 to 10 °C for an hour. The solution was re-cooled to −78 °C and a solution of PhPCl$_2$ (0.18 mL, 1.32 mmol) in THF (5 mL) was added dropwise. The mixture was left to warm up to room temperature overnight with stirring. The volatiles were removed in vacuo and replaced with dichloromethane (30 mL) and the resulting suspension was filtered. Removal of the volatiles in vacuo once again gave **8** as a yellow solid (0.48 g, ca. 70%). Crystals suitable for X-ray diffraction were grown from dichloromethane. **Elemental analysis:** The recrystallised material did not give satisfactory elemental analysis, presumably because of the contamination of the material with inorganic salts; the data fitted **8**·1.45LiBr: calcd. (%) for C$_{32}$H$_{26}$AsP·(LiBr)$_{1.45}$: C 59.83, H 4.08; found: C 60.03, H 3.99. **^{1}H NMR:** δ$_H$ (300.1 MHz, CD$_2$Cl$_2$) 7.96 (2H, dd, $^3J_{HH}$ = 7.2, $^3J_{HP}$ = 4.5 Hz, H-8), 7.81–7.74 (2H, m, o-Ph), 7.59 (2H, d, $^3J_{HH}$ = 7.3 Hz, H-2), 7.46–7.38 (2H, m, m-Ph), 7.38–7.31 (1H, m, p-Ph), 7.20 (4H, $^3J_{HH}$ = 7.2 Hz, H-3 and H-7), 3.39–3.16 (8H, m, H-11 and H-12), 2.28 (2H, q, $^3J_{HH}$ = 7.7 Hz, As-CH$_2$), 1.54 (3H, t, $^3J_{HH}$ = 7.7 Hz, CH$_3$). **^{13}C{^{1}H} NMR** was not acquired due to low solubility of **8**. **^{31}P{^{1}H} NMR:** δ$_P$ (121.5 MHz, CD$_2$Cl$_2$) δ$_P$ = −26.5 (s). **IR** (KBr disc, cm^{-1}): ν = 3071 w (ν$_{Ar-H}$), 2962 s (ν$_{C-H}$), 1325 s, 1260 vs, 1095 vs, br, 1022 vs, 802 vs, 735 s, 701 m. **Raman** (glass capillary, cm^{-1}) ν = 3068 m (ν$_{Ar-H}$), 2919 m (ν$_{C-H}$), 1603 m, 1583 vs, 1565 s, 1330 vs, 997 s, 537 s, 253 m. **MS (CI+):** m/z (%) 517.1 (58) [M + H], 487.1 (100) [M−Et]. **HRMS (CI+):** m/z Calcd. for C$_{32}$H$_{27}$AsP (M + H): 517.1066, found 517.1059.

4.2.6. (Ph$_2$PAcenap)$_2$AsEt (9)

TMEDA (0.51 mL, 3.4 mmol) was added to a suspension of **7** (750 mg, 1.32 mmol) in THF (35 mL) and the mixture was cooled to −78 °C. A solution of n-butyllithium (1.26 mL, 2.5 M in hexanes, 3.2 mmol) was added dropwise with stirring over 1 h. The solution was maintained at −78 °C for 30 min before being warmed up and maintained at 0 to 10 °C for an hour. The solution was re-cooled to −78 °C and a solution of chlorodiphenylphosphine (0.49 mL, 2.64 mmol) in THF (5 mL) was added dropwise over 45 min. The solution was left to warm to room temperature overnight with stirring. The volatiles were removed in vacuo and replaced with dichloromethane (30 mL), and the resulting mixture was filtered. Removal of the volatiles in vacuo once again gave **9** as a pale yellow solid (0.52 g, ~52%). Crystals suitable for X-ray diffraction were obtained in the form of monoxide **9a** from dichloromethane. **^{1}H NMR:** δ$_H$ (500.1 MHz, CD$_2$Cl$_2$) 7.51 (2H, d, $^3J_{HH}$ = 7.0 Hz, H-2), 7.26–7.13 (12H, br m, H-7 and H-8 and m-Ph), 7.10–7.03 (2H, br m, H-3), 7.02–6.95 (4H, br m, p-Ph), 6.76 (8H, br s, o-Ph), 3.37–3.24 (8H, m, H-11 and H-12), 2.19 (2H, q, $^3J_{HH}$ = 7.6 Hz, As-CH$_2$), 1.32 (3H, t, $^3J_{HH}$ = 7.6 Hz, CH$_3$). **^{13}C{^{1}H} NMR:** δ$_C$ NMR was not acquired due to low solubility of **9**. **^{31}P{^{1}H} NMR:** δ$_P$ (121.5 MHz, CD$_2$Cl$_2$) δ$_P$ = −19.9 (s). **IR** (KBr disc, cm^{-1}): ν = 3044 m (ν$_{Ar-H}$), 2962 vs, (ν$_{C-H}$), 1601 s, 1477 s, 1261 vs, 1020 vs, br, 800 vs, 693 vs. **Raman** (glass capillary, cm^{-1}) ν = 3047 vs, (ν$_{Ar-H}$), 2930 s (ν$_{C-H}$), 1604 s, 1583 s, 1325 vs, 998 s, 536 vs. **MS (CI+):** m/z (%) 779.1 (8) [M + H], 749.1 (16) [M−Et], 412.1 (32) [Ph$_2$PAcenapAs]. **HRMS (CI+)** m/z Calcd. for C$_{50}$H$_{42}$P$_2$As (M + H): 779.1978, found 779.1977.

5. Conclusions

Utilising *peri*-substitution, novel dative complexes bearing pnictogen dibromide acceptor groups have been isolated and fully characterised. While their stability resembles that of their chlorine analogues, compounds **3** and **4** are thermally stable up to well above 100 °C and can be stored as solids under an inert atmosphere indefinitely, which makes them interesting for further synthetic use. Using excess $AsBr_3$ gave species **5**, in which a bromine atom of molecule **4** is ionically separated through interactions with co-crystallised molecules of $AsBr_3$. The structures of **3–5** illustrate that in phosphine−pnictine complexes the local polarity (packing) effects within the crystal lattice result in rather different Pn−X bond distances due to the fine balance of the polar covalent and ionic bonding around the (formally) negatively charged pnictoranido motifs. This is further corroborated by computational evaluation of the close As···As interaction observed in **4**, which is found to consist mostly of dispersion forces.

Isolation of compound **6** indicates disproportionationative redox instability of **5** at elevated temperatures, resulting in an unusual heterocubane dianion $[As_6Br_8]^{2-}$, containing As(I) motifs.

Reaction of BrAcenapLi with $EtAsI_2$ afforded no singly substituted species, only geminally disubstituted species **7**. This compound has been shown to be a useful synthon, with the As–C bonds being tolerant to the strong base nBuLi. Thus, treatment of **7** with nBuLi proceeds through lithium-halogen exchange; subsequent reactions with chlorophosphine electrophiles afforded the new species, cyclic **8** and acyclic **9**.

Supplementary Materials: The following are available online at, File S1: Experimental data, including general considerations, preparation of arsenic tribromide, X-ray diffraction and computational details [28,35,37–57]. Accession Codes CCDC 2116902-2116908 contain the supplementary crystallographic data for this paper. These data can be obtained free of charge via www.ccdc.cam.ac.uk/data_request/cif, accessed on 28 October 2021, or by emailing data_request@ccdc.cam.ac.uk, or by contacting The Cambridge Crystallographic Data Centre, 12 Union Road, Cambridge CB2 1EZ, UK; Fax: +44-1223-336033. The research data underpinning this publication can be accessed at DOI https://doi.org/10.17630/7bae9495-1422-4bae-aa2b-d160ce8c1d3f.

Author Contributions: P.K., B.A.C. and D.M.U.K.S. conceived and designed the project; B.A.C., D.M.U.K.S. and B.A.S. performed the experiments; A.M.Z.S. and K.S.A.A. did the crystallography; P.K., B.A.C. and J.D.W. contributed to writing the manuscript, M.B. did the computations. All authors have read and agreed to the published version of the manuscript.

Funding: This research was funded by the Engineering and Physical Sciences Research Council (EPSRC, Grant EP/J500549/1 and EP/M506631/1), COST action CM1302 SIPs and CM0802, and the EaStCHEM School of Chemistry.

Data Availability Statement: The research data underpinning this publication can be accessed at DOI https://doi.org/10.17630/7bae9495-1422-4bae-aa2b-d160ce8c1d3f.

Acknowledgments: The authors thank the Engineering and Physical Sciences Research Council (EPSRC, Grant EP/J500549/1 and EP/M506631/1), COST action CM1302 SIPs and CM0802, and the EaStCHEM School of Chemistry for support. M.B. thanks H. Früchtl for technical assistance with the computations.

Conflicts of Interest: The authors declare no conflict of interest.

Sample Availability: Samples of all compounds are not available from the authors.

References

1. Peruzzini, M.; Gonsalvi, L. *Phosphorus Compounds*; Springer: Berlin/Heidelberg, Germany, 2011; Volume 37.
2. Greenwood, N.N.; Earnshaw, A. *Chemistry of the Elements*, 2nd ed.; Elsevier: Oxford, UK, 1997.
3. Crabtree, R.H. *The Organometallic Chemistry of the Transition Metals*, 5th ed.; John Wiley and Sons Inc.: Hoboken NJ, USA, 2009.
4. Holmes, R.R.; Bertaut, E.F. The Reduction of Phosphorus and Antimony Chlorides by Trimethylarsine and Trimethylstibine. *J. Am. Chem. Soc.* **1958**, *80*, 2983–2985. [CrossRef]

5. Holmes, R.R.; Bertaut, E.F. The Reactions of Phosphorus and Antimony Chlorides with Trimethylamine, Triethylamine and Trimethylphosphine1. *J. Am. Chem. Soc.* **1958**, *80*, 2980–2983. [CrossRef]
6. Spangenberg, S.F.; Sisler, H.H. Reactions of some tri-n-alkylphosphines with some chlorophosphines. *Inorg. Chem.* **1969**, *8*, 1006–1010. [CrossRef]
7. Muller, G.; Matheus, H.J.; Winkler, M. Donor-Acceptor Complexes between Simple Phosphines. First Structural Data for an Almost Forgotten Class of Compounds. *Z. Nat. B J. Chem. Sci.* **2001**, *56*, 1155–1162. [CrossRef]
8. Wawrzyniak, P.; Fuller, A.L.; Slawin, A.M.Z.; Kilian, P. Intramolecular Phosphine-Phosphine Donor-Acceptor Complexes. *Inorg. Chem.* **2009**, *48*, 2500–2506. [CrossRef] [PubMed]
9. Summers, J.C.; Sisler, H.H. Reactions of some trialkyls of phosphorus, arsenic, or antimony with some organohalophosphines, -arsines, or -stibines. *Inorg. Chem.* **1970**, *9*, 862–869. [CrossRef]
10. Chitnis, S.S.; Burford, N.; McDonald, R.; Ferguson, M.J. Prototypical Phosphine Complexes of Antimony(III). *Inorg. Chem.* **2014**, *53*, 5359–5372. [CrossRef] [PubMed]
11. Hill, N.J.; Levason, W.; Reid, G. Arsenic(iii) halide complexes with phosphine and arsine co-ligands: Synthesis, spectroscopic and structural properties. *J. Chem. Soc. Dalton Trans.* **2002**, 1188–1192. [CrossRef]
12. Burt, J.; Levason, W.; Reid, G. Coordination chemistry of the main group elements with phosphine, arsine and stibine ligands. *Coord. Chem. Rev.* **2014**, *260*, 65–115. [CrossRef]
13. Robertson, A.P.M.; Gray, P.A.; Burford, N. Interpnictogen Cations: Exploring New Vistas in Coordination Chemistry. *Angew. Chem. Int. Ed.* **2014**, *53*, 6050–6069. [CrossRef] [PubMed]
14. Chitnis, S.S.; Burford, N. Phosphine complexes of lone pair bearing Lewis acceptors. *Dalton Trans.* **2015**, *44*, 17–29. [CrossRef]
15. Chalmers, B.A.; Bühl, M.; Athukorala Arachchige, K.S.; Slawin, A.M.Z.; Kilian, P. Geometrically Enforced Donor-Facilitated Dehydrocoupling Leading to an Isolable Arsanylidine-Phosphorane. *J. Am. Chem. Soc.* **2014**, *136*, 6247–6250. [CrossRef]
16. Chalmers, B.A.; Bühl, M.; Athukorala Arachchige, K.S.; Slawin, A.M.Z.; Kilian, P. Structural, Spectroscopic and Computational Examination of the Dative Interaction in Constrained Phosphine–Stibines and Phosphine–Stiboranes. *Chem. Eur. J.* **2015**, *21*, 7520–7531. [CrossRef] [PubMed]
17. Hupf, E.; Lork, E.; Mebs, S.; Chęcińska, L.; Beckmann, J. Probing Donor–Acceptor Interactions in peri-Substituted Diphenylphosphinoacenaphthyl–Element Dichlorides of Group 13 and 15 Elements. *Organometallics* **2014**, *33*, 7247–7259. [CrossRef]
18. Carmalt, C.J.; Cowley, A.H.; Culp, R.D.; Jones, R.A.; Kamepalli, S.; Norman, N.C. Synthesis and Structures of Intramolecularly Base-Coordinated Group 15 Aryl Halides. *Inorg. Chem.* **1997**, *36*, 2770–2776. [CrossRef] [PubMed]
19. Taylor, L.J.; Bühl, M.; Wawrzyniak, P.; Chalmers, B.A.; Woollins, J.D.; Slawin, A.M.Z.; Fuller, A.L.; Kilian, P. Hydride Abstraction and Deprotonation—An Efficient Route to Low Coordinate Phosphorus and Arsenic Species. *Eur. J. Inorg. Chem.* **2016**, *2016*, 659–666. [CrossRef]
20. Ray, M.J.; Slawin, A.M.Z.; Buehl, M.; Kilian, P. peri-Substituted Phosphino-Phosphonium Salts: Synthesis and Reactivity. *Organometallics* **2013**, *32*, 3481–3492. [CrossRef]
21. Taylor, L.J.; Buhl, M.; Chalmers, B.A.; Ray, M.J.; Wawrzyniak, P.; Walton, J.C.; Cordes, D.B.; Slawin, A.M.Z.; Woollins, J.D.; Kilian, P. Dealkanative Main Group Couplings across the peri-Gap. *J. Am. Chem. Soc.* **2017**, *139*, 18545–18551. [CrossRef]
22. Nejman, P.S.; Curzon, T.E.; Bühl, M.; McKay, D.; Woollins, J.D.; Ashbrook, S.E.; Cordes, D.B.; Slawin, A.M.Z.; Kilian, P. Phosphorus–Bismuth Peri-Substituted Acenaphthenes: A Synthetic, Structural, and Computational Study. *Inorg. Chem.* **2020**, *59*, 5616–5625. [CrossRef]
23. Kilian, P.; Slawin, A.M.Z.; Woollins, J.D. Preparation and structures of 1,2-dihydro-1,2-diphosphaacenaphthylenes and rigid backbone stabilized triphosphenium cation. *Dalton Trans.* **2006**, 2175–2183. [CrossRef] [PubMed]
24. Chalmers, B.A.; Meigh, C.B.E.; Nejman, P.S.; Bühl, M.; Lébl, T.; Woollins, J.D.; Slawin, A.M.Z.; Kilian, P. Geminally Substituted Tris(acenaphthyl) and Bis(acenaphthyl) Arsines, Stibines, and Bismuthine: A Structural and Nuclear Magnetic Resonance Investigation. *Inorg. Chem.* **2016**, *55*, 7117–7125. [CrossRef] [PubMed]
25. Furan, S.; Hupf, E.; Boidol, J.; Brunig, J.; Lork, E.; Mebs, S.; Beckmann, J. Transition metal complexes of antimony centered ligands based upon acenaphthyl scaffolds. Coordination non-innocent or not? *Dalton Trans.* **2019**, *48*, 4504–4513. [CrossRef]
26. Gabes, W.; Olie, K. Refinement of the crystal structure of phosphorus pentabromide, PBr5. *Acta Crystallogr. Sect. B Struct. Crystallogr. Cryst. Chem.* **1970**, *26*, 443–444. [CrossRef]
27. Hubert, S.; Wolfgang, B.; Brigitte, H.; Gerhard, M. Arene Adducts with Weak Interactions: Hexaethylbenzene-bis(tribromoarsane). *Angew. Chem. Int. Ed. Engl.* **1987**, *26*, 234–236. [CrossRef]
28. The WBI is a measure for the covalent character of a bond and adopts values close to 1 and 2 for true single and double bonds, respectively: Wiberg, K.B. Application of the pople-santry-segal CNDO method to the cyclopropylcarbinyl and cyclobutyl cation and to bicyclobutane. *Tetrahedron* **1968**, *24*, 1083–1096. [CrossRef]
29. Müller, U.; Sinning, H. Octabromocyclohexaarsenate, [As6Br8]$^{2\ominus}$. *Angew. Chem. Int. Ed. Engl.* **1989**, *28*, 185–186. [CrossRef]
30. Ellis, B.D.; Macdonald, C.L.B. Stabilized Arsenic(I) Iodide: A Ready Source of Arsenic Iodide Fragments and a Useful Reagent for the Generation of Clusters. *Inorg. Chem.* **2004**, *43*, 5981–5986. [CrossRef] [PubMed]
31. Batsanov, S.S. Van der Waals Radii of Elements. *Inorg. Mater.* **2001**, *37*, 871–885. [CrossRef]
32. Tanaka, N.; Kasai, T. Reactions of 5,6-Dilithioacenaphthene-*N*,*N*,*N'*,*N'*-Tetramethyl-1,2-ethanediamine Complex with α-Diketones. I.cis-Directing 1:1 Cyclic Additions with Acyclic and Cyclic α-Diketones and Related Compounds. *Bull. Chem. Soc. Jpn.* **1981**, *54*, 3020–3025. [CrossRef]

33. Armarego, W.L.F.; Chai, C.L.L. *Purification of Laboratory Chemicals*, 6th ed.; Elsevier: Burlington, VT, USA, 2009.
34. Neudorff, W.D.; Lentz, D.; Anibarro, M.; Schlüter, A.D. The Carbon Skeleton of the Belt Region of Fullerene C84 (D2). *Chem. Eur. J.* **2003**, *9*, 2745–2757. [CrossRef]
35. Arnaiz, F.J.; Miranda, M.J.; Rheingold, A.L. Arsenic (III) Bromide. *Inorg. Synth.* **2002**, *33*, 203. [CrossRef]
36. Matuska, V.; Slawin, A.M.Z.; Woollins, J.D. Five-Membered Arsenic–Sulfur–Nitrogen Heterocycles, RAs(S2N2) (R = Me, Et, iPr, tBu, Ph, Mes). *Inorg. Chem.* **2010**, *49*, 3064–3069. [CrossRef] [PubMed]
37. Chalmers, B.A.; Somisara, D.M.U.K.; Surgenor, B.A.; Athukorala Arachchige, K.S.; Woollins, J.D.; Buehl, M.; Slawin, A.M.Z.; Kilian, P. *Synthetic and Structural Study of peri-Substituted Phosphine-Arsines (Dataset)*; University of St Andrews Research Portal: St Andrews, UK, 2021. [CrossRef]
38. *CrystalClear 2.0*; Rigaku Americas: The Woodlands, TX, USA; Rigaku Corporation: Tokyo, Japan, 2007.
39. *CrystalClear Software User's Guide*; Molecular Structure Corportation, 2017.
40. Sheldrick, G. Crystal structure refinement with SHELXL. *Acta Crystallogr. Sect. C* **2015**, *71*, 3–8. [CrossRef] [PubMed]
41. Sheldrick, G. SHELXT-Integrated space-group and crystal-structure determination. *Acta Crystallogr. Sect. A* **2015**, *71*, 3–8. [CrossRef] [PubMed]
42. *CrystalStructure 4.3.0*; Rigaku Americas: The Woodlands, TX, USA; Rigaku Corporation: Tokyo, Japan, 2018.
43. Bruno, I.J.; Cole, J.C.; Edgington, P.R.; Kessler, M.; Macrae, C.F.; McCabe, P.; Pearson, J.; Taylor, R. New software for searching the Cambridge Structural Database and visualizing crystal structures. *Acta Crystallogr. Sect. B* **2002**, *58*, 389–397. [CrossRef] [PubMed]
44. Thomas, I.R.; Bruno, I.J.; Cole, J.C.; Macrae, C.F.; Pidcock, E.; Wood, P.A. WebCSD: The online portal to the Cambridge Structural Database. *J. Appl. Crystallogr.* **2010**, *43*, 362–366. [CrossRef]
45. Dolomanov, O.V.; Bourhis, L.J.; Gildea, R.J.; Howard, J.A.K.; Puschmann, H. OLEX2: A complete structure solution, refinement and analysis program. *J. Appl. Crystallogr.* **2009**, *42*, 339–341. [CrossRef]
46. Macrae, C.F.; Bruno, I.J.; Chisholm, J.A.; Edgington, P.R.; McCabe, P.; Pidcock, E.; Rodriguez-Monge, L.; Taylor, R.; van de Streek, J.; Wood, P.A. Mercury CSD 2.0—New features for the visualization and investigation of crystal structures. *J. Appl. Crystallogr.* **2008**, *41*, 466–470. [CrossRef]
47. Becke, A.D. Density-functional thermochemistry. III. The role of exact exchange. *J. Chem. Phys.* **1993**, *98*, 5648–5652. [CrossRef]
48. Lee, C.; Yang, W.; Parr, R.G. Development of the Colle-Salvetti correlation-energy formula into a functional of the electron density. *Phys. Rev. B* **1988**, *37*, 785–789. [CrossRef]
49. Grimme, S.; Antony, J.; Ehrlich, S.; Krieg, H. A consistent and accurate ab initio parametrization of density functional dispersion correction (DFT-D) for the 94 elements H-Pu. *J. Chem. Phys.* **2010**, *132*, 154104. [CrossRef]
50. Grimme, S.; Ehrlich, S.; Goerigk, L. Effect of the damping function in dispersion corrected density functional theory. *J. Comput. Chem.* **2011**, *32*, 1456–1465. [CrossRef]
51. Risthaus, T.; Grimme, S. Benchmarking of London Dispersion-Accounting Density Functional Theory Methods on Very Large Molecular Complexes. *J. Chem. Theory Comput.* **2013**, *9*, 1580–1591. [CrossRef] [PubMed]
52. Becke, A.D.; Johnson, E.R. Exchange-hole dipole moment and the dispersion interaction. *J. Chem. Phys.* **2005**, *122*, 154104. [CrossRef]
53. Johnson, E.R.; Becke, A.D. A post-Hartree-Fock model of intermolecular interactions: Inclusion of higher-order corrections. *J. Chem. Phys.* **2006**, *124*, 174104. [CrossRef] [PubMed]
54. Binning, R.C.; Curtiss, L.A. Compact contracted basis sets for third-row atoms: Ga–Kr. *J. Comput. Chem.* **1990**, *11*, 1206–1216. [CrossRef]
55. Boys, S.F.; Bernardi, F. The calculation of small molecular interactions by the differences of separate total energies. Some procedures with reduced errors. *Mol. Phys.* **1970**, *19*, 553–566. [CrossRef]
56. Reed, A.E.; Curtiss, L.A.; Weinhold, F. Intermolecular interactions from a natural bond orbital, donor-acceptor viewpoint. *Chem. Rev.* **1988**, *88*, 899–926. [CrossRef]
57. Frisch, M.J.; Trucks, G.W.; Schlegel, H.B.; Scuseria, G.E.; Robb, M.A.; Cheeseman, J.R.; Scalmani, G.; Barone, V.; Mennucci, B.; Petersson, G.A. *Gaussian 09*; Gaussian, Inc.: Wallingford, CT, USA, 2009.

molecules

Article

A Comparison of Cysteine-Conjugated Nitroxide Spin Labels for Pulse Dipolar EPR Spectroscopy

Katrin Ackermann, Alexandra Chapman and Bela E. Bode *

EaStCHEM School of Chemistry, Biomedical Sciences Research Complex, and Centre of Magnetic Resonance, University of St Andrews, North Haugh, St Andrews KY16 9ST, UK; ka44@st-andrews.ac.uk (K.A.); ac377@st-andrews.ac.uk (A.C.)
* Correspondence: beb2@st-andrews.ac.uk

Abstract: The structure-function and materials paradigms drive research on the understanding of structures and structural heterogeneity of molecules and solids from materials science to structural biology. Functional insights into complex architectures are often gained from a suite of complementary physicochemical methods. In the context of biomacromolecular structures, the use of pulse dipolar electron paramagnetic resonance spectroscopy (PDS) has become increasingly popular. The main interest in PDS is providing long-range nanometre distance distributions that allow for identifying macromolecular topologies, validating structural models and conformational transitions as well as docking of quaternary complexes. Most commonly, cysteines are introduced into protein structures by site-directed mutagenesis and modified site-specifically to a spin-labelled side-chain such as a stable nitroxide radical. In this contribution, we investigate labelling by four different commercial labelling agents that react through different sulfur-specific reactions. Further, the distance distributions obtained are between spin-bearing moieties and need to be related to the protein structure via modelling approaches. Here, we compare two different approaches to modelling these distributions for all four side-chains. The results indicate that there are significant differences in the optimum labelling procedure. All four spin-labels show differences in the ease of labelling and purification. Further challenges arise from the different tether lengths and rotamers of spin-labelled side-chains; both influence the modelling and translation into structures. Our comparison indicates that the spin-label with the shortest tether in the spin-labelled side-group, (bis-(2,2,5,5-Tetramethyl-3-imidazoline-1-oxyl-4-yl) disulfide, may be underappreciated and could increase the resolution of structural studies by PDS if labelling conditions are optimised accordingly.

Keywords: PDS; PELDOR; DEER; nitroxide spin label; Comparative DEER Analyzer; mtsslSuite; MMM

1. Introduction

Research into the functional characteristics of molecules and materials is underpinned by the fundamental dogma that the molecular structure determines properties. Thus, structure determination lies at the core of chemistry, and virtually every undergraduate will learn Bragg's Law describing X-ray diffraction [1]. The prevalence of structure as a determinant of all properties persists when studying the molecules of life. The fundamental hypothesis that all biomolecular functions are encoded in the structure [2] remains the central dogma of structural biology. The ever-increasing complexity of biological systems under study has been accompanied by a rise in awareness that structural context is of crucial relevance, and integrative structural biology is becoming increasingly important for consolidating information from a variety of methods into a holistic model. Similar approaches allow integrating results from multiple methods for materials characterisation.

X-ray crystallography, especially in its high-throughput forms [3], remains the gold standard for structure determination of crystals from small molecules to solid-state materials. The amorphous nature of polymers and their composites with other components requires a more involved approach. While diverse forms of microscopy (including cryo-electron microscopy), diffraction and scattering (such as small-angle X-ray scattering) and spectroscopic methods provide a plethora of structural data, magnetic resonance can provide structural information with atomic resolution and within native context based on labelling with stable isotopes or exploiting the low natural abundance of unpaired electron spins. Here, we focus on the use of electron paramagnetic resonance (EPR) spectroscopy in determining precise nanometre distances between selected sites in biomolecules [4,5] to validate structural models [6,7] and establish conformational topologies [8–10].

For pulse dipolar EPR spectroscopy (PDS) [11–13], specific labelling sites within the fold of the protein of choice are subjected to site-directed mutagenesis to establish cysteines at the sites of interest (requiring knockout of other accessible cysteines). These cysteines are site-specifically spin labelled with sulfide-specific labelling reagents to introduce a stable spin bearing moiety, most commonly a nitroxide radical [14,15]. While this arguably provides limited information, yielding merely a single distance distribution per label pair, this can be extremely powerful, especially in combination with complementary methods. Importantly, knowledge of possible spin label conformations is crucial to predict corresponding distance distributions [16–20]. In addition to distance distributions encoding conformational flexibility [21,22], potential weak exchange interactions between the spin centres can be quantified [23–25] as well as the number of coupled electron spins interacting in one structural object [12,26,27] and their distribution within nano-confinements [28,29]. Initially informed by a plethora of chemical model systems [30] designed for proof-of-principle and benchmark studies, PDS has allowed significant contributions to the understanding of complex protein systems [24,31]. Illustrative examples are homo-multimeric membrane channels where insights from simulations [32,33] and model systems [34,35] could be translated to significantly improved structural resolution [36,37], and these optimised conditions ultimately yielded functional insights into channel gating [38]. Other examples include the identification of physiologically relevant dimer interfaces in viral proteins [6] and the investigation of the self-assembly of archaeal single-stranded DNA binding proteins [39].

The aim of this study is to investigate four commercially available nitroxide labels (Scheme 1) based on cysteine-mediated conjugation to the protein of interest. Here, we aim to compare labelling efficiency as well as measurement sensitivity and accuracy based on the immunoglobulin-binding B1 domain of group G streptococcal protein G (GB1) that has been extensively used for nitroxide and copper(II) spin labelling [40–45]. While numerous detailed characterisations [46,47] and reviews [48–50] exist, there is, to our best knowledge, no published study directly comparing these four spin labels. The results indicate a surprisingly large breadth in terms of ease of labelling and purification, and agreement with structural modelling.

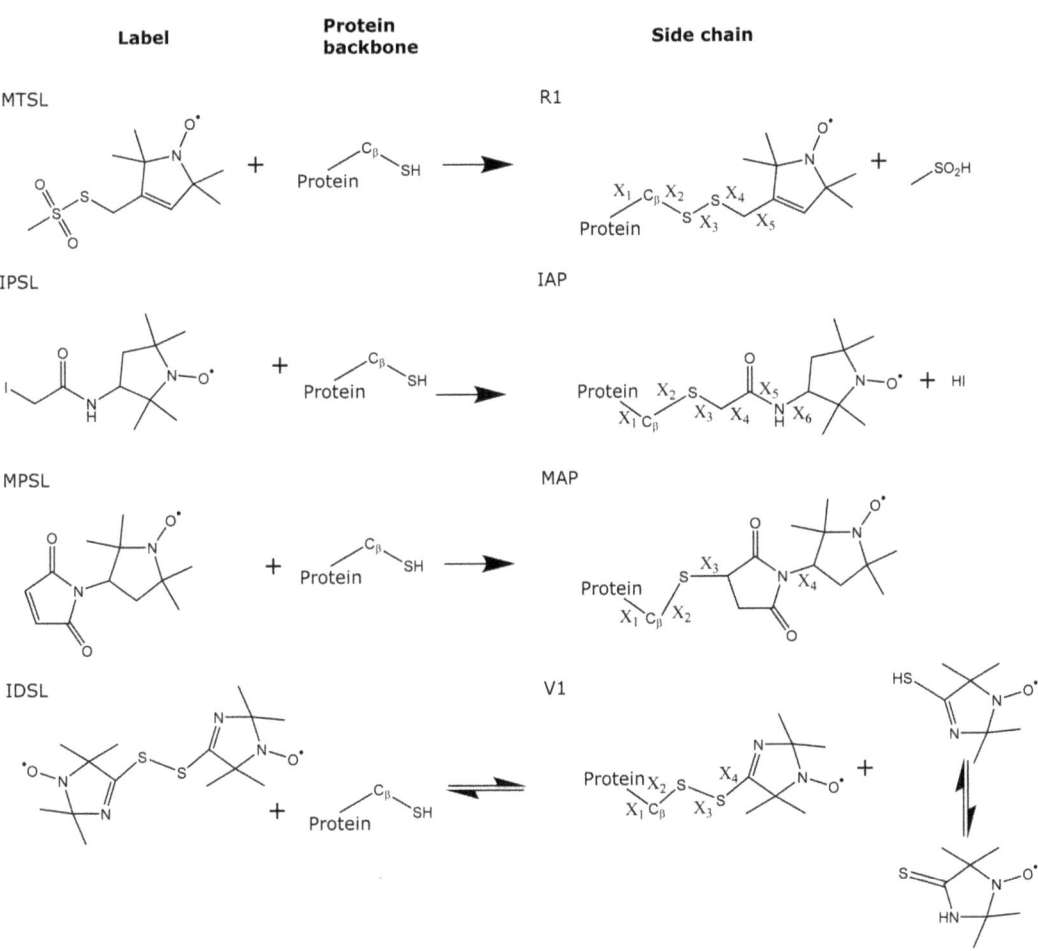

Scheme 1. Scheme displaying the four spin labelling compounds (MTSL, IPSL, MPSL and IDSL) used in this study and the resulting cysteine conjugates and leaving groups. Rotatable bonds are indicated (X).

2. Materials and Methods

2.1. Protein Construct and Spin Labelling

The following nitroxide spin labels were employed in this study: MTSL [4,51] ((1-Oxyl-2,2,5,5-tetramethyl-3-pyrroline-3-methyl)methanethiosulfonate; Santa Cruz Biotechnology), IPSL [52–54] (3-(2-Iodoacetamido)-2,2,5,5-tetramethyl-1-pyrrolidinyloxy; Sigma-Aldrich), MPSL [55–57] (3-Maleimido-2,2,5,5-tetramethyl-1-pyrrolidinyloxy; Santa Cruz Biotechnology), and the biradical IDSL [58–61] (bis-(2,2,5,5-Tetramethyl-3-imidazoline-1-oxyl-4-yl)disulfide; Noxygen). MTSL and IDSL stock solutions were prepared at 100 mM concentration in dimethyl sulfoxide (DMSO) and stored at −80 °C; IDSL required sonication for full dissolution. An MPSL stock solution was prepared at a 100 mM concentration in 100% ethanol; an IPSL stock solution was prepared at a 77 mM concentration in 100% methanol. MPSL and IPSL stock solutions were aliquoted at 1 mg label per aliquot (40 and 42 µL per aliquot for IPSL and MPSL, respectively), dried in a SpeedVac and stored at −20 °C.

A double-cysteine mutant (I6C/K28C) of the immunoglobulin-binding B1 domain of group G streptococcal protein G (GB1) was used as the model protein. In this mutant, one cysteine is introduced into an α-helix (K28C) while the second cysteine is located in a β-sheet (I6C) [41]. Expression and purification were performed as described previously [40,43]. 12 mg of GB1 I6C/K28C in 2 mL of phosphate buffer (150 mM NaCl, 42.4 mM Na_2HPO_4, 7.6 mM KH_2PO_4, pH 7.4) were freshly reduced with dithiothreitol (DTT) using a 5-fold DTT concentration per cysteine (10-fold per protein molecule) overnight at 4 °C. DTT was removed using a desalting PD10 column, and the eluted protein solution was split into 8 equal parts for labelling.

Two labelling reactions per spin label were set up, each adding a 3-fold molar concentration of label per cysteine (6-fold per protein molecule). MTSL and IDSL were added from the DMSO stock, while MPSL and IPSL were redissolved in ethanol and methanol before use, respectively. One sample set was kept at room temperature in the dark for 2 h while the second set was kept in the dark at 4 °C overnight. After their respective incubation periods, aliquots were taken and immediately frozen before submission to mass spectrometry (MALDI-TOF) using the in-house facility to confirm labelling. The residual free label was removed via PD10 columns, and aliquots for spin counting to determine labelling efficiencies via continuous wave (CW) EPR, aliquots for mass spectrometry (MALDI-TOF) of the purified samples, as well as the remaining samples, were frozen until use.

For IDSL, an additional labelling reaction was performed using a 20-fold molar concentration of the label per cysteine, with incubation overnight at 4 °C.

2.2. Continuous Wave (CW) EPR Spectroscopy

Room-temperature CW EPR measurements to determine labelling efficiencies were performed using a Bruker EMX 10/12 spectrometer equipped with an ELEXSYS Super Hi-Q resonator at an operating frequency of ~9.9 GHz (X-band) with 100 kHz modulation. 50 μL samples in micro capillaries (Brand; one end flame-sealed) were recorded using a 120 G field sweep centred at 3445 G, a time constant of 20.48 ms, a conversion time of 20.10 ms and 2048 points resolution. An attenuation of 20 dB (2 mW power), 50 dB receiver gain and a modulation amplitude of 0.7 G were used for all samples. GB1 samples were measured at a ~50 μM protein (~100 μM spin) concentration, and double integrals (corrected for the actual protein concentration and the number of scans) were compared to 100 μM MTSL as a standard.

2.3. Pulse Dipolar EPR—Sample Preparation and Measurement

EPR samples from the overnight incubation for each spin label were prepared at a 24 μM final protein concentration with 50% ethylene glycol for cryoprotection. For buffer exchange into the deuterated solvent, 100 μL of each protonated labelled sample were freeze-dried and reconstituted in 100 μL D_2O. Samples were prepared at a 24 μM final protein concentration with 50% fully deuterated ethylene glycol for cryoprotection. All samples had a final volume of 65 μL and were transferred to 3 mm quartz EPR tubes, which were immediately frozen by immersion into liquid nitrogen.

PDS experiments were performed at Q-band frequency (34 GHz) operating on a Bruker ELEXSYS E580 spectrometer with a 3 mm cylindrical resonator (ER 5106QT-2w in TE012 mode) using a second frequency option (E580-400U). The temperature was controlled via a cryogen-free variable temperature cryostat (Cryogenic Ltd.) operating in the 3.5 to 300 K temperature range. Pulses were amplified by a pulse travelling wave tube (TWT) amplifier (Applied Systems Engineering) with a nominal output of 150 W.

Specifically, pulsed electron–electron double resonance (PELDOR/DEER) experiments were performed with the 4-pulse DEER [13,62,63] pulse sequence $(\pi/2(\nu_A) - \tau_1 - \pi(\nu_A) - (\tau_1 + t) - \pi(\nu_B) - (\tau_2 - t) - \pi(\nu_A) - \tau_2 - $ echo) at 50 K as described previously, [6] with a frequency offset (pump—detection frequency) of +80 MHz (~3 mT). The shot repetition time (SRT) was set to 4 ms (deuterated samples) or 3 to 4.5 ms (protonated samples); τ_1 was set to 380 ns, and τ_2 was set to 8000 ns for the deuterated samples and to 2400 ns for the protonated samples apart from the IDSL-labelled GB1 sample where 3200 ns were used to allow sufficient resolution for the detection of longer distances. Pulse lengths were 16 and 32 ns for $\pi/2$ and π detection, and 12 ns for the ELDOR pump π pulse. The pump pulse was placed on the resonance frequency of the resonator and applied to the maximum of the nitroxide field-swept spectrum.

PELDOR data were subjected to the Comparative DEER Analyzer (CDA) within DeerAnalysis2021b [64] for unbiased data processing and analysis according to recent recommendations [65], employing DEERNet [66] neural network processing and Deer-Lab [67] Tikhonov regularisation. Full reports of the CDA analysis are provided in the supplementary information.

2.4. Modelling

Distance distributions were modelled based on the I6H/N8H/K28H/Q32H construct (PDB ID: 4WH4) [41]; histidine residues at positions 6 and 28 were mutated to cysteine residues, while histidine residues at positions 8 and 32 were mutated to asparagine and glutamine residues, respectively.

Modelling was performed using the MATLAB plugin MMM 2018 [19] under ambient (298 K) and cryogenic (175 K) temperature conditions and compared with modelling using the mtsslWizard [17] within the mtsslSuite [68] server-based modelling software (Table 1) using 'tight' (vdW-restraint 0 clashes, 3.4 Å cutoff) and 'loose' (vdW-restraint 5 clashes, 2.5 Å cutoff) settings. Cartoon structural representations of spin-labelled GB1 were generated using Pymol [69].

Table 1. List of nitroxide spin labels used in this study and respective names of modifications in modelling software.

Spin Label Name This Study	Side-Chain in MMM	Side-Chain in MtsslSuite
MTSL	MTSL	R1
IPSL	IA-Proxyl	Proxyl
MPSL	MA-Proxyl	malR1
IDSL	V1	sR1

3. Results and Discussion

3.1. Spin Labelling

A deliberately small ratio of spin label to cysteine of 3 to 1 was chosen to allow for assessing differences in the ease of labelling for the different nitroxide labels. In addition, two different incubation conditions were tested for each label, a quick labelling reaction of 2 h at room temperature and an overnight labelling reaction at 4 °C. Successful labelling and labelling efficiencies were determined using mass spectrometry and continuous wave (CW) EPR spectroscopy, respectively.

3.1.1. CW EPR

Individual CW EPR spectra are shown in Figure 1, and a summary of labelling efficiencies is given in Table 2. MTSL labelling efficiency was around 100% after just two hours of labelling. MPSL labelling was determined at about 125%, indicating more label present than available cysteines, already after two hours. The sharp component (especially visible in the high-field line) in MPSL-labelled GB1 spectra suggests that some free label might be present in the samples despite the PD10 column used to remove the free label. This would explain the determined labelling efficiency of well above 100% and indicates purification protocols that were empirically optimised for MTSL may not be sufficient for MPSL. For IPSL, quantitative labelling was achieved after the overnight incubation. The biradical IDSL is attached by substituting one sulfide of its disulfide bond. In contrast to substitutions with good leaving groups (MTSL and IPSL) or addition reactions (MPSL), this disulfide exchange has an equilibrium constant closer to unity, thus incomplete labelling and free dimeric label are very likely. It should be noted that the free thiolate released as leaving group can attack another disulfide bond and this can, at least in theory, result in the complete scrambling of disulfide bonds that could also entail disulfide-linked protein dimers, if steric demand around the cysteine residues permits. IDSL did not yield more than two-thirds of labelling efficiency even after the overnight incubation time. Therefore, a second overnight labelling reaction using a 20-fold concentration of IDSL with respect to cysteine was performed, which yielded above 90% efficiency. Additional lines in IDSL-labelled GB1 spectra are attributed to the intact (free) label [60].

CW data suggest that MTSL, IPSL and MPSL can provide quantitative labelling at relatively small (here: 3:1) label-to-cysteine ratios for both secondary structure elements, α-helix and β-sheet. It should be noted that both labelling sites in the GB1 construct are easily accessible, thus higher ratios might be required if sites are buried. MTSL and IPSL could easily be removed using a PD10 desalting column, while MPSL labelling might need additional chromatographic steps to remove the residual free label. IDSL, presumably due to the equilibrium reaction, was shown to require a higher excess of the label to approach quantitative labelling of the cysteine residues.

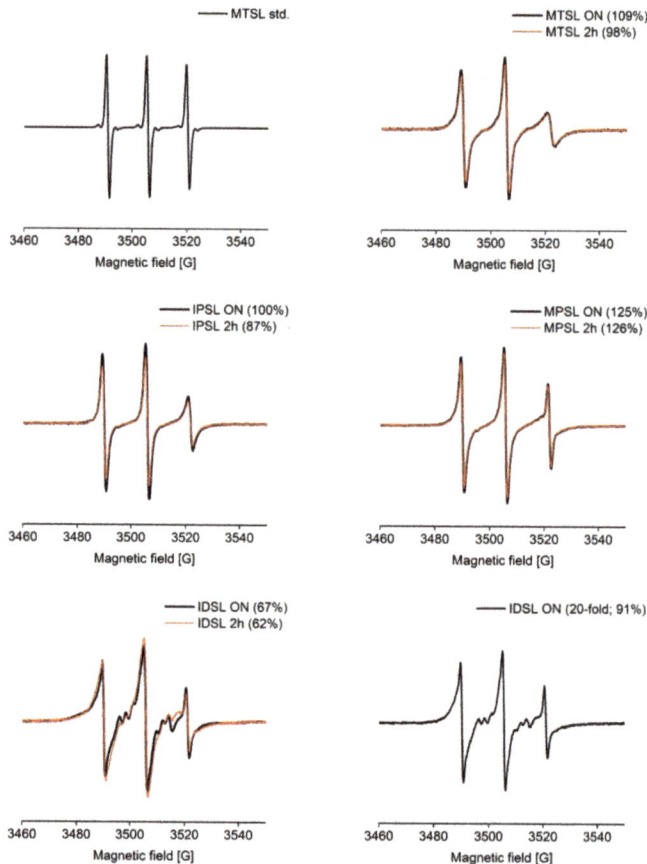

Figure 1. Continuous wave (CW) EPR spectra for GB1 labelled with MTSL, IPSL, MPSL and IDSL and respective labelling efficiencies relative to MTSL standard. All spectra were recorded with the same experimental parameters and normalised to the number of scans. 2 h = two hours incubation time; ON = overnight incubation time. For IDSL-labelled GB1, due to the low labelling efficiencies for a 3-fold label concentration, an additional labelling experiment was performed at a 20-fold label concentration incubating overnight.

Table 2. Summary of determined labelling efficiencies per cysteine relative to MTSL standard.

Incubation	MTSL	IPSL	MPSL	IDSL
2 h, 3-fold	98%	87%	126%	62%
ON, 3-fold	109%	100%	125%	67%
ON, 20-fold	-	-	-	91%

3.1.2. Mass Spectrometry

Samples were analysed by MALDI-TOF after 2 h and overnight incubation time and a 3:1 label-to-cysteine ratio confirming successful spin labelling before the removal of excess free label. Since overnight reactions generally showed better labelling, these were taken forward for PDS and the purified samples were re-analysed by MALDI-TOF, showing excellent agreement with results from the unpurified labelling reactions. In the case of IDSL-labelled GB1, MALDI-TOF was also performed after overnight reaction with 20-fold IDSL. Details and individual mass spectrometry results are shown in the supplementary information (Figures S1–S10). Overall, MALDI-TOF spectra are in line with the results obtained from CW EPR; although for IPSL- and especially IDSL-labelled GB1, less of the fully labelled protein is seen with MALDI-TOF than would be expected from CW EPR. This could be due to residual free label (not all labels attached to the protein), label-specific differences in the ionisation for labelled and unlabelled protein or the laser could lead to partial label detachment. Interestingly, upon measuring the second distance in PELDOR for IDSL-labelled GB1 (see below), MALDI could also confirm the presence of a small amount of a species with a mass corresponding to a GB1 dimer, which was no longer present in the sample with a 20-fold ratio of label-to-cysteine. This highlights the need for a larger excess of the IDSL label—not only to drive the equilibrium towards quantitative cysteine labelling, but also to avoid significant equilibrium concentrations of the disulfide-linked protein dimer formed by thiolate exchanges.

3.2. PDS Distance Measurements (PELDOR/DEER) and Comparison to Modelling

Initially, protonated samples from the overnight labelling reaction were prepared for PDS after removal of the free spin label. We rationalised that echo dephasing (T_m) would be sufficient to resolve the expected short distance (below 3 nm) in the GB1 I6C/K28C construct. Distance distributions obtained from PDS (PELDOR/DEER) primary data on the protonated samples are shown in Figure 2.

MTSL and, to a lesser extent, IPSL labelling results suggest a bimodal distance distribution within the range of 2–3 nm, while MPSL and IDSL labelling results do not. An interesting finding is the appearance of a larger distance (~4.5 nm) for IDSL-labelled GB1, further supporting the hypothesis of dimer formation based on label chemistry and mass spectrometry, which is further discussed below. However, distance resolution is not sufficient in the protonated samples, therefore the second set of samples was prepared with reconstitution of the freeze-dried protein into D_2O and a fully deuterated cryoprotectant. A comparison of respective refocused echo decay experiments demonstrating the gain in the distance resolution with deuteration is given in Figure S13. In addition, IDSL labelling was repeated using a much larger excess of the label (ratio 20:1) to assess changes in the labelling efficiency and the presence of the disulfide-linked protein dimer.

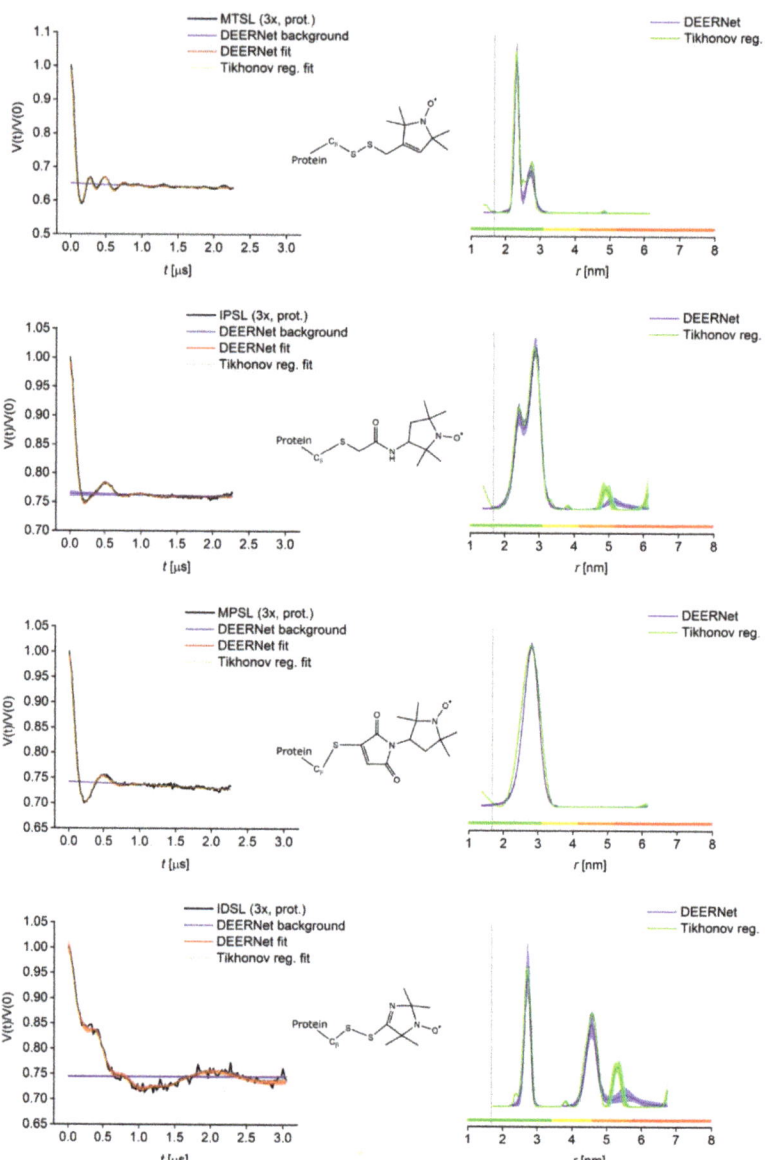

Figure 2. PELDOR results for protonated GB1 I6C/K28C labelled with MTSL, IPSL, MPSL and IDSL. Left: Superposition of primary data (black) with Tikhonov regularisation fit (green) and DEERNet separated background (blue) and fit (red) with uncertainty bounds. Right: Superposition of distance distributions obtained from DEERNet (blue) and Tikhonov regularisation (green) with uncertainty bands. Vertical lines indicate the C_β-C_β distance 6C–28C (1.67 nm). Colour bars represent reliability ranges (green: Shape reliable; yellow: Mean and width reliable; orange: Mean reliable; red: No quantification possible).

Distance distributions obtained from PDS (PELDOR/DEER) primary data on the deuterated samples are shown below (Figures 3 and 4), and a superposition of distributions obtained from protonated and deuterated samples is given Figure S12. MTSL labelling still suggests a bimodal distance distribution, but to a lesser extent than observed for the protonated sample. This bimodality with MTSL has been observed previously [40] and can be rationalised with additional label conformations. Another potential explanation would be the presence of different conformational states of the protein itself. However, GB1 is not known to exist in a conformational equilibrium but is a very rigid small protein model. If there were additional protein conformations, these would be expected to also be visible with the other labels, which is not the case.

For IDSL-labelled GB1, the larger distance (~4.5 nm) at the 3:1 label-to-cysteine ratio is confirmed with the deuterated sample with a sufficient distance resolution. If the thiolate exchange is dynamic, then a certain fraction of the disulfide-bridged protein dimer will equilibrate based on its proportion in the mixture. A disulfide-bridged GB1 protein dimer will still have one cysteine per monomer available for labelling. Considering a distance of just over 2.5 nm for IDSL-labelled GB1 and a distance of 1.7 nm between C_β atoms of the labelled residues, a disulfide-bridged dimer of singly IDSL-labelled GB1 monomers will have a distance shorter than two times 2.5 nm but substantially longer than two times 1.7 nm, in good agreement with the 4.5 nm observed. SDS-PAGE for spin-labelled GB1 I6C/K28C with non-reduced samples to preserve disulfide linkages demonstrates that disulfides do indeed form for IDSL at the 3:1 label-to-cysteine ratio (see Figure S11). As expected, this dimer peak vanished after increasing the amount of IDSL label to a 20:1 ratio, indicating the equilibrium labelling reaction can be driven towards quantitative binding by a larger excess of the label. In this case, a labelling efficiency of >90% was determined using CW EPR. Notably, due to the shortest linker of all labels tested, the IDSL provides the highest precision with significantly narrower distributions, similar to those observed for Cu^{II}-labelling of double-histidine sites [41,43].

As our samples may contain free label, it is important to note that the residual free label will add to the unmodulated part of the echo, thereby reducing modulation depth. However, it should not alter the distance distribution or influence the overall signal-to-noise. Unspecific labelling (i.e., with the label attached to the protein at a non-cysteine residue) might lead to added modulation depth (and potentially multi-spin effects), and additional distances affecting the overall distance distribution. We did not observe these effects in our PDS data nor indications of unspecific labelling from mass spectrometry.

Modelling of expected distance distributions for the four nitroxide labels was performed using MMM and mtsslWizard; overlays of the resulting simulated and experimental distributions are shown in Figure 5 for recommended settings (MMM at ambient temperature, mtsslWizard with 'tight' vdW-restraint setting) and in Figure S14 with additional settings (MMM at cryogenic temperature, mtsslWizard with 'loose' vdW-restraint setting). Overall, consistent models were obtained. The bimodal distance distribution observed with the protonated sample of MTSL-labelled GB1 is in contrast to modelling results. This difference could be rationalised with an additional label conformation induced by the interaction with the protein surface escaping both modelling approaches [16,20]. Interestingly, this seems to be much less pronounced in the deuterated sample even though all other conditions are nominally identical. For IPSL-labelled GB1, the modelled distributions matched well with the experimentally obtained distribution, while for MPSL-labelled GB1, some deviation between the two modelling approaches was observed, with the mtsslWizard model matching experimental distributions more closely. As seen in Figure 5, the spin-labelled side-chain of IDSL-labelled GB1 occupies the most restricted rotameric space of all four labels tested. Notably, experimental distance distributions were even narrower than the simulated ones. Expectedly, the distance corresponding to the disulfide-bridged GB1 dimer could not be modelled as it is not present in the coordinate file.

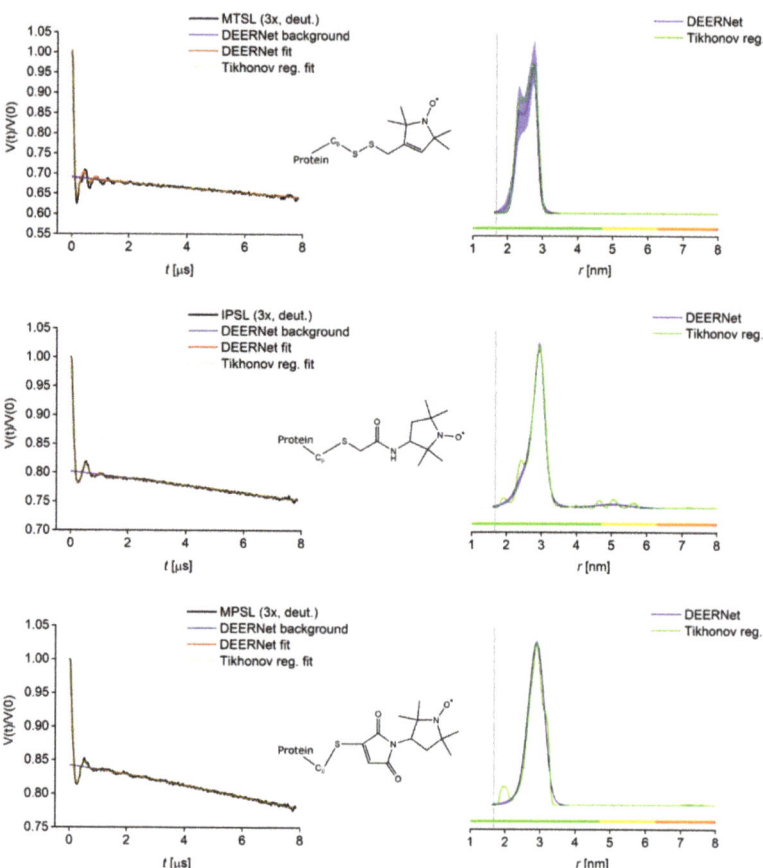

Figure 3. PELDOR results for GB1 I6C/K28C labelled with MTSL, IPSL and MPSL and buffer exchanged into deuterated solvent. Left: Superposition of primary data (black) with Tikhonov regularisation fit (green) and DEERNet separated background (blue) and fit (red) with uncertainty bounds. Right: Superposition of distance distributions obtained from DEERNet (blue) and Tikhonov regularisation (green) with uncertainty bands. Vertical lines indicate the C_β-C_β distance 6C–28C (1.67 nm). Colour bars represent reliability ranges (green: Shape reliable; yellow: Mean and width reliable; orange: Mean reliable; red: No quantification possible).

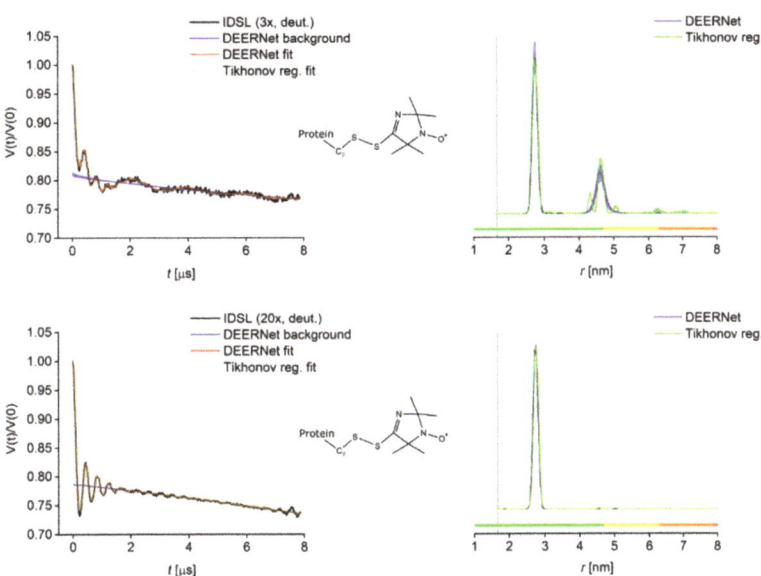

Figure 4. PELDOR results for GB1 I6C/K28C labelled with IDSL at 3:1 and 20:1 label-to-cysteine ratio and buffer exchanged into deuterated solvent. Left: Superposition of primary data (black) with Tikhonov regularisation fit (green) and DEERNet separated background (blue) and fit (red) with uncertainty bounds. Right: Superposition of distance distributions obtained from DEERNet (blue) and Tikhonov regularisation (green) with uncertainty bands. Vertical lines indicate the C_β-C_β distance 6C–28C (1.67 nm). Colour bars represent reliability ranges (green: Shape reliable; yellow: Mean and width reliable; orange: Mean reliable; red: No quantification possible).

We further investigated the agreement between the two modelling approaches (MMM and mtsslWizard) and their agreement with experimental distance distributions obtained from two different processing methods (neuronal network analysis and Tikhonov regularisation). Comparing deviations of the distributions obtained from protonated and deuterated buffer samples reveals that the agreement between different modelling approaches and the corresponding experiments is not much worse than between different samples (Table 3). MTSL labelling leads to different populations of conformers and IDSL labelling leads to a different population of dimers dominating the rmsd between experiments. For the latter, both modelling approaches are in excellent agreement, but the width of the distribution is found to be narrower than modelled and this dominates the rmsd. Interestingly, the model deviation is worst for MPSL-labelled GB1, where this seems largely down to modelled short-distance conformers in MMM at 298 K that do not manifest in the experimental distribution with the agreement between experiment and mtsslWizard being significantly better. Here, the agreement between experiments for protonated and deuterated samples is best for all labels. Interestingly, MMM at 175 K substantially shifts the modelled distance distribution for MPSL-labelled GB1, giving much better agreement than at ambient temperatures, while 'loose' settings for the mtsslWizard result in broadened and slightly shortened distributions for all labels (Figure S14). In contrast, for MTSL-labelled GB1, both models agree better with the experiment than with each other. While there are significant differences between protonated and deuterated samples, MMM agrees much better with the latter than mtsslWizard. On the other hand, the distance distributions for IPSL-labelled GB1 are predicted remarkably well, especially by mtsslWizard. Finally, it is obvious that the prediction of spin-label conformations remains problematic for highly accurate and precise distance measurements, and although IDSL and IPSL labelling appear more robust in this example, it remains to be seen if this holds for other scenarios. However, these results reveal that there is significant promise in systematic comparisons between labels both in silico and in experiments to reveal systematic trends with respect to the reliability of predictions and specific advantages and disadvantages of individual labels.

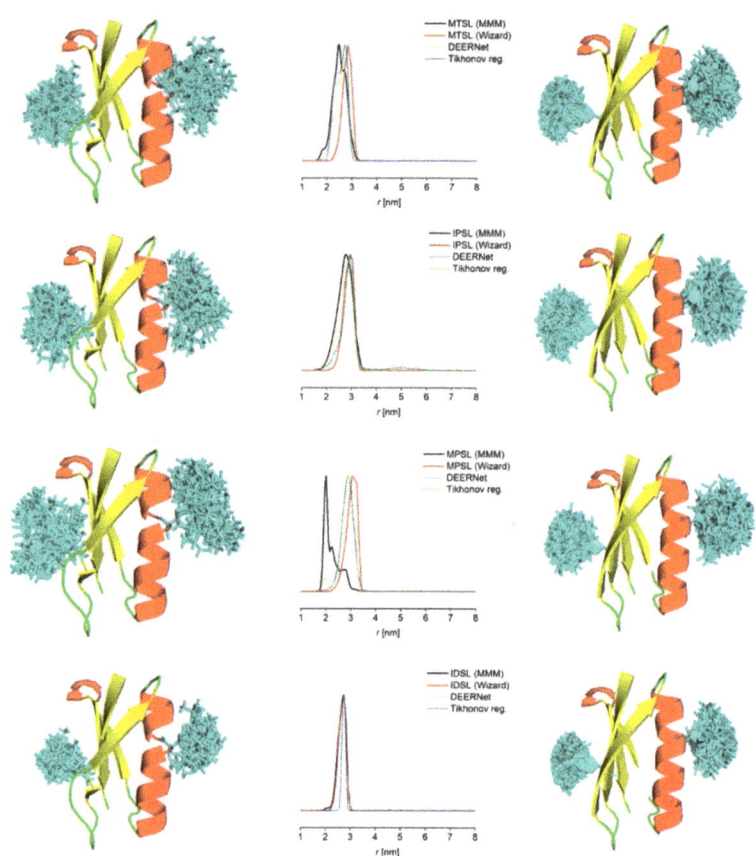

Figure 5. Modelled distance distributions for GB1 I6C/K28C labelled with MTSL, IPSL, MPSL and IDSL. Left: Cartoon representation of GB1 and modelled spin label conformations (cyan sticks) as obtained from MMM. Middle: Superposition of predicted and experimental (buffer exchanged into deuterated solvent, 3:1 ratio except IDSL 20:1 ratio) distance distributions. Right: Cartoon representation of GB1 and modelled spin label conformations (cyan sticks) as obtained from mtsslWizard.

Table 3. Root-mean-square-deviations of intensity-normalised distance distributions based on various experiments and modelling approaches.

	Wizard/MMM	Deuterated/Protonated		DEERNet		Regularised	
	RMSD	DEERNet	Regularised	Deuterated/Wizard	Deuterated/MMM	Deuterated/Wizard	Deuterated/MMM
MTSL	0.1363	0.1118	0.1161	0.1165	0.0459	0.1237	0.0421
IPSL	0.0803	0.0568	0.0628	0.0311	0.0750	0.0340	0.0732
MPSL	0.1866	0.0432	0.0613	0.0981	0.1741	0.0917	0.1624
IDSL	0.0156	0.0907	0.1099	0.0771	0.0819	0.0714	0.0769

For IDSL values from the deuterated sample using the 20:1 label-to-cysteine ratio were used. RMSD = root mean square deviation; Wizard = mtsslWizard; Regularised = Tikhonov regularisation.

4. Conclusions and Outlook

While MTSL is arguably the current 'work-horse' and best-established nitroxide spin label for protein labelling, this study shows that other commercially available cysteine-reactive nitroxide labels are attractive alternatives. Both MPSL and IPSL are supposedly less prone to suffering from cleavage in reducing environments and thus might be better options for more native environments such as in-cell studies, presuming the samples can be frozen before the nitroxide itself is reduced. In our hands, even at a low label-to-cysteine ratio, excess MPSL was not fully removed using a simple PD10 desalting column, suggesting that this label is more demanding in downstream purification steps. In contrast, IPSL delivered quantitative labelling after overnight incubation at the same low excess and was fully removed in the PD10 clean-up step. A very interesting alternative spin label is the biradical IDSL, which has the most restricted rotameric space of all labels compared in this study. However, it requires a large excess of the label to achieve near quantitative labelling. PELDOR results agreed well with modelling based on the crystal structure for all four labels and the differences between modelling approaches and between analysis methods were generally similar to the deviations between the models and experiments. Notably, distance distributions for IDSL-labelled GB1 exhibited significantly enhanced precision due to the short linker that is truer to the protein backbone. This makes IDSL particularly attractive to investigate small conformational changes requiring a high precision with narrow distributions. It will be interesting to further explore the specific advantages of the individual labels for distinct applications.

The research data underpinning this publication can be accessed at https://doi.org/10.17630/70739623-1bb5-40e6-b326-5fff1e22c289 [70].

Supplementary Materials: The following are available online, Mass spectrometry data (Figures S1–S10); SDS-PAGE gel of spin-labelled GB1 6C/28C (Figure S11); superposition of distance distributions of protonated and deuterated samples (Figure S12); refocused echo decay (Hahn echo) experiments (Figure S13); additional modelling (Figure S14), and full CDA reports.

Author Contributions: Conceptualization, methodology, validation, formal analysis, writing—review and editing, visualization, K.A. and B.E.B.; investigation, K.A. and A.C.; writing—original draft preparation, K.A.; resources, supervision, project administration, funding acquisition, B.E.B. All authors have read and agreed to the published version of the manuscript.

Funding: This research was funded, in whole or in part, by the Wellcome Trust (099149/Z/12/Z and 204821/Z/16/Z). B.E.B. and K.A. acknowledge support from the Leverhulme Trust (RPG-2018–397). B.E.B. acknowledges equipment funding from BBSRC (BB/R013780/1 and BB/T017740/1).

Institutional Review Board Statement: Not applicable.

Informed Consent Statement: Not applicable.

Data Availability Statement: Digital data underpinning the results presented in this manuscript is freely available [70].

Acknowledgments: We thank the mass spectrometry facility at the Biomedical Sciences Research Complex in St Andrews and colleagues from the StAnD (St Andrews and Dundee) EPR grouping.

Conflicts of Interest: The authors declare no conflict of interest.

Sample Availability: Samples of the compounds are not available from the authors.

References

1. Bragg, W.H.; Bragg, W.L. The Reflection of X-rays by Crystals. *Proc. R. Soc. Lond. A* **1913**, *88*, 428–438. [CrossRef]
2. Anfinsen, C.B. Principles that Govern the Folding of Protein Chains. *Science* **1973**, *181*, 223–230. [CrossRef]
3. Fuller, A.L.; Scott-Hayward, L.A.S.; Li, Y.; Bühl, M.; Slawin, A.M.Z.; Woollins, J.D. Automated Chemical Crystallography. *J. Am. Chem. Soc.* **2010**, *132*, 5799–5802. [CrossRef] [PubMed]
4. Jeschke, G. DEER Distance Measurements on Proteins. *Annu. Rev. Phys. Chem.* **2012**, *63*, 419–446. [CrossRef] [PubMed]
5. Schiemann, O.; Prisner, T.F. Long-Range Distance Determinations in Biomacromolecules by EPR Spectroscopy. *Q. Rev. Biophys.* **2007**, *40*, 1–53. [CrossRef]

6. Kerry, P.S.; Turkington, H.L.; Ackermann, K.; Jameison, S.A.; Bode, B.E. Analysis of Influenza A Virus NS1 Dimer Interfaces in Solution by Pulse EPR Distance Measurements. *J. Phys. Chem. B* **2014**, *118*, 10882–10888. [CrossRef] [PubMed]
7. Theillet, F.-X.; Binolfi, A.; Bekei, B.; Martorana, A.; Rose, H.M.; Stuiver, M.; Verzini, S.; Lorenz, D.; van Rossum, M.; Goldfarb, D.; et al. Structural Disorder of Monomeric α-Synuclein Persists in Mammalian Cells. *Nature* **2016**, *530*, 45–50. [CrossRef]
8. Bleicken, S.; Assafa, T.E.; Stegmueller, C.; Wittig, A.; Garcia-Saez, A.J.; Bordignon, E. Topology of Active, Membrane-Embedded Bax in the Context of a Toroidal Pore. *Cell Death Differ.* **2018**, *25*, 1717–1731. [CrossRef]
9. Joseph, B.; Sikora, A.; Cafiso, D.S. Ligand Induced Conformational Changes of a Membrane Transporter in *E. coli* Cells Observed with DEER/PELDOR. *J. Am. Chem. Soc.* **2016**, *138*, 1844–1847. [CrossRef] [PubMed]
10. Sameach, H.; Narunsky, A.; Azoulay-Ginsburg, S.; Gevorkyan-Aiapetov, L.; Zehavi, Y.; Moskovitz, Y.; Juven-Gershon, T.; Ben-Tal, N.; Ruthstein, S. Structural and Dynamics Characterization of the MerR Family Metalloregulator CueR in its Repression and Activation States. *Structure* **2017**, *25*, 988–996.e3. [CrossRef] [PubMed]
11. Kulik, L.V.; Dzuba, S.A.; Grigoryev, I.A.; Tsvetkov, Y.D. Electron Dipole–Dipole Interaction in ESEEM of Nitroxide Biradicals. *Chem. Phys. Lett.* **2001**, *343*, 315–324. [CrossRef]
12. Milov, A.D.; Ponomarev, A.B.; Tsvetkov, Y.D. Electron-Electron Double Resonance in Electron Spin Echo: Model Biradical Systems and the Sensitized Photolysis of Decalin. *Chem. Phys. Lett.* **1984**, *110*, 67–72. [CrossRef]
13. Milov, A.D.; Salikov, K.M.; Shirov, M.D. Application of ELDOR in Electron-Spin Echo for Paramagnetic Center Space Distribution in Solids. *Fiz. Tverd. Tela* **1981**, *23*, 975–982.
14. Altenbach, C.; Marti, T.; Khorana, H.G.; Hubbell Wayne, L. Transmembrane Protein Structure: Spin Labeling of Bacteriorhodopsin Mutants. *Science* **1990**, *248*, 1088–1092. [CrossRef] [PubMed]
15. Hubbell, W.L.; Altenbach, C. Investigation of Structure and Dynamics in Membrane Proteins Using Site-Directed Spin Labeling. *Curr. Opin. Struct. Biol.* **1994**, *4*, 566–573. [CrossRef]
16. Abdullin, D.; Hagelueken, G.; Schiemann, O. Determination of Nitroxide Spin Label Conformations via PELDOR and X-ray Crystallography. *Phys. Chem. Chem. Phys.* **2016**, *18*, 10428–10437. [CrossRef] [PubMed]
17. Hagelueken, G.; Ward, R.; Naismith, J.H.; Schiemann, O. MtsslWizard: In Silico Spin-Labeling and Generation of Distance Distributions in PyMOL. *Appl. Magn. Reson.* **2012**, *42*, 377–391. [CrossRef] [PubMed]
18. Jeschke, G. Conformational Dynamics and Distribution of Nitroxide Spin Labels. *Prog. Nucl. Magn. Reson. Spectrosc.* **2013**, *72*, 42–60. [CrossRef] [PubMed]
19. Polyhach, Y.; Bordignon, E.; Jeschke, G. Rotamer Libraries of Spin Labelled Cysteines for Protein Studies. *Phys. Chem. Chem. Phys.* **2011**, *13*, 2356–2366. [CrossRef] [PubMed]
20. Spicher, S.; Abdullin, D.; Grimme, S.; Schiemann, O. Modeling of Spin–Spin Distance Distributions for Nitroxide Labeled Biomacromolecules. *Phys. Chem. Chem. Phys.* **2020**, *22*, 24282–24290. [CrossRef] [PubMed]
21. Jeschke, G. The Contribution of Modern EPR to Structural Biology. *Emerg. Top. Life Sci.* **2018**, *2*, 9–18.
22. Margraf, D.; Bode, B.E.; Marko, A.; Schiemann, O.; Prisner, T.F. Conformational Flexibility of Nitroxide Biradicals Determined by X-Band PELDOR Experiments. *Mol. Phys.* **2007**, *105*, 2153–2160. [CrossRef]
23. Bode, B.E.; Plackmeyer, J.; Bolte, M.; Prisner, T.F.; Schiemann, O. PELDOR on an Exchange Coupled Nitroxide Copper(II) Spin Pair. *J. Organomet. Chem.* **2009**, *694*, 1172–1179. [CrossRef]
24. Richert, S.; Cremers, J.; Kuprov, I.; Peeks, M.D.; Anderson, H.L.; Timmel, C.R. Constructive Quantum Interference in a Bis-Copper Six-Porphyrin Nanoring. *Nat Commun* **2017**, *8*, 14842. [CrossRef] [PubMed]
25. Weber, A.; Schiemann, O.; Bode, B.; Prisner, T.F. PELDOR at S- and X-Band Frequencies and the Separation of Exchange Coupling from Dipolar Coupling. *J. Magn. Reson.* **2002**, *157*, 277–285. [CrossRef] [PubMed]
26. Ackermann, K.; Giannoulis, A.; Cordes, D.B.; Slawin, A.M.Z.; Bode, B.E. Assessing Dimerisation Degree and Cooperativity in a Biomimetic Small-Molecule Model by Pulsed EPR. *Chem. Commun.* **2015**, *51*, 5257–5260. [CrossRef] [PubMed]
27. Bode, B.E.; Margraf, D.; Plackmeyer, J.; Dürner, G.; Prisner, T.F.; Schiemann, O. Counting the Monomers in Nanometer-Sized Oligomers by Pulsed Electron−Electron Double Resonance. *J. Am. Chem. Soc.* **2007**, *129*, 6736–6745. [CrossRef] [PubMed]
28. Bode, B.E.; Dastvan, R.; Prisner, T.F. Pulsed Electron–Electron Double Resonance (PELDOR) Distance Measurements in Detergent Micelles. *J. Magn. Reson.* **2011**, *211*, 11–17. [CrossRef] [PubMed]
29. Dastvan, R.; Bode, B.E.; Karuppiah, M.P.R.; Marko, A.; Lyubenova, S.; Schwalbe, H.; Prisner, T.F. Optimization of Transversal Relaxation of Nitroxides for Pulsed Electron−Electron Double Resonance Spectroscopy in Phospholipid Membranes. *J. Phys. Chem. B* **2010**, *114*, 13507–13516. [CrossRef] [PubMed]
30. Valera, S.; Bode, B.E. Strategies for the Synthesis of Yardsticks and Abaci for Nanometre Distance Measurements by Pulsed EPR. *Molecules* **2014**, *19*, 20227–20256. [CrossRef] [PubMed]
31. Assafa, T.E.; Nandi, S.; Śmiłowicz, D.; Galazzo, L.; Teucher, M.; Elsner, C.; Pütz, S.; Bleicken, S.; Robin, A.Y.; Westphal, D.; et al. Biophysical Characterization of Pro-apoptotic BimBH3 Peptides Reveals an Unexpected Capacity for Self-Association. *Structure* **2020**, *29*, 114–124. [CrossRef] [PubMed]
32. Giannoulis, A.; Ward, R.; Branigan, E.; Naismith, J.H.; Bode, B.E. PELDOR in Rotationally Symmetric Homo-Oligomers. *Mol. Phys.* **2013**, *111*, 2845–2854. [CrossRef] [PubMed]
33. Jeschke, G.; Sajid, M.; Schulte, M.; Godt, A. Three-Spin Correlations in Double Electron–Electron Resonance. *Phys. Chem. Chem. Phys.* **2009**, *11*, 6580–6591. [CrossRef] [PubMed]

34. Valera, S.; Taylor, J.E.; Daniels, D.S.B.; Dawson, D.M.; Athukorala Arachchige, K.S.; Ashbrook, S.E.; Slawin, A.M.Z.; Bode, B.E. A Modular Approach for the Synthesis of Nanometer-Sized Polynitroxide Multi-Spin Systems. *J. Org. Chem.* **2014**, *79*, 8313–8323. [CrossRef]
35. von Hagens, T.; Polyhach, Y.; Sajid, M.; Godt, A.; Jeschke, G. Suppression of Ghost Distances in Multiple-Spin Double Electron–Electron Resonance. *Phys. Chem. Chem. Phys.* **2013**, *15*, 5854–5866. [CrossRef]
36. Ackermann, K.; Pliotas, C.; Valera, S.; Naismith, J.H.; Bode, B.E. Sparse Labeling PELDOR Spectroscopy on Multimeric Mechanosensitive Membrane Channels. *Biophys. J.* **2017**, *113*, 1968–1978. [CrossRef] [PubMed]
37. Valera, S.; Ackermann, K.; Pliotas, C.; Huang, H.; Naismith, J.H.; Bode, B.E. Accurate Extraction of Nanometer Distances in Multimers by Pulse EPR. *Chem. Eur. J.* **2016**, *22*, 4700–4703. [CrossRef] [PubMed]
38. Kapsalis, C.; Wang, B.; El Mkami, H.; Pitt, S.J.; Schnell, J.R.; Smith, T.K.; Lippiat, J.D.; Bode, B.E.; Pliotas, C. Allosteric Activation of an Ion Channel Triggered by Modification of Mechanosensitive Nano-Pockets. *Nat. Commun.* **2019**, *10*, 4619. [CrossRef]
39. Morten, M.J.; Peregrina, J.R.; Figueira-Gonzalez, M.; Ackermann, K.; Bode, B.E.; White, M.F.; Penedo, J.C. Binding Dynamics of a Monomeric SSB Protein to DNA: A Single-Molecule Multi-Process Approach. *Nucleic Acids Res.* **2015**, *43*, 10907–10924. [CrossRef]
40. Ackermann, K.; Wort, J.L.; Bode, B.E. Nanomolar Pulse Dipolar EPR Spectroscopy in Proteins: Cu^{II}–Cu^{II} and Nitroxide–Nitroxide Cases. *J. Phys. Chem. B* **2021**, *125*, 5358–5364. [CrossRef] [PubMed]
41. Cunningham, T.F.; Putterman, M.R.; Desai, A.; Horne, W.S.; Saxena, S. The Double-Histidine Cu^{2+}-Binding Motif: A Highly Rigid, Site-Specific Spin Probe for Electron Spin Resonance Distance Measurements. *Angew. Chem. Int. Ed.* **2015**, *54*, 6330–6334. [CrossRef] [PubMed]
42. Singewald, K.; Bogetti, X.; Sinha, K.; Rule, G.S.; Saxena, S. Double Histidine Based EPR Measurements at Physiological Temperatures Permit Site-Specific Elucidation of Hidden Dynamics in Enzymes. *Angew. Chem. Int. Ed.* **2020**, *59*, 23040–23044. [CrossRef] [PubMed]
43. Wort, J.L.; Ackermann, K.; Giannoulis, A.; Stewart, A.J.; Norman, D.G.; Bode, B.E. Sub-Micromolar Pulse Dipolar EPR Spectroscopy Reveals Increasing Cu^{II}-labelling of Double-Histidine Motifs with Lower Temperature. *Angew. Chem. Int. Ed.* **2019**, *58*, 11681–11685. [CrossRef]
44. Wort, J.L.; Ackermann, K.; Norman, D.G.; Bode, B.E. A General Model to Optimise Cu^{II} Labelling Efficiency of Double-Histidine Motifs for Pulse Dipolar EPR Applications. *Phys. Chem. Chem. Phys.* **2021**, *23*, 3810–3819. [CrossRef] [PubMed]
45. Wort, J.L.; Arya, S.; Ackermann, K.; Stewart, A.J.; Bode, B.E. Pulse Dipolar EPR Reveals Double-Histidine Motif Cu^{II}–NTA Spin-Labeling Robustness against Competitor Ions. *J. Phys. Chem. Lett.* **2021**, *12*, 2815–2819. [CrossRef] [PubMed]
46. Bordignon, E.; Polyhach, Y. EPR Techniques to Probe Insertion and Conformation of Spin-Labeled Proteins in Lipid Bilayers. In *Lipid-Protein Interactions: Methods and Protocols*; Kleinschmidt, J.H., Ed.; Humana Press: Totowa, NJ, USA, 2013; pp. 329–355.
47. Joseph, B.; Jaumann, E.A.; Sikora, A.; Barth, K.; Prisner, T.F.; Cafiso, D.S. In Situ Observation of Conformational Dynamics and Protein Ligand–Substrate Interactions in Outer-Membrane Proteins with DEER/PELDOR Spectroscopy. *Nat. Protoc.* **2019**, *14*, 2344–2369. [CrossRef] [PubMed]
48. *Structural Information from Spin-Labels and Intrinsic Paramagnetic Centres in the Biosciences*; Timmel, C.R.; Harmer, J.R. (Eds.) Springer: Berlin/Heidelberg, Germany, 2013.
49. Bordignon, E. EPR Spectroscopy of Nitroxide Spin Probes. *eMagRes* **2017**, *6*, 235–254.
50. Fanucci, G.E.; Cafiso, D.S. Recent Advances and Applications of Site-Directed Spin Labeling. *Curr. Opin. Struct. Biol.* **2006**, *16*, 644–653. [CrossRef] [PubMed]
51. Berliner, L.J.; Grunwald, J.; Hankovszky, H.O.; Hideg, K. A Novel Reversible Thiol-Specific Spin Label: Papain Active Site Labeling and Inhibition. *Anal. Biochem.* **1982**, *119*, 450–455. [CrossRef]
52. Gmeiner, C.; Dorn, G.; Allain, F.H.T.; Jeschke, G.; Yulikov, M. Spin Labelling for Integrative Structure Modelling: A Case Study of the Polypyrimidine-Tract Binding Protein 1 Domains in Complexes with Short RNAs. *Phys. Chem. Chem. Phys.* **2017**, *19*, 28360–28380. [CrossRef]
53. Hellmich, U.A.; Lyubenova, S.; Kaltenborn, E.; Doshi, R.; van Veen, H.W.; Prisner, T.F.; Glaubitz, C. Probing the ATP Hydrolysis Cycle of the ABC Multidrug Transporter LmrA by Pulsed EPR Spectroscopy. *J. Am. Chem. Soc.* **2012**, *134*, 5857–5862. [CrossRef] [PubMed]
54. Volkov, A.; Dockter, C.; Bund, T.; Paulsen, H.; Jeschke, G. Pulsed EPR Determination of Water Accessibility to Spin-Labeled Amino Acid Residues in LHCIIb. *Biophys. J.* **2009**, *96*, 1124–1141. [CrossRef]
55. Páli, T.; Finbow, M.E.; Marsh, D. Membrane Assembly of the 16-kDa Proteolipid Channel from Nephrops norvegicus Studied by Relaxation Enhancements in Spin-Label ESR. *Biochemistry* **1999**, *38*, 14311–14319. [CrossRef] [PubMed]
56. Pantusa, M.; Sportelli, L.; Bartucci, R. Spectroscopic and Calorimetric Studies on the Interaction of Human Serum Albumin with DPPC/PEG:2000-DPPE Membranes. *Eur. Biophys. J.* **2008**, *37*, 961–973. [CrossRef] [PubMed]
57. Shenberger, Y.; Gottlieb, H.E.; Ruthstein, S. EPR and NMR Spectroscopies Provide Input on the Coordination of Cu(I) and Ag(I) to a Disordered Methionine Segment. *JBIC J. Biol. Inorg. Chem.* **2015**, *20*, 719–727. [CrossRef] [PubMed]
58. Balo, A.R.; Feyrer, H.; Ernst, O.P. Toward Precise Interpretation of DEER-Based Distance Distributions: Insights from Structural Characterization of V1 Spin-Labeled Side Chains. *Biochemistry* **2016**, *55*, 5256–5263. [CrossRef] [PubMed]
59. Toledo Warshaviak, D.; Khramtsov, V.V.; Cascio, D.; Altenbach, C.; Hubbell, W.L. Structure and Dynamics of an Imidazoline Nitroxide Side Chain with Strongly Hindered Internal Motion in Proteins. *J. Magn. Reson.* **2013**, *232*, 53–61. [CrossRef] [PubMed]

60. Weiner, L.M. [8] Quantitative Determination of Thiol Groups in Low and High Molecular Weight Compounds by Electron Paramagnetic Resonance. In *Methods in Enzymology*; Elsevier: Amsterdam, The Netherlands, 1995; Volume 251, pp. 87–105.
61. Wingler, L.M.; Elgeti, M.; Hilger, D.; Latorraca, N.R.; Lerch, M.T.; Staus, D.P.; Dror, R.O.; Kobilka, B.K.; Hubbell, W.L.; Lefkowitz, R.J. Angiotensin Analogs with Divergent Bias Stabilize Distinct Receptor Conformations. *Cell* **2019**, *176*, 468–478.e11. [CrossRef]
62. Larsen, R.G.; Singel, D.J. Double Electron-Electron Resonance Spin-Echo Modulation—Spectroscopic Measurement of Electron-Spin Pair Separations in Orientationally Disordered Solids. *J. Chem. Phys.* **1993**, *98*, 5134–5146. [CrossRef]
63. Pannier, M.; Veit, S.; Godt, A.; Jeschke, G.; Spiess, H.W. Dead-Time Free Measurement of Dipole-Dipole Interactions between Spins. *J. Magn. Reson.* **2000**, *142*, 331–340. [CrossRef]
64. Jeschke, G.; Chechik, V.; Ionita, P.; Godt, A.; Zimmermann, H.; Banham, J.; Timmel, C.R.; Hilger, D.; Jung, H. DeerAnalysis2006—A Comprehensive Software Package for Analyzing Pulsed ELDOR Data. *Appl. Magn. Reson.* **2006**, *30*, 473–498. [CrossRef]
65. Schiemann, O.; Heubach, C.A.; Abdullin, D.; Ackermann, K.; Azarkh, M.; Bagryanskaya, E.G.; Drescher, M.; Endeward, B.; Freed, J.H.; Galazzo, L.; et al. Benchmark Test and Guidelines for DEER/PELDOR Experiments on Nitroxide-Labeled Biomolecules. *J. Am. Chem. Soc.* **2021**, *143*, 17875–17890. [CrossRef] [PubMed]
66. Worswick, S.G.; Spencer, J.A.; Jeschke, G.; Kuprov, I. Deep Neural Network Processing of DEER Data. *Sci. Adv.* **2018**, *4*, eaat5218. [CrossRef] [PubMed]
67. Fábregas Ibáñez, L.; Jeschke, G.; Stoll, S. DeerLab: A Comprehensive Software Package for Analyzing Dipolar Electron Paramagnetic Resonance Spectroscopy Data. *Magn. Reson.* **2020**, *1*, 209–224. [CrossRef] [PubMed]
68. Hagelueken, G.; Abdullin, D.; Schiemann, O. mtsslSuite: Probing Biomolecular Conformation by Spin-Labeling Studies. *Methods Enzymol.* **2015**, *563*, 595–622. [PubMed]
69. *The PyMOL Molecular Graphics System*; Version 1.2r3pre; Schrödinger, LLC: New York, NY, USA, 2011.
70. Ackermann, K.; Chapman, A.; Bode, B.E. A Comparison of Cysteine-Conjugated Nitroxide Spin Labels for Pulse Dipolar EPR Spectroscopy (Dataset), University of St Andrews Research Portal. Available online: https://doi.org/10.17630/70739623-1bb5-40e6-b326-5fff1e22c289. [CrossRef]

Article

Rationalisation of Patterns of Competing Reactivity by X-ray Structure Determination: Reaction of Isomeric (Benzyloxythienyl)oxazolines with a Base

R. Alan Aitken *, Andrew D. Harper and Alexandra M. Z. Slawin

EaStCHEM School of Chemistry, University of St Andrews, North Haugh, St Andrews KY16 9ST, Fife, UK; andydharper@hotmail.com (A.D.H.); amzs@st-and.ac.uk (A.M.Z.S.)
* Correspondence: raa@st-and.ac.uk; Tel.: +44-1334-463865

Abstract: Three isomeric (benzyloxythienyl)oxazolines **9**, **11** and **13** have been prepared and are found, upon treatment with a strong base, to undergo either Wittig rearrangement or intramolecular attack of the benzylic anion on the oxazoline function to give products derived from cleavage of the initially formed 3-aminothienofuran products. This pattern of reactivity is directly linked to the distance between the two reactive groups as determined by X-ray diffraction, with the greatest distance in **11** leading to exclusive Wittig rearrangement, the shortest distance in **13** giving exclusively cyclisation-derived products, and the intermediate distance in **9** leading to both processes being observed. The corresponding N-butyl amides were also obtained in two cases and one of these undergoes efficient Wittig rearrangement leading to a thieno[2,3-c]pyrrolone product.

Keywords: oxazoline; Wittig rearrangement; thiophene; thieno[2,3-c]pyrrolone; X-ray structure

1. Introduction

Some time ago we described the reaction of 2-(2-benzyloxyphenyl)oxazoline **1** with a strong base to give either the 3-aminobenzofuran product **2** resulting from intramolecular nucleophilic ring-opening of the oxazoline by the benzyl anion, or the oxazoline **3** in which the benzyloxy group has undergone a Wittig rearrangement (Scheme 1) [1]. Heterocycle formation by cyclisation of an aryloxy carbanion onto an *ortho* functional group is rather uncommon but formation of 3-aminobenzofurans by the so-called Gewald reaction of benzonitriles provides one example [2]. Similarly, although the Wittig rearrangement has been known for almost a century [3], it is not commonly used in synthesis and a recent review shows rather limited developments over the last 20 years [4]. While the aminobenzofuran formation could be optimised by using 3.3 equiv. of Schlosser's base (*n*-BuLi/*t*-BuOK) and applied to a number of substituted examples [1], the Wittig rearrangement process was not so favourable and, under optimal conditions of 2.2 equiv. butyllithium (*n*-BuLi) in THF, an isolated yield of just 29% was obtained. As will shortly be reported elsewhere, the N-butyl amide group is a more effective promoter of the Wittig rearrangement and treating compound **4** with 3.3 equiv. *n*-BuLi in THF gives almost entirely the rearranged product **5**, conveniently isolated as the phthalide **6** after acid-mediated cyclisation in 90% yield. However in the latter study, simply changing to the N,N-diisopropyl amide **7** and treating with 2.2 equiv. *n*-BuLi in toluene again resulted in cyclisation to give the 3-aminobenzofuran **8**. It is clear from these studies that there is a delicate balance between Wittig rearrangement of the benzyloxy group without affecting the adjacent activating group, and interaction of the two groups with the formation of a furan ring.

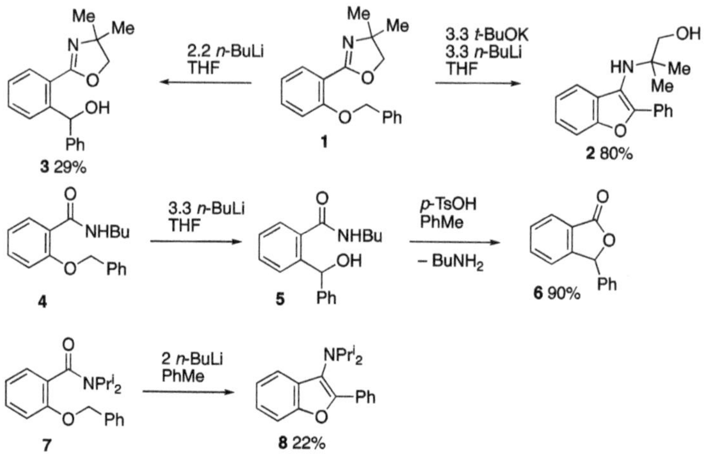

Scheme 1. Competition between Wittig rearrangement and cyclisation in benzene-based systems.

In contrast to the symmetrical benzene ring, the different adjacent positions on a heterocycle such as thiophene are not equivalent and so a more interesting pattern of reactivity can be expected, which would also lead to some unusual and novel heterocyclic products. In this paper, we report the synthesis, characterisation and reactivity upon treatment with a strong base, of the three isomeric (benzyloxythienyl)oxazolines **9**, **11** and **13** (Scheme 2) as well as the corresponding (benzyloxythienyl)-*N*-butylcarboxamides **10**, **12** and **14**. As well as examining the pattern of reactivity we were interested to discover whether there was any correlation between this and the distance between the two adjacent groups as determined by X-ray diffraction.

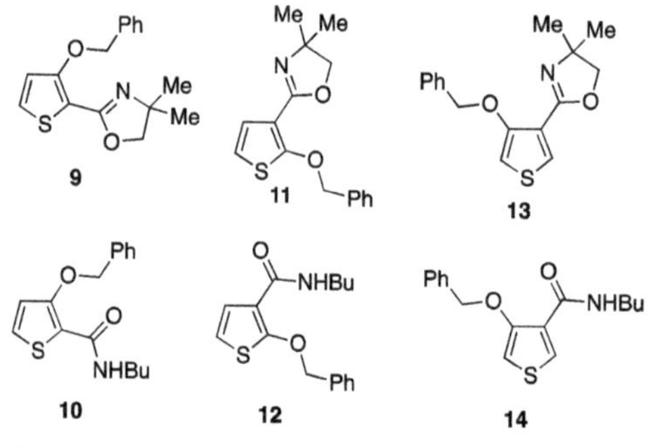

Scheme 2. The six isomeric thiophene compounds targeted for reactivity studies.

2. Results

2.1. Synthesis of 3-Benzyloxy-2-thienyl Systems **9** and **10**

O-Benzylation of the commercially available methyl ester **15** followed by ester hydrolysis of **16** gave the carboxylic acid **17** (Scheme 3). This was readily converted into the corresponding acid chloride which was reacted immediately with 2-amino-2-methylpropan-1-ol to give the hydroxy amide **18**, which was cyclised using thionyl chloride to give the target oxazoline **9** in good overall yield. Alternatively, treating acid **17** with thionyl chloride followed by an excess of butylamine gave the target amide **10** also in high yield.

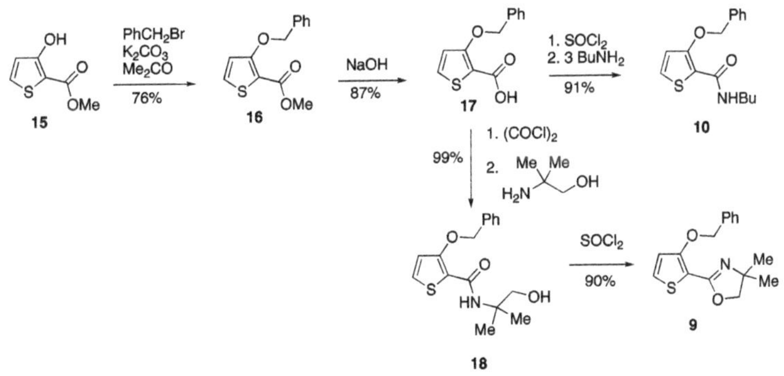

Scheme 3. Synthesis of 3-benzyloxy-2-thienyl compounds **9** and **10**.

2.2. Reaction 3-Benzyloxy-2-thienyl Systems 9 and 10 with Base

When compound **9** was subjected to the same conditions used to convert **1** into **2**, a mixture of two products was formed which were separated by preparative thin-layer chromatography (TLC) on alumina and identified as the expected Wittig rearrangement product **19**, formed in 8% yield, and a second more major product (43%) which was initially thought to be the expected cyclisation product **21** (Scheme 4). However certain features of the spectra were not consistent with this, notably the non-equivalence of the two methyl groups and CH$_2$O hydrogens which suggested the presence of a stereogenic centre. After further evidence from ^{13}C and 2D NMR studies suggested the presence of a 2-alkylidenethiophen-3(2H)-one structure, and the HRMS result showed the presence of an extra oxygen atom as compared to **21**, the actual structure was finally confirmed as the unusual morpholine-containing thiophenone **20** by X-ray diffraction.

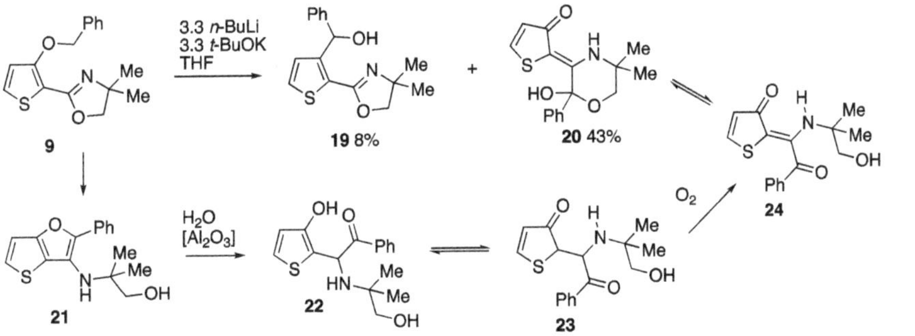

Scheme 4. Reaction of oxazoline **9** with n-BuLi/t-BuOK.

The molecular structure of **20** features two independent molecules in the unit cell in addition to one molecule each of CH$_2$Cl$_2$ and acetone. The two molecules are actually enantiomers and they both take up half chair conformations with the ring oxygen out of the plane, the NH in the plane and one methyl axial and one equatorial (Figure 1). Where they differ is that one has phenyl axial and OH equatorial while for the other it is the opposite way round.

In the crystal, there is hydrogen bonding both intramolecularly between the NH and C=O and intermolecularly between C=O and OH; in terms of the Etter–Bernstein graph-set descriptors [5] C$^1_1$(7) [S(6)]. The intermolecular interaction involves the two enantiomeric molecules alternating and the pattern is shown schematically in Figure 2 with parameters in Table 1.

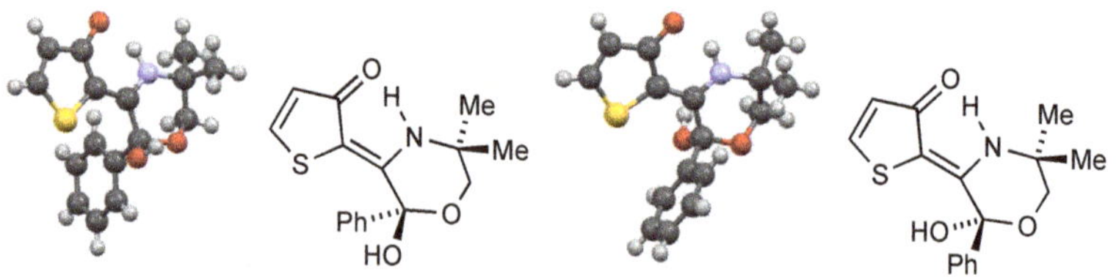

Figure 1. Conformations of the two enantiomeric molecules of **20** in the crystal.

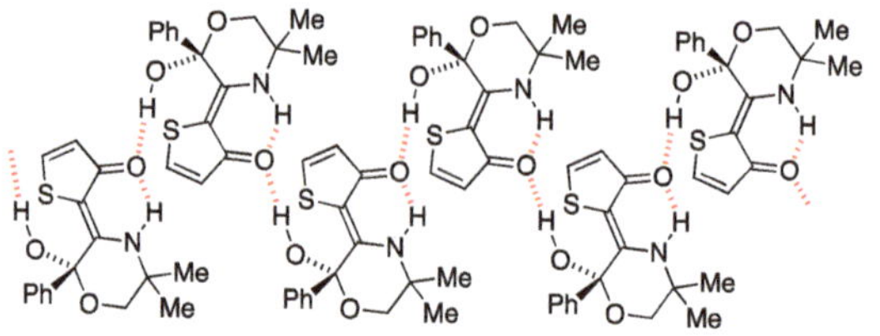

Figure 2. Schematic representation of the hydrogen bonding pattern for **20**.

Table 1. Hydrogen bonding parameters for **20** (Å, °).

D—H...A	D—H	H...A	D...A	D—H...A
O(7)–H(7)...O(23)	0.98(7)	1.61(7)	2.592(6)	177(7)
N(9)–H(9)...O(3)	0.98(6)	1.74(7)	2.649(6)	153(7)
O(27)–H(27)...O(3)	0.98(5)	1.72(5)	2.694(6)	173(6)
N(29)–H(29)...O(23)	0.98(5)	1.70(5)	2.584(6)	147(5)

Once the structure of **20** was clear, its formation could be rationalised as shown in Scheme 4 by initial cyclisation of **9** to give the thieno[3,2-*b*]furan product **21** and hydrolysis of this on the alumina with opening of the furan ring to give **22**, which in its thiophen-3-one tautomeric form can be oxidised by air to form the favourable fully conjugated ene-dione structure **24**, which then cyclises to form the cyclic hemiketal **20**. It should be noted that compound **20** was found to be quite unstable and although it was isolated in small amount with sufficient purity for identification, further attempts at purification resulted in decomposition. As described in a recent review [6], thiophene-based *o*-quinomethane analogues have a rich and varied chemistry, however, the formation of such a structure by hydrolysis then oxidation of a thieno[3,2-*b*]furan is unprecedented.

We now turned to the *N*-butyl amide **10** and found a much simpler pattern of reactivity. Treatment of **10** with *n*-BuLi under the standard conditions developed in the benzene series, resulted in exclusive Wittig rearrangement to give secondary alcohol **25** which could be characterised spectroscopically but was converted for isolation into the stable thieno[2,3-*c*]pyrrolone product **26** by treatment with *p*-toluenesulfonic acid in boiling toluene (Scheme 5) [7]. This method was also used to obtain stable cyclic products in the benzene-based systems, however, note that while **5** and analogues cyclise to lactones **6** with loss of butylamine, here we have a loss of water to form the lactam in excellent yield. The synthesis and chemistry of thieno[*c*]pyrrolones and their dihydro analogues has been recently reviewed [8].

Since such fused-ring heterocycles are rather uncommon we took the chance to determine the X-ray structure of compound **26** (Scheme 5). Only one previous X-ray structure of a compound with this ring system appears to have been published, that of the compound with butyl replaced by quinolin-8-yl [9], but the molecular dimensions are very similar. Interestingly the *tert*-butyl amide **27** isomeric with **25** has been prepared by *ortho*-directed metalation of *N*-*tert*-butylthiophene-2-carboxamide with *n*-BuLi followed by reaction with benzaldehyde [10].

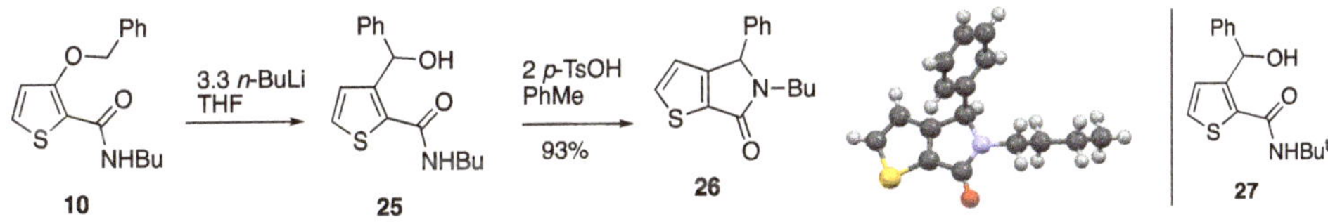

Scheme 5. Reaction of 10 with *n*-BuLi to give 25 and 26 and structure of an analogous product 27.

2.3. Synthesis of 2-Benzyloxy-3-thienyl Systems 11 and 12

Entry to this system was gained by starting with the 3-thienyloxazoline 28 and introducing oxygen functionality at the 2-position by lithiation and treatment with bis(trimethylsilyl) peroxide (Scheme 6). As we have described in detail elsewhere [11], the resulting product had the 3-(oxazolidin-2-ylidene)thiophen-2-one structure 29 which exhibited an interesting and varied pattern of reactivity. However, for the present purpose, it could be cleanly *O*-benzylated in moderate yield to give the desired compound 11.

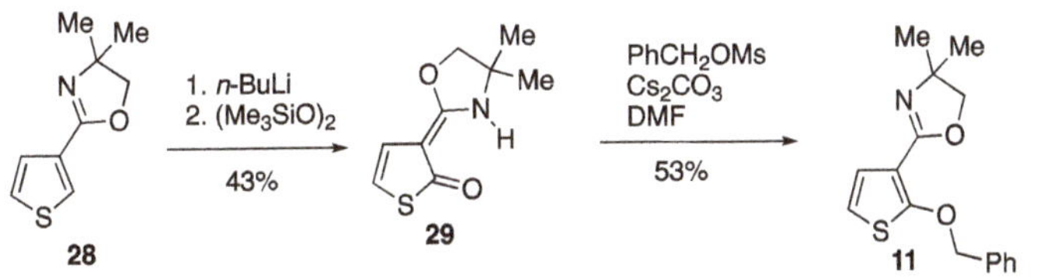

Scheme 6. Synthesis of oxazoline 11.

As described below, attempted application of a similar method to the formation of the amide 12 failed since lithiation of the corresponding *N*-butyl amide 32 followed by treatment with bis(trimethylsilyl) peroxide instead gave the silyl compound 33.

2.4. Reaction of 2-Benzyloxy-3-thienyl Systems 11 and 12 with Base

Treatment of oxazoline 11 with *n*-BuLi under the standard conditions developed for ring closure of 1 to give 2 gave exclusively the Wittig rearrangement product 30 in good yield (Scheme 7).

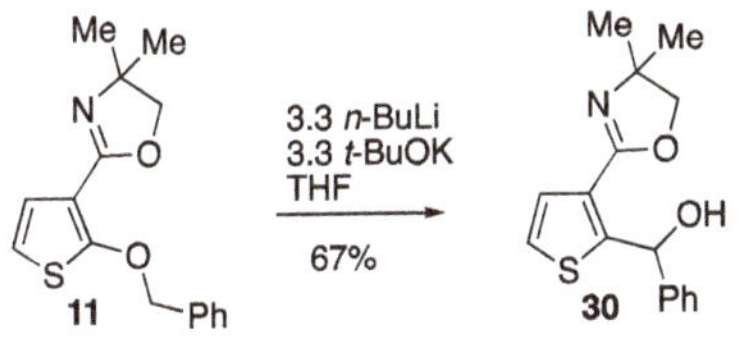

Scheme 7. Wittig rearrangement of oxazoline 11.

Although attempts to prepare the amide 12 by lithiation and bis(trimethylsilyl) peroxide treatment failed, instead giving the new silane 33 (Scheme 8), the expected Wittig rearrangement product from 12, compound 34, was prepared by lithiation and benzaldehyde treatment of 32. It was not isolated, however, the reaction product was directly treated with *p*-toluenesulfonic acid giving the thieno[2,3-*c*]pyrrolone [8] product 35 isomeric with 26 together with a low yield of the oxidation product 36. This last product showed extra signals in the ^{13}C NMR spectrum due to amide rotamers (Supplementary Materials).

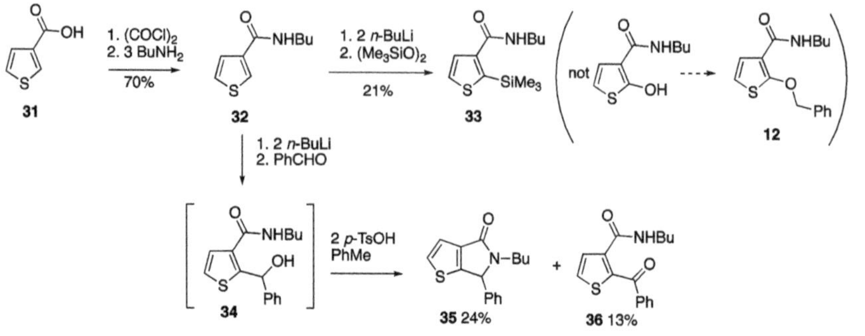

Scheme 8. Formation and cyclisation of secondary alcohol **34**.

2.5. Synthesis of 4-Benzyloxy-3-thienyl Systems 13 and 14

Synthesis of the required compounds in this series was more challenging since suitably substituted thiophene starting materials are not commercially available. Instead, we had to resort to a ring-synthesis of a thiophene with the desired functionality in place. This started from the sulfide-containing diester **37** prepared by conjugate addition of methyl thioglycolate to methyl acrylate [12], which underwent base-induced ring closure [13] to give compound **38** in low yield (Scheme 9). Aromatisation of this was achieved using sulfuryl chloride [14] to give the thiophene ester **39**. Conversion of this into the required benzyl ether **41** proved to be more difficult than expected. Simple alkylation using benzyl bromide and either potassium carbonate or sodium hydride resulted in polymerisation and reaction with phenyldiazomethane [15] also failed.

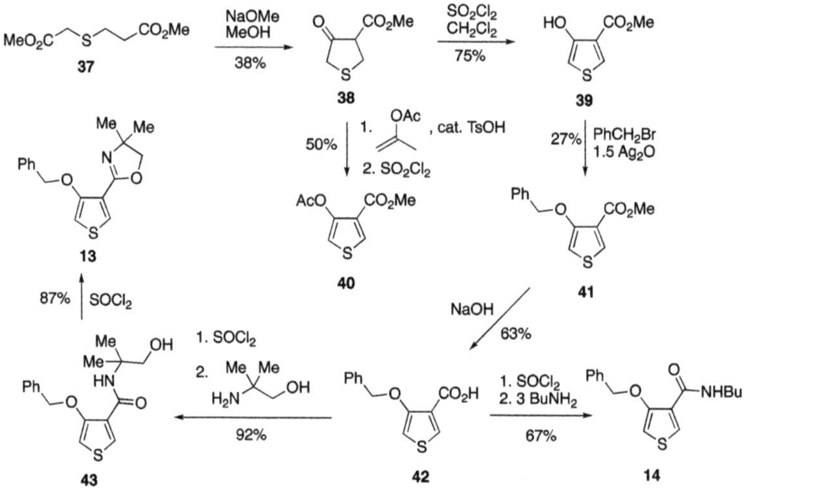

Scheme 9. Synthesis of 4-benyloxy-3-thienyl compounds **13** and **14**.

Following a literature report that reaction of the 4-acetoxy compound **40** with ethanol and sulfuric acid gave the 4-ethoxy compound [15], this compound was prepared by reaction of **38** with isopropenyl acetate followed by sulfuryl chloride, but the treatment of this with benzyl alcohol and sulfuric acid again resulted in polymerisation. Access to **41** was finally achieved, albeit in low yield, by resorting to treatment with benzyl bromide in the presence of silver oxide in a process reminiscent of the Purdie–Irvine method for methylation of sugars developed in St Andrews over 100 years ago [16]. With the key intermediate **41** in hand, the remaining synthetic steps proceeded without incident: hydrolysis gave the acid **42** which was converted into its acid chloride and then reacted either with 2-amino-2-methylpropan-1-ol to give amide **43** which was cyclised with thionyl chloride to oxazoline **13**, or with butylamine to directly afford the amide **14**.

2.6. Reaction of 4-Benzyloxy-3-thienyl Systems **13** and **14** with Base

Treatment of oxazoline **13** with 3.3 equiv. of Schlosser's base gave largely unreacted starting material, however, increasing this to 4.4 equiv. did give a reaction and after chromatographic purification, the 4-benzoyloxy-3-thienyl amide **45** was isolated in moderate yield (Scheme 10). This is evidently formed by air oxidation of the expected cyclisation product, the 3-aminothieno[3,4-b]furan **44**. As shown in our previous work [1], such ring-fused 3-aminofuran products are susceptible to oxidative ring-cleavage.

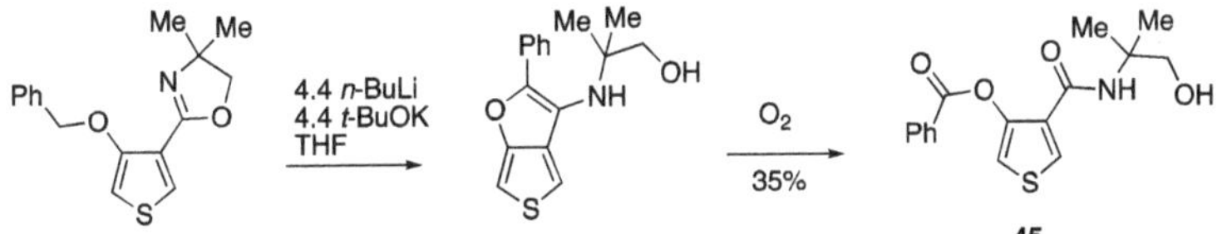

Scheme 10. Base-induced cyclisation and oxidative ring opening of **13**.

The reaction of the corresponding N-butyl amide **14** with n-BuLi under the conditions required for Wittig rearrangement gave largely unreacted starting material and the only new products isolated in low yield after chromatographic purification (Scheme 11) were the 2,3,4-trisubstituted thiophene **46** together with the debenzylated compound **47** which was found to exist in solution as a mixture with the thiophen-3(2H)-one tautomer **47a** (see Section 3). It seems likely that the products have resulted from the intermolecular reaction between two carbanions derived from **14** but in view of their very low yield this process was not investigated further. Products **45**, **46** and **47** which were isolated in low amounts following one or two stages of chromatography were found to decompose upon attempted further purification.

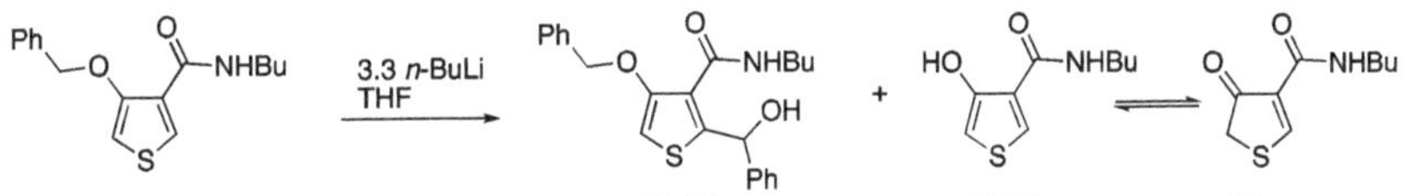

Scheme 11. Reaction of amide **14** with strong base.

To summarise the reactivity of the isomeric systems, oxazoline **11** underwent exclusive Wittig rearrangement and oxazoline **13** gave products derived from cyclisation, while for **9** Wittig rearrangement was observed as a minor process with the major product derived from cyclisation. The N-butyl amides gave a less complete picture with **10** undergoing exclusive Wittig rearrangement in high yield, **12** not being available for investigation (although its expected Wittig rearrangement product was obtained by other means), and **14** remaining largely unreacted under the conditions. In the case of the three isomeric oxazolines, each compound was obtained as good quality crystals suitable for X-ray diffraction and so it was decided to determine their molecular structures to examine whether there might be a direct link between the distance between the benzyloxy and oxazoline groups and the observed reactivity. All three compounds gave structures with the monoclinic $P2_1/c$ space group and these are shown in a similar orientation in Figure 3.

For the intramolecular cyclisation to compete with Wittig rearrangement, the key distance is that between benzyloxy carbanionic carbon and C-2 of the oxazoline. Since the benzyloxy groups have rotated to place this carbon pointing away from the oxazoline in each case, the benzyloxy oxygen is taken as a reference point and it can be seen that the molecular geometry correlates well with the observed reactivity. Thus, for **11**, the benzyloxy group is too far away (3.046(1) Å) for cyclisation and we observe exclusively a Wittig rearrangement, for **13** the benzyloxy group is much closer (2.945(1) Å) and only products derived from cyclisation are observed, while in **9** we have an intermediate situation (3.006(3) Å) and mainly cyclisation-derived products are observed but with a little Wittig rearrangement.

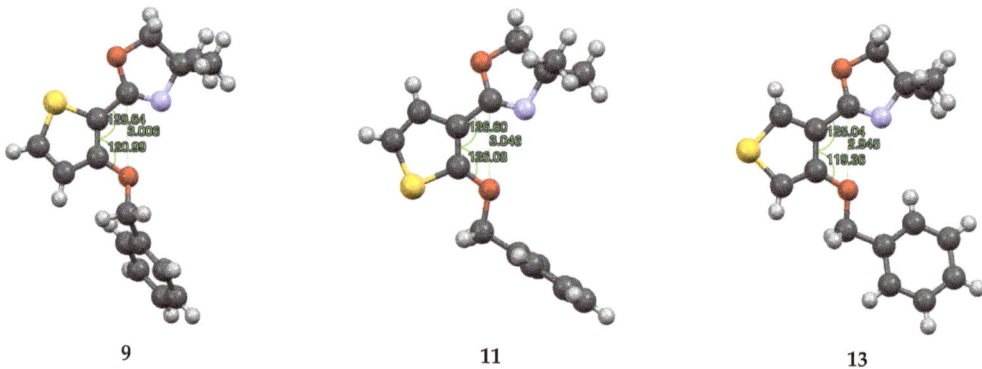

Figure 3. Molecular structures of **9, 11** and **13** showing angles (°) and benzyloxy O to oxazoline C(2) distance (Å).

3. Experimental

3.1. General Experimental Details

NMR spectra were recorded on solutions in CDCl$_3$ unless otherwise stated using Bruker instruments and chemical shifts are given in ppm to high frequency from Me$_4$Si with coupling constants J in Hz. IR spectra were recorded using the ATR technique on a Shimadzu IRAffinity 1S instrument. The ionisation method used for high-resolution mass spectra is noted in each case. Column chromatography was carried out using silica gel of 40–63 mm particle size and preparative TLC was carried out using 1.0 mm layers of Merck alumina 60G containing 0.5% Woelm fluorescent green indicator on glass plates. Melting points were recorded on a Gallenkamp 50W melting point apparatus or a Reichert hot-stage microscope.

3.2. Preparation and Reactions of 3-Benzyloxy-2-thienyl Systems

3.2.1. Methyl 3-(Benzyloxy)thiophene-2-carboxylate 16

A literature procedure [17] was modified as follows: benzyl bromide (11.9 cm^3, 17.11 g, 0.100 mol) was added to a stirred mixture of methyl 3-hydroxythiophene-2-carboxylate **15** (15.85 g, 0.100 mol) and potassium carbonate (27.60 g, 0.200 mol) in acetone (50 cm^3) and the reaction mixture was heated at reflux for 18 h. After cooling to rt, the inorganic salts were removed by filtration and the filtrate was concentrated *in vacuo*. The residue was dissolved in CH$_2$Cl$_2$ (150 cm^3) and washed with water (100 cm^3) before being dried and evaporated. The crude residue was recrystallised (aq. MeOH) to give **16** (18.94 g, 76%) as pale yellow crystals; mp 69–72 °C; (lit. [17] 66–67 °C); δ_H (500 MHz) 7.46–7.44 (2H, m, Ph), 7.39–7.35 (2H, m, Ph), 7.35 (1H, d, J 5.5, 5-H), 7.32–7.29 (1H, m, Ph), 6.82 (1H, d, J 5.5, 4-H), 5.24 (2H, s, CH$_2$) and 3.85 (3H, s, CH$_3$); δ_C (125 MHz) 162.1 (C), 160.8 (C), 136.4 (C), 130.4 (CH), 128.6 (2CH), 127.9 (CH), 126.8 (2CH), 117.5 (CH), 110.4 (C), 73.2 (CH$_2$) and 51.6 (CH$_3$). The ^{1}H NMR spectral data were in accordance with those previously reported [17]. ^{13}C NMR data are reported for the first time.

3.2.2. 3-(Benzyloxy)thiophene-2-carboxylic Acid 17

Following a literature procedure [17], a mixture of methyl 3-(benzyloxy)thiophene-2-carboxylate **16** (18.40 g, 74.1 mmol) and sodium hydroxide (5.99 g, 0.150 mol) in water (150 cm^3) was heated at reflux for 2.5 h. After cooling to rt, the aqueous layer was washed with CH$_2$Cl$_2$ (2 × 50 cm^3) before being acidified to pH 1 by the addition of 2 M HCl. The resultant suspension was extracted with CH$_2$Cl$_2$ (2 × 100 cm^3) and the combined organic layers were dried and evaporated. The crude residue was recrystallised (aq. MeOH) to give **17** (15.07 g, 87%) as tan-coloured crystals; mp 122–125 °C; (lit. [17] 125–126 °C); δ_H (400 MHz, CD$_3$SOCD$_3$) 12.49 (1H, br s, CO$_2$H), 7.74 (1H, d, J 5.6, 5-H), 7.48–7.45 (2H, m, Ph), 7.41–7.37 (2H, m, Ph), 7.35–7.30 (1H, m, Ph), 7.14 (1H, d, J 5.6, 4-H) and 5.25 (2H, s, CH$_2$); δ_C (100 MHz, CD$_3$SOCD$_3$) 162.4 (C), 160.1 (C), 136.8 (C), 131.2 (CH), 128.4 (2CH), 127.9 (CH), 127.3 (2CH), 118.5 (Ar CH), 110.5 (C) and 72.4 (CH$_2$). The ^{1}H NMR spectral data were in accordance with those previously reported [17]. ^{13}C NMR data are reported for the first time.

3.2.3. 3-(Benzyloxy)-N-(1-hydroxy-2-methylpropan-2-yl)thiophene-2-carboxamide 18

Oxalyl chloride (3.0 cm^3, 4.50 g, 35.5 mmol) was added to a suspension of 3-(benzyloxy)thiophene-2-carboxylic acid **17** (4.00 g, 17.1 mmol) and in Et$_2$O (25 cm^3) and the mixture was stirred for 18 h. Evaporation gave 3-(benzyloxy)thiophene-2-carbonyl chloride as a brown oil which was used without further purification.

A solution of 3-(benzyloxy)thiophene-2-carbonyl chloride (assuming 17.1 mmol) in CH$_2$Cl$_2$ (40 cm^3) was added dropwise to a solution of 2-amino-2-methylpropan-1-ol (3.10 g, 34.8 mmol) in CH$_2$Cl$_2$ (40 cm^3) stirred at 0 °C. After the addition, the mixture was allowed to warm to rt and stirred for 18 h before being poured into water. The organic layer was separated and the aqueous layer was extracted with CH$_2$Cl$_2$ (2 × 20 cm^3). The combined organic layers were washed successively with 2M HCl, 2M NaOH and water before being dried and evaporated to give **18** (5.14 g, 99%) as a pale yellow solid which was used without further purification; mp 112–115 °C; ν_{max}/cm^{-1} 3358, 3065, 2968, 1626, 1531, 1425, 1240, 1057, 777, 704 and 600; δ_H (500 MHz) 7.43–7.40 (6H, m, NH and Ph), 7.41 (1H, d, J 5.5, 5-H), 6.93 (1H, d, J 5.5, 4-H), 5.18 (2H, s, OCH$_2$Ph), 3.58 (2H, s, CH$_2$OH) and 1.19 (6H, s, CH$_3$); δ_C (125 MHz) 162.5 (C), 155.3 (C), 135.2 (C), 129.2 (CH), 129.0 (CH), 128.8 (2CH), 128.0 (2CH), 117.6 (C), 116.2 (CH), 74.1 (CH$_2$), 70.9 (CH$_2$), 56.3 (CMe$_2$) and 24.8 (2CH$_3$); HRMS (NSI$^+$): found 306.1150. C$_{16}$H$_{20}$NO$_3$S (M + H) requires 306.1158.

3.2.4. 2-(3-(Benzyloxy)thiophen-2-yl)-4,4-dimethyl-4,5-dihydrooxazole 9

Thionyl chloride (1.4 cm^3, 2.28 g, 19.2 mmol) was added to a solution of 3-(benzyloxy)-N-(1-hydroxy-2-methylpropan-2-yl)thiophene-2-carboxamide **18** (4.71 g, 15.4 mmol) in CH$_2$Cl$_2$ (50 cm^3) and the mixture was stirred at room temperature for 18 h. The mixture was washed with 2M NaOH and water before being dried and evaporated to give **9** (4.01 g, 90%) as a pale brown oil which solidified on standing as a tan-coloured solid; mp 65–68 °C; ν_{max}/cm^{-1} 3080, 2965, 1632, 1545, 1260, 1231, 1200, 1069, 1026, 766 and 745; δ_H (500 MHz) 7.45–7.42 (2H, m, Ph), 7.37–7.33 (2H, m, Ph), 7.31–7.27 (1H, m, Ph), 7.24 (1H, d, J 5.5, 5-H), 6.78 (1H, d, J 5.5, 4-H), 5.24 (2H, s, OCH$_2$Ph), 4.09 (2H, s, OCH$_2$) and 1.38 (6H, s, CH$_3$) δ_C (125 MHz) 157.6 (C), 157.3 (C), 136.8 (C), 128.4 (2CH), 127.8 (2CH), 126.9 (2CH), 118.1 (CH), 108.9 (C), 79.2 (CH$_2$), 73.4 (CH$_2$), 67.1 (CMe$_2$) and 28.3 (2CH$_3$); HRMS (ESI$^+$): found 288.1047. C$_{16}$H$_{18}$NO$_2$S (M + H) requires 288.1053.

3.2.5. 3-(Benzyloxy)-N-butylthiophene-2-carboxamide 10

Thionyl chloride (2.5 cm^3, 4.08 g, 34.3 mmol) was added to a suspension of 3-(benzyloxy)thiophene-2-carboxylic acid **17** (4.00 g, 17.1 mmol) in toluene (30 cm^3) and the mixture was heated under reflux for 3 h. After cooling to room temperature, the mixture was evaporated to give 3-(benzyloxy)thiophene-2-carbonyl chloride as a brown oil which was used without further purification.

A solution of 3-(benzyloxy)thiophene-2-carbonyl chloride (assuming 17.1 mmol) in toluene (30 cm^3) was added dropwise to a solution of n-butylamine (5.1 cm^3, 3.77 g, 51.6 mmol) in toluene (10 cm^3) stirred at 0 °C. Once the addition was complete, the reaction mixture was allowed to warm to room temperature over 1 h before being poured into water. The organic layer was separated and washed with 2M NaOH and brine, dried and evaporated to give, after purification by column chromatography (SiO$_2$, Et$_2$O/hexane 7:3), at R$_f$ 0.65, **10** (4.49 g, 91%) as an orange oil; ν_{max}/cm^{-1} 3364, 2961, 1628, 1558, 1435, 1364,

1310, 1074, 976, 773 and 606; δ_H (500 MHz) 7.43–7.38 (5H, m, Ph), 7.36 (1H, d, J 5.5, 5-H), 7.19 (1H, br s, NH), 6.89 (1H, d, J 5.5, 4-H), 5.19 (2H, s, OCH_2), 3.36 (2H, td, J 7.0, 5.5, NCH_2), 1.47–1.41 (2H, m, NCH$_2$$CH_2$), 1.28–1.20 (2H, m, CH_2CH$_3$) and 0.85 (3H, t, J 7.5, CH$_3$); δ_C (125 MHz) 161.7 (C), 154.9 (C), 135.6 (C), 128.84 (2CH), 128.76 (CH), 128.5 (CH), 127.7 (2CH), 118.1 (C), 116.2 (CH), 73.9 (OCH$_2$), 38.9 (NCH$_2$), 31.5 (CH$_2$), 20.0 (CH$_2$) and 13.7 (CH$_3$); HRMS (ESI$^+$): found 312.1017. $C_{16}H_{19}NaNO_2S$ (M + Na) requires 312.1029.

3.2.6. (2-(4,4-Dimethyl-4,5-dihydrooxazol-2-yl)thiophen-3-yl)(phenyl)methanol 19 and (E)-2-(2-Hydroxy-5,5-dimethyl-2-phenylmorpholin-3-ylidene)thiophen-3(2H)-one 20

Under a nitrogen atmosphere, n-butyllithium (2.5 M in hexanes, 0.66 cm^3, 1.65 mmol) was added to a stirred mixture of 2-(3-(benzyloxy)thiophen-2-yl)-4,4-dimethyl-4,5-dihydrooxazole 9 (0.1440 g, 0.50 mmol) and potassium *tert*-butoxide (0.1850 g, 1.65 mmol) in dry THF (5 cm^3). The mixture was stirred at rt for 2 h before being quenched by the addition of saturated aq. NH$_4$Cl and extracted with Et$_2$O (3 × 10 cm^3). The combined extracts were dried and evaporated to give, after purification by preparative TLC (Al$_2$O$_3$, Et$_2$O/hexane 7:3), at R$_f$ 0.65, 19 (12 mg, 8%) as an orange oil; δ_H (400 MHz) 7.42–7.38 (2H, m, Ph), 7.34–7.30 (3H, m, ArH and Ph), 7.28–7.23 (1H, m, Ph), 6.71 (1H, d, J 5.2, ArH), 6.02 (1H, s, CHOH), 4.11 and 4.09 (2H, AB pattern, J 8.2, CH$_2$), 1.40 (3H, s, CH$_3$) and 1.27 (3H, s, CH$_3$); δ_C (125 MHz) 158.6 (C), 150.4 (C), 142.8 (C), 130.1 (CH), 128.2 (CH), 128.0 (2CH), 127.1 (CH), 126.5 (2CH), 124.7 (C), 79.5 (CH$_2$), 70.9 (CHOH), 68.1 (*C*Me$_2$), 28.3 (CH$_3$) and 28.1 (CH$_3$). The ^{1}H NMR spectral data were consistent with those previously reported [18]. ^{13}C NMR data are reported for the first time.

This was followed by a second fraction, at R$_f$ 0.15, to give 20 (64.5 mg, 42%) in slightly impure form as brown crystals; mp 103–105 °C; ν_{max}/cm^{-1} 1582, 1537, 1449, 1317, 1260, 1221, 1067, 768, 698 and 669; δ_H (400 MHz, CD$_3$COCD$_3$) 7.65–7.62 (2H, m, Ph), 7.52 (1H, d, J 5.6, 5-H), 7.40–7.35 (3H, m, Ph), 6.32 (1H, d, J 5.6, 4-H), 4.12 and 3.69 (2H, AB pattern, J 11.6, CH$_2$), 3.06 (2H, br s, OH and NH), 1.53 (3H, s, CH$_3$) and 1.39 (3H, s, CH$_3$); δ_C (125 MHz, CD$_3$COCD$_3$) 182.1 (C=O), 162.7 (C)), 141.5 (C), 138.5 (CH), 129.5 (CH), 128.6 (2CH), 127.6 (2CH), 122.1 (CH), 103.5 (C), 95.0 (C), 68.3 (CH$_2$), 51.4 (*C*Me$_2$), 26.8 (CH$_3$) and 26.5 (CH$_3$); HRMS (NSI$^+$): found 304.1004. $C_{16}H_{18}NO_3S$ (M + H) requires 304.1002.

3.2.7. N-Butyl-3-(hydroxy(phenyl)methyl)thiophene-2-carboxamide 25 and 5-Butyl-4-phenyl-4,5-dihydro-6H-thieno[2,3-c]pyrrol-6-one 26

Under a nitrogen atmosphere, n-butyllithium (2.5 M in hexanes, 6.6 cm^3, 16.5 mmol) was added dropwise to a stirred solution of 3-(benzyloxy)-N-butylthiophene-2-carboxamide 10 (1.45 g, 5.01 mmol) in dry THF (50 cm^3). After stirring at room temperature for 2 h, the reaction mixture was quenched by the addition of saturated aq. NH$_4$Cl and extracted with Et$_2$O (3 × 30 cm^3). The combined organic extracts were washed with NaOH and water before being dried and evaporated to give 25 as a pale brown oil which was used without further purification; ν_{max}/cm^{-1} 3256, 3086, 2957, 2930, 1612, 1545, 1450, 1302, 1026, 698 and 669; δ_H (400 MHz) 7.37–7.29 (4H, m, Ph), 7.27–7.23 (1H, m, Ph), 7.21 (1H, d, J 5.0, 5-H), 6.96 (1H, t, J 5.6, NH), 6.71 (1H, d, J 5.0, 4-H), 6.02 (1H, s, CHOH), 5.87 (1H, br s, OH), 3.30 (2H, td, J 7.2, 5.6, NCH$_2$), 1.52–1.45 (2H, m, NCH$_2$$CH_2$), 1.35–1.26 (2H, m, CH_2CH$_3$) and 0.89 (3H, t, J 7.2, CH$_3$); δ_C (75 MHz) 163.1 (C=O), 147.8 (C), 142.3 (C), 133.4 (C), 130.4 (CH), 128.2 (2CH), 127.3 (CH), 126.7 (CH), 126.2 (2CH), 70.9 (CHOH), 39.9 (NCH$_2$), 31.3 (CH$_2$), 20.0 (CH$_2$) and 13.7 (CH$_3$); HRMS (ESI$^+$): found 312.1023. $C_{16}H_{19}NaNO_2S$ (M + Na) requires 312.1029.

A mixture of N-butyl-3-(hydroxy(phenyl)methyl)thiophene-2-carboxamide 25 (assuming 5.01 mmol) and p-toluenesulfonic acid monohydrate (1.90 g, 9.99 mmol) in toluene (50 cm^3) was heated at reflux for 1 h. After cooling to room temperature, the reaction mixture was washed with water (50 cm^3), 2 M NaOH (50 cm^3) and brine (50 cm^3) before being dried and evaporated. The crude residue was purified by filtration through a silica plug (Et$_2$O) to give 26 (1.26 g, 93%) as a tan-coloured solid; mp 90–93 °C; ν_{max}/cm^{-1} 2955, 1668, 1441, 1398, 1310, 1069, 781, 743, 698 and 637; δ_H (400 MHz) 7.55 (1H, d, J 4.8, 5-H), 7.38–7.32 (3H, m, Ph), 7.16–7.13 (2H, m, Ph), 6.79 (1H, d, J 4.8, 4-H), 5.39 (1H, s, CHPh), 3.84

(1H, dt, *J* 14.4, 7.8, NCH), 2.86–2.79 (1H, m, NCH), 1.54–1.46 (2H, m, NCH$_2$C*H*$_2$), 1.34–1.24 (2H, m, C*H*$_2$CH$_3$) and 0.88 (3H, t, *J* 7.2, CH$_3$); δ_C (125 MHz) 164.4 (C=O), 155.7 (C), 136.0 (C), 134.8 (C), 134.5 (CH), 129.0 (2CH), 128.6 (CH), 127.3 (2CH), 120.7 (CH), 62.9 (CHPh), 40.3 (NCH$_2$), 30.6 (CH$_2$), 19.9 (CH$_2$) and 13.7 (CH$_3$); HRMS (NSI$^+$): found 272.1103. C$_{16}$H$_{18}$NOS (M + H) requires 272.1104.

3.3. Preparation and Reactions of 2-Benzyloxy-3-thienyl Systems

3.3.1. Attempted Cyclisation of 2-(2-(Benzyloxy)thiophen-3-yl)-4,4-dimethyl-4,5-dihydrooxazole 11

Under a nitrogen atmosphere, *n*-butyllithium (2.5 M in hexanes, 0.66 cm^3, 1.65 mmol) was added to a stirred mixture of 2-(2-(benzyloxy)thiophen-3-yl)-4,4-dimethyl-4,5-dihydrooxazole 11 [11] (0.1437 g, 0.50 mmol) and potassium *tert*-butoxide (0.1875 g, 1.67 mmol) in dry THF (5 cm^3). The mixture was stirred at rt for 2 h before being quenched by the addition of saturated aq. NH$_4$Cl and extracted with Et$_2$O (3 × 10 cm^3). The combined extracts were dried and evaporated to give, after purification by preparative TLC (Al$_2$O$_3$, Et$_2$O/hexane 1:1), at R$_f$ 0.50, (3-(4,4-Dimethyl-4,5-dihydrooxazol-2-yl)thiophen-2-yl)(phenyl)methanol 30 (96.7 mg, 67%) as an orange oil; ν_{max}/cm^{-1} 3177, 2965, 1636, 1535, 1452, 1288, 1194, 1148, 974 and 698; δ_H (500 MHz) 8.04 (1H, br s, OH), 7.49 (2H, d, *J* 7.0, Ph), 7.36–7.28 (4H, m, ArH and Ph), 7.06 (1H, d, *J* 5.0, ArH), 6.10 (1H, s, C*H*OH), 4.09 and 4.06 (2H, AB pattern, *J* 6.8, CH$_2$), 1.38 (3H, s, CH$_3$) and 1.28 (3H, s, CH$_3$); δ_C (125 MHz) 159.6 (C), 153.7 (C), 141.8 (C), 128.6 (CH), 127.9 (2CH), 127.7 (CH), 126.8 (2CH), 125.5 (C), 123.2 (CH), 79.0 (CH$_2$), 69.5 (CHOH), 67.3 (4ry, *C*Me$_2$), 28.5 (CH$_3$) and 28.1 (CH$_3$); HRMS (NSI$^+$): found 288.1052. C$_{16}$H$_{18}$NO$_2$S (M + H) requires 288.1053.

3.3.2. *N*-Butylthiophene-3-carboxamide 32

Oxalyl chloride (2.0 cm^3, 3.00 g, 23.6 mmol) was added to a solution of thiophene-3-carboxylic acid 31 (2.52 g, 19.7 mmol) in CH$_2$Cl$_2$ (30 cm^3) and the mixture was stirred for 18 h. Evaporation gave thiophene-3-carbonyl chloride as a pale-yellow solid which was used immediately without further purification.

A solution of thiophene-3-carbonyl chloride (assuming 19.7 mmol) in toluene (30 cm^3) was added dropwise to a solution of *n*-butylamine (5.8 cm^3, 4.29 g, 58.7 mmol) in toluene (30 cm^3) stirred at 0 °C. Once the addition was complete, the reaction mixture was allowed to warm to rt over 1 h before being poured into water. The organic layer was separated and washed with 2M NaOH and brine, dried and evaporated to give, after recrystallisation (EtOAc/hexane), 32 (2.54 g, 70%) as colourless crystals; mp 66–68 °C; (lit. [19] 53–55 °C); ν_{max}/cm^{-1} 3253, 3085, 2921, 1617, 1555, 1301, 1220, 1127, 881, 831, 741 and 707; δ_H (500 MHz) 7.84 (1H, dd, *J* 3.0, 1.5, ArH), 7.37 (1H, dd, *J* 5.0, 1.5, ArH), 7.33 (1H, dd, *J* 5.0, 3.0, ArH), 6.02 (1H, br s, NH), 3.43 (2H, td, *J* 7.0, 6.0, NCH$_2$), 1.62–1.56 (2H, m, NCH$_2$C*H*$_2$), 1.44–1.37 (2H, m, C*H*$_2$CH$_3$) and 0.95 (3H, t, *J* 7.5, CH$_3$). The ^{1}H NMR spectral data were in accordance with those previously reported [19]. IR data are reported for the first time.

3.3.3. *N*-Butyl-2-(trimethylsilyl)thiophene-3-carboxamide 33

Under a nitrogen atmosphere, *n*-butyllithium (2.5 M in hexane, 5.2 cm^3, 13.0 mmol) was added dropwise to a stirred −78 °C solution of *N*-butylthiophene-3-carboxamide 32 (1.10 g, 6.00 mmol) in dry THF (30 cm^3). After stirring at −78 °C for 5 min, the reaction mixture was allowed to warm to rt for 1 h, before being cooled to −78 °C and treated with bis(trimethylsilyl) peroxide (1.28 g, 7.18 mmol). The reaction mixture was allowed to warm to rt over 18 h before being poured into sat. aq. NH$_4$Cl (100 cm^3) and extracted with Et$_2$O (3 × 50 cm^3). The combined organic extracts were dried and evaporated and the crude residue was purified by column chromatography (SiO$_2$, Et$_2$O/hexane 3:2) to give, at R$_f$ 0.90, 33 (0.32 g, 21%) as tan-coloured crystals; mp 86–89 °C; ν_{max}/cm^{-1} 3285, 2957, 1620, 1558, 1402, 1296, 1240, 1005, 833, 746, 704 and 604; δ_H (500 MHz) 7.50 (1H, d, *J* 5.0, ArH), 7.29 (1H, d, *J* 5.0, ArH), 5.94 (1H, br s, NH), 3.41 (2H, td, *J* 7.0, 6.0, NCH$_2$), 1.61–1.55 (2H, m, NCH$_2$C*H*$_2$), 1.43–1.36 (2H, m, C*H*$_2$CH$_3$), 0.94 (3H, t, *J* 7.5, CH$_2$C*H*$_3$) and

0.39 (9H, s, SiMe$_3$); δ_C (125 MHz) 164.6 (C=O), 145.7 (C), 143.0 (C), 130.2 (CH), 126.9 (CH), 39.6 (NCH$_2$), 31.7 (CH$_2$), 20.1 (CH$_2$), 13.8 (CH$_3$) and 0.0 (SiMe$_3$); HRMS (NSI$^+$): found 256.1184. C$_{12}$H$_{22}$NOSSi (M + H) requires 256.1186.

3.3.4. 5-Butyl-6-phenyl-5,6-dihydro-4H-thieno[2,3-c]pyrrol-4-one 35 and 2-Benzoyl-N-butylthiophene-3-carboxamide 36

Under a nitrogen atmosphere, n-butyllithium (2.5 M in hexane, 4.2 cm^3, 10.5 mmol) was added dropwise to a stirred −78 °C solution of N-butylthiophene-3-carboxamide 32 (0.9158 g, 5.00 mmol) in dry THF (50 cm^3). After stirring at −78 °C for 5 min, the reaction mixture was allowed to warm to rt for 1 h before benzaldehyde (0.57 cm^3, 0.60 g, 5.61 mmol) was added and stirring was continued for 18 h. The reaction mixture was poured into sat. aq. NH$_4$Cl (100 cm^3) and extracted with Et$_2$O (3 × 50 cm^3) and the combined organic layers were dried and evaporated.

The residue was dissolved in toluene (100 cm^3) and treated with p-toluenesulfonic acid monohydrate (1.90 g, 9.99 mmol) before being heated at reflux for 1 h. After cooling to rt, the reaction mixture was washed with water (50 cm^3), 2 M NaOH (50 cm^3) and brine (50 cm^3) before being dried and evaporated. The crude residue was purified by column chromatography (SiO$_2$, Et$_2$O/hexane 3:2) to give, at R$_f$ 0.50, 35 (0.3223 g, 24%) as a pale yellow solid; mp 96–98 °C; ν_{max}/cm^{-1} 2953, 1668, 1454, 1412, 1375, 1308, 1267, 1070, 760, 700 and 575; δ_H (400 MHz) 7.40–7.34 (4H, m, ArH), 7.27 (1H, d, J 4.8, ArH), 7.17–7.13 (2H, m, ArH), 5.51 (1H, s, CHPh), 3.84 (1H, dt, J 14.0, 8.0, NCH), 2.81 (1H, ddd, J 14.0, 7.6, 6.4, NC), 1.53–1.45 (2H, m, NCH$_2$CH$_2$), 1.35–1.24 (2H, m, CH$_2$CH$_3$) and 0.88 (3H, t, J 7.4, CH$_3$); δ_C (125 MHz) 165.1 (C=O), 155.0 (C), 139.8 (C), 136.7 (C), 130.2 (CH), 129.2 (2CH), 129.0 (CH), 127.3 (2CH), 120.1 (CH), 62.9 (CHPh), 40.3 (NCH$_2$), 30.7 (CH$_2$), 20.0 (CH$_2$) and 13.7 (CH$_3$); HRMS (ASAP$^+$): found 272.1115. C$_{16}$H$_{18}$NOS (M + H) requires 272.1104.

This was followed by a second fraction to give, at R$_f$ 0.30, 36 (0.1817 g, 13%) as an orange oil; ν_{max}/cm^{-1} 3277, 2957, 2870, 1630, 1549, 1449, 1406, 1279, 847 and 691; δ_H (400 MHz) 8.65 (1H, br s, NH), 7.84–7.81 (2H, m, Ph), 7.75 (1H, d, J 5.0, ArH), 7.64–7.60 (1H, m, Ph), 7.55 (1H, d, J 5.0, ArH), 7.51–7.46 (2H, m, Ph), 3.33 (2H, td, J 7.2, 5.6, NCH$_2$), 1.56–1.49 (2H, m, NCH$_2$CH$_2$), 1.42–1.33 (2H, m, CH$_2$CH$_3$) and 0.92 (3H, t, J 7.4, CH$_3$); δ_C (100 MHz, signals for major amide rotamer only) 190.6 (COPh), 162.5 (CONHBu), 142.6 (C), 138.8 (C), 138.1 (C), 133.3 (CH), 132.8 (CH), 130.4 (CH), 129.8 (2CH), 128.3 (2CH), 39.7 (NCH$_2$), 31.2 (CH$_2$), 20.2 (CH$_2$) and 13.7 (CH$_3$); HRMS (NSI$^+$): found 288.1052. C$_{16}$H$_{18}$NO$_2$S (M + H) requires 288.1053.

3.4. Preparation and Reactions of 4-Benzyloxy-3-thienyl Systems

3.4.1. Methyl 3-((2-Methoxy-2-oxoethyl)thio)propanoate 37

Following a literature procedure [12], methyl acrylate (18.37 g, 0.213 mol) was added dropwise to a stirred mixture of methyl thioglycolate (21.18 g, 0.200 mol) and piperidine (0.2 cm^3, 0.17 g, 2.02 mmol). Once approximately half of the methyl acrylate had been added, further piperidine (0.2 cm^3, 0.17 g, 2.02 mmol) was added. Once the addition was complete, the reaction mixture was heated at 80 °C for 30 min. After cooling to rt, the reaction mixture was diluted with Et$_2$O (150 cm^3) and washed with water (5 × 50 cm^3) before being dried and evaporated to give 37 (37.12 g, 97%) as a pale yellow oil which was used without further purification; δ_H (400 MHz) 3.75 (3H, s, CH$_3$), 3.71 (3H, s, CH$_3$), 3.26 (2H, s, MeO$_2$CCH$_2$S), 2.92 (2H, t, J 7.2, SCH$_2$CH$_2$CO$_2$Me) and 2.66 (2H, t, J 7.2, SCH$_2$CH$_2$CO$_2$Me); δ_C (100 MHz) 172.0 (C=O), 170.7 (C=O), 52.4 (CH$_3$), 51.8 (CH$_3$), 34.1 (CH$_2$), 33.4 (CH$_2$) and 27.5 (CH$_2$). The ^{1}H NMR spectral data were in accordance with those previously reported [11]. ^{13}C NMR data are reported for the first time.

3.4.2. Methyl 4-Oxotetrahydrothiophene-3-carboxylate 38

Following a literature procedure [13], sodium methoxide was prepared by the addition of sodium (12.51 g, 0.544 mol) in small portions to methanol (90 cm^3). Once the sodium had fully dissolved, a solution of methyl 3-((2-methoxy-2-oxoethyl)thio)propanoate 37

(37.12 g, 0.193 mol) in methanol (30 cm^3) was added dropwise and the reaction mixture was heated at reflux for 1 h. After cooling to rt, the reaction mixture was poured into a mixture of crushed ice (400 g) and conc. HCl (100 cm^3) before being extracted with CH$_2$Cl$_2$ (2 × 300 cm^3). The combined organic layers were washed with sat. aq. NaHCO$_3$ (250 cm^3) before being dried and evaporated. The crude residue was purified by distillation to give 38 (11.69 g, 38%) as a colourless oil which partially crystallised on standing; bp 103 °C/4.9 Torr; (lit. [20] 109 °C/4 Torr).

3.4.3. Methyl 4-Hydroxythiophene-3-carboxylate 39

Following a literature procedure [14], sulfuryl chloride (9.7 cm^3, 16.15 g, 0.120 mol) was added dropwise to a stirred 0 °C solution of methyl 4-oxotetrahydrothiophene-3-carboxylate 973 (17.41 g, 0.109 mol) in CH$_2$Cl$_2$ (110 cm^3) over a period of 1 h. The reaction mixture was stirred at 0 °C for 30 min before being washed with sat. aq. NaHCO$_3$ (150 cm^3) and water (3 × 50 cm^3). The organic layer was dried and evaporated to give after filtration through a silica plug (Et$_2$O), 39 (12.93 g, 75%) as a light brown low-melting solid; δ_H (300 MHz) 8.72 (1H, s, OH), 7.90 (1H, d, J 3.6, ArH), 6.39 (1H, d, J 3.6, ArH) and 3.92 (3H, s, CH$_3$); δ_C (125 MHz) 165.4 (C=O), 155.3 (C–O), 131.0 (CH), 119.0 (C), 100.0 (CH) and 51.8 (CH$_3$). The ^{1}H NMR spectral data were in accordance with those previously reported [15]. ^{13}C NMR data are reported for the first time.

3.4.4. Methyl 4-Acetoxythiophene-3-carboxylate 40

Following a literature procedure [15], a mixture of methyl 4-oxotetrahydrothiophene-3-carboxylate 38 (30.80 g, 0.192 mol) and p-toluenesulfonic acid monohydrate (0.19 g, 1.00 mmol) in isopropenyl acetate (70 cm^3) was heated at reflux for 18 h. After cooling to rt, the reaction mixture was concentrated in vacuo to give methyl 4-acetoxy-2,5-dihydrothiophene-3-carboxylate as a dark brown oil which was used without further purification.

Following a literature procedure [15], sulfuryl chloride (19.5 cm^3, 32.47 g, 0.241 mol) was added dropwise to a stirred −25 °C solution of methyl 4-acetoxy-2,5-dihydrothiophene-3-carboxylate (assuming 0.192 mol) in CH$_2$Cl$_2$ (80 cm^3) over a period of 1 h. The reaction mixture was allowed to warm to rt for 18 h before being evaporated to give 40 (19.10 g, 50%) as a dark brown oil which was used without further purification; δ_H (500 MHz) 8.07 (1H, d, J 3.8, ArH), 6.98 (1H, d, J 3.8, ArH), 3.83 (3H, s, OCH$_3$) and 2.33 (3H, s, COCH$_3$). The ^{1}H NMR spectral data were in accordance with those previously reported [15].

3.4.5. Silver(I) Oxide

A solution of sodium hydroxide (14.61 g, 0.365 mol) in water (440 cm^3) was added dropwise to a stirred solution of silver nitrate (60.00 g, 0.353 mol) in water (110 cm^3). Once the addition was complete, the precipitate was collected by filtration and washed with water until the washings were neutral before being dried in vacuo to give the title compound (39.60 g, 97%) as a brown solid which was stored in the dark and used without further purification.

3.4.6. Methyl 4-(Benzyloxy)thiophene-3-carboxylate 41

A mixture of methyl 4-hydroxythiophene-3-carboxylate 39 (12.91 g, 81.6 mmol), silver(I) oxide (28.40 g, 0.123 mol) and benzyl bromide (10.7 cm^3, 15.39 g, 90.0 mmol) in CH$_2$Cl$_2$ (500 cm^3) was heated at reflux for 3 d. After cooling to rt, the reaction mixture was filtered and evaporated and the crude residue was purified by column chromatography (SiO$_2$, Et$_2$O/hexane 1:1) to give, at R$_f$ 0.90, 983 (5.47 g, 27%) as a red oil; ν_{max}/cm^{-1} 3113, 2949, 1726, 1541, 1449, 1265, 1082, 770 and 698; δ_H (500 MHz) 8.01 (1H, d, J 3.5, ArH), 7.48 (2H, d, J 7.5, Ph), 7.38 (2H, t, J 7.5, Ph), 7.31 (1H, t, J 7.5, Ph), 6.31 (1H, d, J 3.5, ArH), 5.13 (2H, s, CH$_2$) and 3.86 (3H, s, CH$_3$); δ_C (125 MHz) 162.2 (C=O), 156.1 (C–O), 136.5 (C), 133.2 (CH), 128.5 (2CH), 127.8 (CH), 126.8 (2CH), 123.7 (C), 99.0 (CH), 72.3 (CH$_2$) and 51.6 (CH$_3$); HRMS (ESI$^+$): found 271.0393. C$_{13}$H$_{12}$NaO$_3$S (M + Na) requires 271.0399.

3.4.7. 4-(Benzyloxy)thiophene-3-carboxylic Acid 42

A mixture of methyl 4-(benzyloxy)thiophene-3-carboxylate **41** (5.17 g, 20.8 mmol) and sodium hydroxide (1.74 g, 43.5 mmol) in water (45 cm^3) was heated at reflux for 18 h. After cooling to rt, the reaction mixture was washed with CH_2Cl_2 (30 cm^3) before being adjusted to pH 1 by the addition of 2 M HCl and extracted with CH_2Cl_2 (2 × 50 cm^3). The combined organic extracts were dried and evaporated to give, after recrystallisation (PhMe), **42** (3.08 g, 63%) as brown crystals; mp 101–105 °C; ν_{max}/cm^{-1} 3123, 1667, 1537, 1447, 1285, 1196, 1088, 880 and 764; δ_H (500 MHz) 9.96 (1H, br s, CO_2H), 8.21 (1H, d, J 3.5, ArH), 7.45–7.37 (5H, m, Ph), 6.48 (1H, d, J 3.5, ArH) and 5.20 (2H, s, CH_2); δ_C (125 MHz) 164.7 (C=O), 155.0 (C–O), 135.5 (C), 135.2 (CH), 128.6 (2CH), 128.3 CH), 127.2 (2CH), 122.7 (C), 100.4 (CH) and 72.9 (CH_2); HRMS (ESI$^+$): found 257.0239. $C_{12}H_{10}NaO_3S$ (M + Na) requires 257.0243.

3.4.8. 4-(Benzyloxy)-N-(1-hydroxy-2-methylpropan-2-yl)thiophene-3-carboxamide 43

Thionyl chloride (0.31 cm^3, 0.51 g, 4.25 mmol) was added to a suspension of 4-(benzyloxy)thiophene-3-carboxylic acid **42** (0.50 g, 2.13 mmol) in toluene (5 cm^3) and the mixture was heated under reflux for 3 h. After cooling to rt, the mixture was evaporated to give 4-(benzyloxy)thiophene-3-carbonyl chloride as a red oil which was used immediately without further purification.

A solution of 4-(benzyloxy)thiophene-3-carbonyl chloride (assuming 2.13 mmol) in CH_2Cl_2 (30 cm^3) was added dropwise to a solution of 2-amino-2-methylpropan-1-ol (0.41 g, 4.60 mmol) in CH_2Cl_2 (10 cm^3) stirred at 0 °C. After the addition, the mixture was allowed to warm to rt and stirred for 18 h before being poured into water. The organic layer was separated and the aqueous layer extracted with CH_2Cl_2 (2 × 20 cm^3). The combined organic layers were washed successively with 2M HCl, 2M NaOH and water before being dried and evaporated to give **43** (0.60 g, 92%) as a pale-yellow solid which was used without further purification; mp 90–92 °C; ν_{max}/cm^{-1} 3366, 2970, 1628, 1560, 1435, 1364, 1310, 1074, 976 and 606; δ_H (400 MHz) 8.07 (1H, d, J 3.6, ArH), 7.65 (1H, br s, NH), 7.46–7.37 (5H, m, Ph), 6.46 (1H, d, J 3.6, ArH), 5.22 (1H, br s, OH), 5.08 (2H, s, OCH_2Ph), 3.57 (2H, s, CH_2OH) and 1.16 (6H, s, CH_3); δ_C (100 MHz) 162.1 (C=O), 153.1 (C–O), 135.1 (C), 131.9 (CH), 128.8 (CH), 128.7 (2CH), 128.1 (2CH), 126.6 (C), 99.7 (CH), 73.2 (CH_2), 70.7 (CH_2), 56.0 (CMe_2) and 24.5 (2CH_3); HRMS (NSI$^+$): found 306.1150. $C_{16}H_{20}NO_3S$ (M + H) requires 306.1158.

3.4.9. 2-(4-(Benzyloxy)thiophen-3-yl)-4,4-dimethyl-4,5-dihydrooxazole 13

Thionyl chloride (0.15 cm^3, 0.24 g, 2.06 mmol) was added to a solution of 4-(benzyloxy)-N-(1-hydroxy-2-methylpropan-2-yl)thiophene-3-carboxamide **43** (0.50 g, 1.64 mmol) in CH_2Cl_2 (20 cm^3) and the mixture was stirred at rt for 18 h. The mixture was washed with 2M NaOH and water before being dried and evaporated to give, after purification by column chromatography (SiO$_2$, Et$_2$O/hexane 3:2), at R$_f$ 0.40, **13** (0.41 g, 87%) as a pale yellow solid; mp 77–80 °C; ν_{max}/cm^{-1} 2965, 1651, 1535, 1449, 1371, 1211, 1194, 1042, 733 and 698; δ_H (400 MHz) 7.79 (1H, d, J 3.6, ArH), 7.47 (2H, d, J 7.6, Ph), 7.36 (2H, t, J 7.4, Ph), 7.29 (1H, t, J 7.2, Ph), 6.30 (1H, d, J 3.6, ArH), 5.16 (2H, s, OCH_2Ph), 4.05 (2H, s, OCH_2) and 1.39 (6H, s, CH_3); δ_C (100 MHz) 157.2 (C), 155.3 (C), 136.9 (C), 129.1 (CH), 128.4 (2CH), 127.6 CH), 126.7 (2CH), 121.6 (C), 100.3 (CH), 78.5 (OCH_2), 72.4 (OCH_2), 67.5 (CMe_2) and 28.4 (2CH_3); HRMS (ESI$^+$): found 288.1047. $C_{16}H_{18}NO_2S$ (M + H) requires 288.1053.

3.4.10. 4-((1-Hydroxy-2-methylpropan-2-yl)carbamoyl)thiophen-3-yl Benzoate 45

Under a nitrogen atmosphere, n-butyllithium (2.5 M in hexanes, 0.88 cm^3, 2.20 mmol) was added to a stirred mixture of 2-(4-(benzyloxy)thiophen-3-yl)-4,4-dimethyl-4,5-dihydrooxazole **13** (0.1443 g, 0.50 mmol) and potassium tert-butoxide (0.2485 g, 2.21 mmol) in dry THF (5 cm^3). The mixture was stirred at rt for 2 h before being quenched by the addition of saturated aq. NH$_4$Cl and extracted with Et$_2$O (3 × 10 cm^3). The combined extracts were dried and evaporated to give, after purification by preparative TLC (Al$_2$O$_3$,

Et$_2$O/hexane 1:1), at R$_f$ 0.35, **45** (55.8 mg, 35%) in slightly impure form as a brown oil; ν_{max}/cm^{-1} 3335, 2972, 1744, 1638, 1545, 1450, 1246, 1177, 1047, 908, 766 and 700; δ_H (400 MHz) 8.19–8.16 (2H, m, Ph), 7.99 (1H, d, J 3.6, ArH), 7.71–7.67 (1H, m, Ph), 7.55 (2H, t, J 7.8, Ph), 7.34 (1H, d, J 3.6, ArH), 6.61 (1H, br s, NH), 4.58 (1H, br s, OH), 3.59 (2H, s, CH$_2$) and 1.24 (6H, s, CH$_3$); δ_C (125 MHz) 163.7 (CO), 162.2 (CO), 143.2 (C), 134.4 (CH), 130.0 (2CH), 129.9 (CH), 129.3 (C), 128.9 (2CH), 128.4 (C), 113.7 (CH), 69.9 (OCH$_2$), 56.3 (CMe$_2$) and 24.6 (2CH$_3$); HRMS (NSI$^+$): found 320.0953. C$_{16}$H$_{18}$NO$_4$S (M + H) requires 320.0951.

3.4.11. 4-(Benzyloxy)-N-butylthiophene-3-carboxamide **14**

Thionyl chloride (1.0 cm^3, 1.63 g, 13.7 mmol) was added to a suspension of 4-(benzyloxy)thiophene-3-carboxylic acid **42** (1.50 g, 6.40 mmol) in toluene (15 cm^3) and the mixture was heated under reflux for 3 h. After cooling to rt, the mixture was evaporated to give 4-(benzyloxy)thiophene-3-carbonyl chloride as a red oil which was used immediately without further purification.

A solution of 4-(benzyloxy)thiophene-3-carbonyl chloride (assuming 6.40 mmol) in toluene (30 cm^3) was added dropwise to a solution of n-butylamine (1.9 cm^3, 1.41 g, 19.2 mmol) in toluene (10 cm^3) stirred at 0 °C. Once the addition was complete, the reaction mixture was allowed to warm to rt over 1 h before being poured into water. The organic layer was separated and washed with 2M NaOH and brine, dried and evaporated to give, after purification by column chromatography (SiO$_2$, Et$_2$O/hexane 3:2), at R$_f$ 0.50, **14** (1.24 g, 67%) as a tan-coloured solid; mp 51–53 °C; ν_{max}/cm^{-1} 3385, 3111, 2970, 1630, 1557, 1435, 1364, 1265, 1184, 1074, 988, 714 and 579; δ_H (400 MHz) 8.10 (1H, d, J 3.6, ArH), 7.45–7.37 (6H, m, NH and Ph), 6.43 (1H, d, J 3.6, ArH), 5.10 (2H, s, OCH$_2$), 3.34 (2H, td, J 6.8, 5.6, NCH$_2$), 1.45–1.38 (2H, m, NCH$_2$CH$_2$), 1.25–1.15 (2H, m, CH$_2$CH$_3$) and 0.81 (3H, t, J 7.2, CH$_3$); δ_C (100 MHz) 161.5 (C=O), 153.5 (C–O), 135.5 (C), 131.5 (CH), 128.8 (2CH), 128.7 (CH), 127.9 (2CH), 127.1 (C), 99.5 (CH), 73.1 (OCH$_2$), 38.7 (NCH$_2$), 31.3 (CH$_2$), 20.0 (CH$_2$) and 13.7 (CH$_3$); HRMS (NSI$^+$): found 290.1207. C$_{16}$H$_{20}$NO$_2$S (M + H) requires 290.1209.

3.4.12. Attempted [1,2]-Wittig Rearrangement of 4-(Benzyloxy)-N-butylthiophene-3-carboxamide **14**

Under a nitrogen atmosphere, n-butyllithium (2.5 M, 2.7 cm^3, 6.75 mmol) was added dropwise to a stirred solution of 4-(benzyloxy)-N-butylthiophene-3-carboxamide **14** (0.5787 g, 2.00 mmol) in dry THF (20 cm^3). After stirring at rt for 2 h, the reaction mixture was quenched by the addition of saturated aq. NH$_4$Cl and extracted with Et$_2$O (3 × 30 cm^3). The combined organic extracts were washed with NaOH and water before being dried and evaporated to give, after purification by column chromatography (SiO$_2$, Et$_2$O/hexane 3:2), at R$_f$ 0.65, 4-(benzyloxy)-N-butyl-2-(hydroxy(phenyl)methyl)thiophene-3-carboxamide **46** (27.8 mg, 4%) in slightly impure form as a tan-coloured solid; mp 74–76 °C; ν_{max}/cm^{-1} 3393, 3084, 2930, 2870, 1624, 1557, 1443, 1217, 1173, 1016 and 696; δ_H (400 MHz) 7.73 (1H, t, J 5.0, NH), 7.52–7.48 (2H, m, Ph), 7.44–7.30 (8H, m, Ph), 6.77 (1H, br s, OH), 6.26 (1H, s, ArH), 6.21 (1H, s, CHOH), 5.04 (2H, s, OCH$_2$), 3.41–3.25 (2H, m, NCH$_2$), 1.42–1.35 (2H, m, NCH$_2$CH$_2$), 1.21–1.12 (2H, m, CH$_2$CH$_3$) and 0.80 (3H, t, J 7.2, CH$_3$); δ_C (100 MHz) 163.6 (C=O), 155.9 (C), 154.3 (C), 140.9 (C), 135.2 (C), 128.8 (3CH), 128.03 (2CH), 128.01 (2CH), 127.97 (CH), 127.3 (2CH), 121.9 (C), 97.5 (CH), 73.0 (OCH$_2$), 70.8 (CHOH), 39.0 (NCH$_2$), 31.0 (CH$_2$), 19.9 (CH$_2$) and 13.7 (CH$_3$); HRMS (ESI$^+$): found 418.1439. C$_{23}$H$_{25}$NaNO$_3$S (M + Na) requires 418.1447.

This was followed by a second fraction at R$_f$ 0.50 which was further purified by preparative TLC (SiO$_2$, CH$_2$Cl$_2$) to give, at R$_f$ 0.35 N-butyl-4-hydroxythiophene-3-carboxamide **47** (30.9 mg, 8%) in slightly impure form as a brown oil; ν_{max}/cm^{-1} 3327, 2957, 2930, 1634, 1557, 1441, 1273, 739 and 698; ^{1}H NMR revealed a 3:2 mixture of enol and keto tautomers; δ_H (500 MHz, enol tautomer **47**) 10.36 (1H, br s, OH), 7.54 (1H, d, J 3.3, ArH), 6.36 (1H, d, J 3.3, ArH), 6.35 (1H, br s, NH), 3.41 (2H, td, J 7.3, 6.0, NCH$_2$), 1.62–1.51 (2H, m, NCH$_2$CH$_2$), 1.44–1.33 (2H, m, CH$_2$CH$_3$) and 0.95 (3H, t, J 7.5, CH$_3$); δ_C (125 MHz, enol tautomer **47**) 165.1 (C=O), 156.1 (C–O), 124.6 (CH), 121.7 (C), 100.1 (CH), 39.1 (NCH$_2$), 31.6 (CH$_2$), 20.08 (CH$_2$) and 13.7 (CH$_3$); δ_H (500 MHz, keto tautomer **47a**) 9.33 (1H, s, CH), 8.06 (1H, br s, NH), 3.89 (2H, s, SCH$_2$), 3.36 (2H, td, J 7.0, 6.0, NCH$_2$), 1.62–1.51 (2H, m, NCH$_2$CH$_2$), 1.44–1.33 (2H, m, CH$_2$CH$_3$) and 0.93 (3H, t, J 7.3, CH$_3$); δ$_C$ (125 MHz, keto tautomer **47a**) 199.9 (C=O), 174.7 (CH), 159.9 (CONH), 128.2 (C), 42.7 (SCH$_2$), 38.7 (NCH$_2$), 31.5 (CH$_2$), 20.11 (CH$_2$) and 13.7 (CH$_3$); HRMS (NSI$^+$): found 200.0740. C$_9$H$_{14}$NO$_2$S (M + H) requires 200.0740.

3.5. X-ray Structure Determination

Data have been deposited at the Cambridge Crystallographic Data Centre as CCDC 2111424 (**9**), 2111425 (**20**), 2111426 (**11**), 2111427 (**26**) and 2111428 (**13**). The data can be obtained free of charge from the Cambridge Crystallographic Data Centre via http://www.ccdc.cam.ac.uk/getstructures. In all cases, data were collected on a Rigaku XtaLAB 200 diffractometer using graphite monochromated Mo-Kα radiation, λ = 0.71075 Å and the structures were solved by direct methods and refined by full-matrix least-squares against F2 (SHELXL Version 2014/7 [21]).

Compound 20

Slow evaporation of an acetone/CH$_2$Cl$_2$ solution gave tan-coloured crystals suitable for X-ray structure determination. Crystal data for **20**: 2C$_{16}$H$_{17}$NO$_3$S•CH$_2$Cl$_2$•Me$_2$CO, M = 749.76, yellow prism, crystal dimensions 0.10 × 0.10 × 0.10 mm, monoclinic, space group P2$_1$/n, a = 17.4520, b = 9.9840, c = 21.1050 Å, β = 92.3760°, V = 3674.1899 Å^3, Z = 4, D$_c$ = 1.355 Mg m^{-3}, T = 93 K, R = 0.0929, R$_W$ = 0.2464 for 4473 reflections with I > 2σ(I) and 458 variables.

Compound 26

Slow evaporation of a CH$_2$Cl$_2$ solution gave crystals suitable for X-ray structure determination. Crystal data for **26**: C$_{16}$H$_{17}$NOS, M = 271.38, colourless needle, crystal dimensions 0.12 × 0.02 × 0.02 mm, monoclinic, space group P2$_1$, a = 8.237(3), b = 5.717(2), c = 15.097(6) Å, β = 90.539(10)°, V = 710.9(5) Å^3, Z = 2, D$_c$ = 1.268 Mg m^{-3}, T = 93 K, R = 0.0481, R$_W$ = 0.1086 for 2072 reflections with I > 2σ(I) and 172 variables.

Compound 9

Slow evaporation of an MeCN solution gave crystals suitable for X-ray structure determination. Crystal data for **9**: C$_{16}$H$_{17}$NO$_2$S, M = 287.38, colourless plate, crystal dimensions 0.10 × 0.10 × 0.01 mm, monoclinic, space group P2$_1$/c, a = 15.9970(19), b = 7.5839(6), c = 12.4523(15) Å, β = 111.010(14)°, V = 1410.3(3) Å^3, Z = 4, D$_c$ = 1.353 Mg m^{-3}, T = 93 K, R = 0.0584, R$_W$ = 0.1406 for 2462 reflections with I > 2σ(I) and 181 variables.

Compound 11

Slow evaporation of a CH$_2$Cl$_2$ solution gave crystals suitable for X-ray structure determination. Crystal data for **11**: C$_{16}$H$_{17}$NO$_2$S, M = 287.38, colourless prism, crystal dimensions 0.12 × 0.10 × 0.06 mm, monoclinic, space group P2$_1$/c, a = 14.8176(4), b = 8.72450(17), c = 11.9720(3) Å, β = 113.314(3)°, V = 1421.32(7) Å^3, Z = 4, D$_c$ = 1.343 Mg m^{-3}, T = 93 K, R = 0.0274, R$_W$ = 0.0730 for 2906 reflections with I > 2σ(I) and 181 variables.

Compound 13

Slow evaporation of a toluene solution gave crystals suitable for X-ray structure determination. Crystal data for **13**: C$_{16}$H$_{17}$NO$_2$S, M = 287.38, colourless plate, crystal dimensions 0.20 × 0.20 × 0.01 mm, monoclinic, space group P2$_1$/c, a = 15.6319(4), b = 8.7152(3), c = 10.7572(3) Å, β = 93.909(3)°, V = 1462.10(7) Å^3, Z = 4, D$_c$ = 1.305 Mg m^{-3}, T = 296 K, R = 0.0333, R$_W$ = 0.0888 for 2781 reflections with I > 2σ(I) and 181 variables.

4. Conclusions

The three isomeric thienyloxazolines showed a varied and interesting pattern of reactivity with two of them undergoing Wittig rearrangement and two giving products derived from ring cleavage of an intermediate 3-aminothienofuran. One of the corresponding thienyl amides also underwent Wittig rearrangement and cyclisation of the product, as well as an isomeric one obtained by other means, gave two isomeric thienopyrrolones. The pattern of reactivity in the oxazoline series correlates well with the molecular geometry in the solid state as determined by X-ray diffraction, and the use of this method to explain patterns of competing reactivity between closely similar molecules may be useful more generally.

Supplementary Materials: The following are available online, Figures S1–S41: NMR spectra of new compounds. Cif and check-cif files for X-ray structures of **9, 11, 13, 20** and **26**.

Author Contributions: A.D.H. performed the experiments; A.M.Z.S. collected the X-ray data and solved the structures; R.A.A. designed the experiments, analysed the data and wrote the paper. All authors have read and agreed to the published version of the manuscript.

Funding: We thank EPSRC (UK) for a DTA studentship to ADH (Grant EP/L505079/1) and the EPSRC UK National Mass Spectrometry Facility at Swansea University.

Institutional Review Board Statement: Not applicable.

Informed Consent Statement: Not applicable.

Data Availability Statement: The X-ray data is deposited at CCDC as stated above and all NMR spectroscopic data is in the Supplementary Material.

Conflicts of Interest: The authors declare no conflict of interest.

References

1. Aitken, R.A.; Harper, A.D.; Slawin, A.M.Z. Base-Induced Cyclisation of ortho-Substituted 2-Phenyloxazolines to Give 3-Aminobenzofurans and Related Heterocycles. *Synlett* **2017**, *28*, 1738–1742. [CrossRef]
2. Gewald, K.; Jänsch, H.-J. 3-Amino-benzo[*b*]furane. *J. Prakt. Chem.* **1973**, *315*, 779–785. [CrossRef]
3. Schorigin, P. Über die Carbinol-Umlagerung von Benzyläthern I. *Ber. Dtsch. Chem. Ges.* **1924**, *57*, 1634–1637. [CrossRef]
4. Wang, F.; Wang, J.; Zhang, Y.; Yang, J. The [1,2]- and [1,4]-Wittig rearrangement. *Tetrahedron* **2020**, *76*, 130857. [CrossRef]
5. Etter, M.C.; Macdonald, J.C.; Bernstein, J. Graph-set analysis of hydrogen-bond patterns in organic crystals. *Acta Crystallogr. B.* **1990**, *46*, 256–262. [CrossRef] [PubMed]
6. Aitken, R.A.; Harper, A.D. Thiophene-Based Quinomethane Analogs. *Adv. Heterocycl. Chem.* **2017**, *123*, 169–243. [CrossRef]
7. Kawasaki, T.; Kimachi, T. Sparteine-Mediated Enantioselective [2,3]-Wittig Rearrangement of Allyl ortho-Substituted Benzyl Prenyl Ethers. *Tetrahedron* **1999**, *55*, 6847–6862. [CrossRef]
8. Harper, A.D.; Aitken, R.A. The Chemistry of Thieno[*c*]pyrrolones, Dihydrothieno[*c*]pyrrolones, and Their Fused Derivatives. *Adv. Heterocycl. Chem.* **2018**, *127*, 227–314. [CrossRef]
9. Parella, R.; Babu, S.A. Pd(II)-Catalyzed Arylation and Intramolecular Amidation of γ-C(sp^3)–H Bonds: En Route to Arylheteroarylmethane and Pyrrolidone Ring Annulated Furan/Thiophene Scaffolds. *J. Org. Chem.* **2017**, *82*, 7123–7150. [CrossRef] [PubMed]
10. Carpenter, A.J.; Chadwick, D.J. The scope and limitations of carboxamide-induced β-directed metalation of 2-substituted furan, thiophene, and 1-methylpyrrole derivatives. Application of the method to syntheses of 2,3-disubstituted thiophenes and furans. *J. Org. Chem.* **1985**, *50*, 4362–4368. [CrossRef]
11. Aitken, R.A.; Harper, A.D.; Slawin, A.M.Z. Synthesis, Structure, and Unusual Reactivity of a Stable 3-(Oxazolidin-2-ylidene)thiophen-2-one. *J. Org. Chem.* **2016**, *81*, 10527–10531. [CrossRef] [PubMed]
12. Dowd, P.; Choi, S.-C. Free radical ring-expansion leading to novel six- and seven-membered heterocycles. *Tetrahedron* **1991**, *47*, 4847–4860. [CrossRef]
13. Hromatka, O.; Binder, D.; Eichinger, K. Über den Mechanismus der Dieckmann-Reaktion von 3-(Methoxycarbonylmethylthio) propionsäuremethylester. *Monatsh. Chem.* **1973**, *104*, 1520–1525. [CrossRef]
14. Rossy, P.A.; Hoffmann, W.; Mueller, N. Aromatization of dihydrothiophenes. Thiophenesaccharin: A sweet surprise. *J. Org. Chem.* **1980**, *45*, 617–620. [CrossRef]
15. Press, J.B.; Hofmann, C.M.; Safir, S.R. Thiophene systems. 2. Synthesis and chemistry of some 4-alkoxy-3-substituted thiophene derivatives. *J. Org. Chem.* **1979**, *44*, 3292–3296. [CrossRef]
16. Purdie, T.; Irvine, J.C. The Alkylation of Sugars. *J. Chem. Soc. Trans.* **1903**, *83*, 1023–1037. [CrossRef]
17. Plant, A.; Harder, A.; Mencke, N.; Bertram, H.-J. Synthesis and Anthelmintic Activity of 7-Hydroxy-5-oxo-5*H*-thieno[3,2-*b*]pyran-6-carboxanilides and -6-thiocarboxanilides. *Pestic. Sci.* **1996**, *48*, 351–358. [CrossRef]
18. Schöning, A.; Debbaerdemeker, T.; Zander, M.; Friedrichsen, W. 4,6-Diphenylthieno[2,3-c]furan. *Chem. Ber.* **1989**, *122*, 1119–1131. [CrossRef]
19. Gable, R.W.; Laws, M.J.; Schiesser, C.H. *N*-Butylthiophene-3-carboxamide and 2-Benzylseleno-*N*-butylthiophene-3-carboxamide at 130K. *Acta Crystallogr. Sect. C Cryst. Struct. Commun.* **1997**, *53*, 641–644. [CrossRef]
20. Woodward, R.B.; Eastman, R.H. Tetrahydrothiophene ("thiophane") derivatives. *J. Am. Chem. Soc.* **1946**, *68*, 2229–2235. [CrossRef]
21. Sheldrick, G.M. A short history of SHELXL. *Acta Crystallogr. A* **2008**, *64*, 112–122. [CrossRef] [PubMed]

MDPI
St. Alban-Anlage 66
4052 Basel
Switzerland
Tel. +41 61 683 77 34
Fax +41 61 302 89 18
www.mdpi.com

Molecules Editorial Office
E-mail: molecules@mdpi.com
www.mdpi.com/journal/molecules

www.ingramcontent.com/pod-product-compliance
Lightning Source LLC
LaVergne TN
LVHW070156120526
838202LV00013BA/1298